AF355969

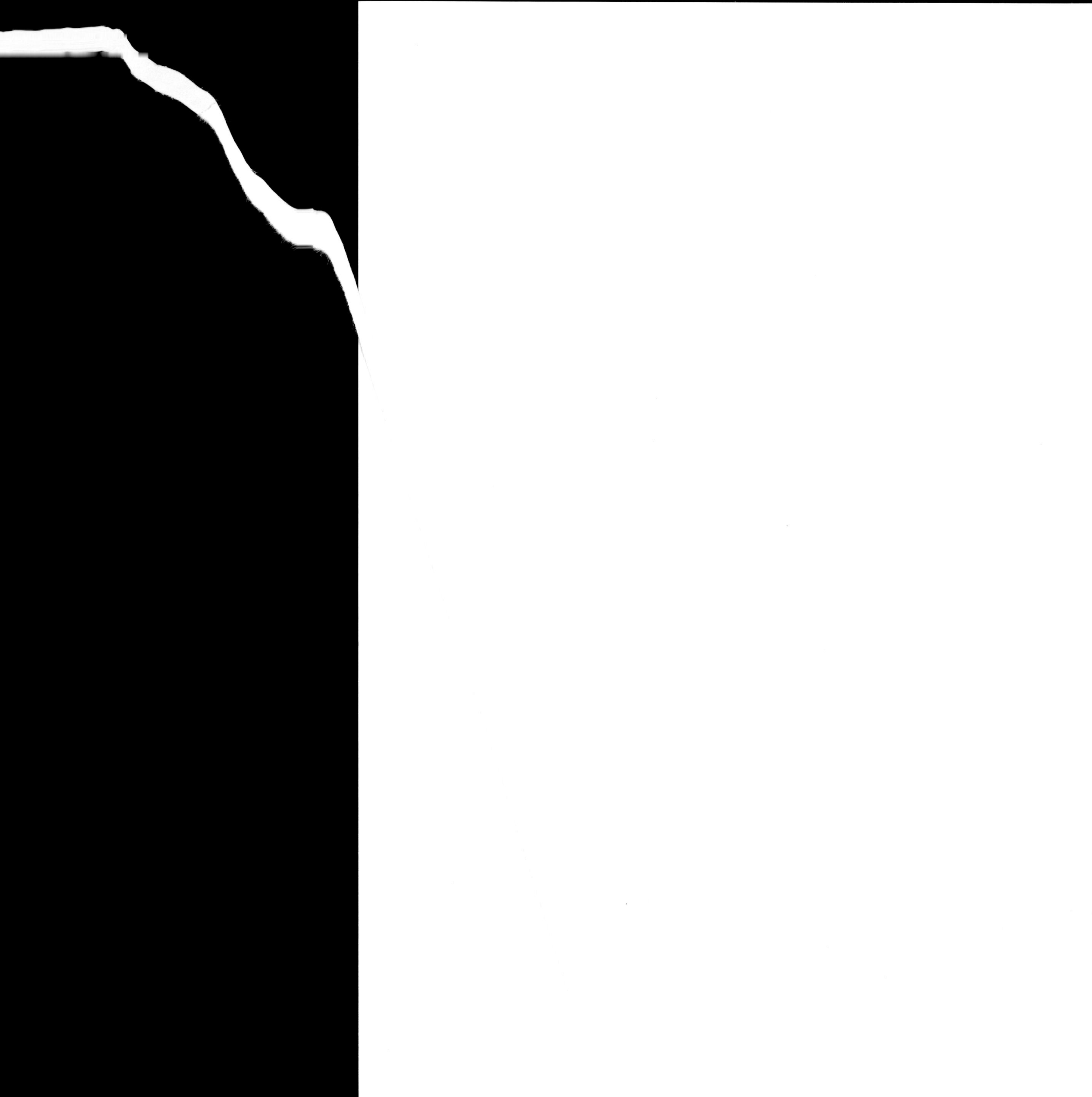

Enzymes in Biomedical, Cosmetic and Food Application

Enzymes in Biomedical, Cosmetic and Food Application

Editors

Chia-Hung Kuo
Chwen-Jen Shieh
Hui-Min David Wang

Basel • Beijing • Wuhan • Barcelona • Belgrade • Novi Sad • Cluj • Manchester

Editors

Chia-Hung Kuo
National Kaohsiung
University of Science and
Technology
Kaohsiung
Taiwan

Chwen-Jen Shieh
National Chung-Hsing
University
Taichung
Taiwan

Hui-Min David Wang
National Chung-Hsing
University
Taichung
Taiwan

Editorial Office
MDPI
St. Alban-Anlage 66
4052 Basel, Switzerland

This is a reprint of articles from the Special Issue published online in the open access journal *Catalysts* (ISSN 2073-4344) (available at: https://www.mdpi.com/journal/catalysts/special_issues/73I51J75AP).

For citation purposes, cite each article independently as indicated on the article page online and as indicated below:

Lastname, A.A.; Lastname, B.B. Article Title. *Journal Name* **Year**, *Volume Number*, Page Range.

ISBN 978-3-7258-0579-2 (Hbk)
ISBN 978-3-7258-0580-8 (PDF)
doi.org/10.3390/books978-3-7258-0580-8

Contents

About the Editors

Chia-Hung Kuo

Chia-Hung Kuo currently holds the positions of Professor of Seafood Science, Chair of the Department of Seafood Science, and Director of the Center for Aquatic Products Inspection Service at the National Kaohsiung University of Science and Technology (NKUST), Taiwan. He earned his MS in Food Science and Technology from the National Taiwan University, and a Ph.D. in Chemical Engineering from the National Taiwan University of Science and Technology. Throughout his career, he has garnered several awards, including the Outstanding New Teacher Award (2016), Outstanding Research Award (2018), Special Outstanding Academia-Industry Collaboration Talent Award (2023), and Special Outstanding Research Talent Award (2018–2023) from NKUST, the second-largest university in Taiwan. He has authored over 90 international SCI papers, 7 books or book chapters, and holds three patents. His H-index is 30, with more than 2800 citations. Additionally, he serves as an Editor for *Catalysts* and a Topical Advisory Panel Member for the *International Journal of Molecular Sciences*, and acts as a Guest Editor for Special Issues in *Catalysts, Sustainability*, and the *International Journal of Molecular Sciences*. His primary research interests revolve around process biochemistry, biocatalysis, food engineering, food analysis, extraction, oil and fat processing, and fermentation biotechnology.

Chwen-Jen Shieh

Dr. Chwen-Jen Shieh holds a Ph.D. in Food Science & Technology from the University of Georgia, USA. He is currently a Distinguished Professor of Biotechnology Center at the National Chung-Hsing University, Taiwan. He was also an Outstanding Research Professor at Dayeh University, and Director of R&D at Taisun Enterprise Company, Taiwan. He has published more than 100 SCI papers in international journals. His main research interests focus on biodiesel, biocatalysis, enzyme technology, bioprocess optimization, supercritical fluid technology, and Chinese herb medicine biotechnology.

Hui-Min David Wang

Hui-Min David Wang, a Full Professor at the Graduate Institute of Biomedical Engineering (National Chung Hsing University), graduated from the Department of Chemical Engineering, National Cheng Kung University, Tainan, Taiwan, in 2003. As a promising young scientist, his focus during the post-doctorate term at the Institute of Biological Chemistry (Academia Sinica, Taiwan) in 2004 was on DNA gene / protein research and SARS virus research. In 2005, he gained a position as a part-time Assistant Professor at the National Taipei University of Technology, Department of Chemical Engineering and Biotechnology and Institute of Chemical Engineering. From 2007 to 2016, he was an Assistant Professor at Kaohsiung Medical University and was promoted to Associate Professor and Full Professor (Division Leader, Office for Operation of Industry and University Cooperation; Director, Division of Academic Collaboration, Office of Global Affairs; Vice Chief, Center of Stem Cell Research, Kaohsiung Medical University). He has published 250 SCI papers on drug discovery and biomaterials on skin and biomass production, and he has also obtained numerous grants and patents for his research. In addition to the above, his current work includes searching for a novel environmental resource in place of our world's depleting natural resources (functional ingredients); melanogenesis inhibitors in skin care; biomedical engineering in microfluidics; and cancer research in vitro and in vivo.

Preface

Enzymes are commonly referred to as biocatalysts that can catalyze many biological and chemical reactions. In this Special Issue, eleven research articles and two review articles summarize current developments of enzymes in biomedical, cosmetic, and food applications. From the synthesis of bioactive compounds to the enhancement of industrial processes, each chapter offers invaluable insights into the versatility and efficacy of enzymatic catalysis. By delving into topics such as enzyme immobilization, metabolic pathway elucidation, and enzyme-mediated transformations, this compilation bridges the gap between fundamental research and practical implementation. We are honored to present this compilation and hope that it will serve as a source of inspiration and knowledge dissemination, fostering further exploration and innovation in the field of biocatalysis related to biomedical, cosmetic, and food applications.

Chia-Hung Kuo, Chwen-Jen Shieh, and Hui-Min David Wang
Editors

Editorial

Enzymes in Biomedical, Cosmetic and Food Application

Chia-Hung Kuo [1,2,*], Hui-Min David Wang [3,*] and Chwen-Jen Shieh [4,*]

[1] Department of Seafood Science, National Kaohsiung University of Science and Technology, Kaohsiung 811, Taiwan
[2] Center for Aquatic Products Inspection Service, National Kaohsiung University of Science and Technology, Kaohsiung 811, Taiwan
[3] Graduate Institute of Biomedical Engineering, National Chung Hsing University, Taichung 402, Taiwan
[4] Biotechnology Center, National Chung Hsing University, Taichung 402, Taiwan
[*] Correspondence: kuoch@nkust.edu.tw (C.-H.K.); davidw@dragon.nchu.edu.tw (H.-M.D.W.); cjshieh@nchu.edu.tw (C.-J.S.); Tel.: +886-7-361-7141 (ext. 23646) (C.-H.K.); +886-4-2284-0733 (ext. 651) (H.-M.D.W.); +886-4-2284-0450 (ext. 5121) (C.-J.S.)

Citation: Kuo, C.-H.; Wang, H.-M.D.; Shieh, C.-J. Enzymes in Biomedical, Cosmetic and Food Application. *Catalysts* **2024**, *14*, 162. https://doi.org/10.3390/catal14030162

Received: 2 February 2024
Accepted: 19 February 2024
Published: 22 February 2024

Enzymes play an important role in biomedical, cosmetic and food applications, and their effects are mainly related to their specific reactions and catalytic activity [1]. Enzymes are important catalysts for promoting the formation of complex organic molecules during the synthesis of medical compounds [2]. Enzymes can be used to synthesize various drugs, including antibiotics, hormones, anticancer agents, etc. In chiral synthesis, enzymes are essential for the preparation of drugs with high chirality, as they can specifically catalyze the synthesis of chiral molecules [3]. Moreover, the highly selective catalysis of specific functional groups and bonding sites by enzymes reduces the occurrence of side reactions and helps improve the purity of a product [4]. Through protein engineering technology, the structure of an enzyme can be modified to improve its activity or give it new catalytic properties to adapt to specific synthetic reaction conditions. Enzyme-mediated synthetic reactions are generally greener and therefore considered more environmentally sustainable than conventional chemical approaches. This is attributed to the ability of enzymes to operate under milder conditions, resulting in less energy consumption and a reduced generation of toxic by-products [5]. High yields are frequently attained in synthesis processes when enzymes serve as catalysts, thereby enhancing overall efficiency. The selective catalysis of an enzyme allows for the generation of high-purity products, reducing the need for subsequent purification steps [4]. Various enzymes are known for their applications in cosmetics and the food industry. Proteases can remove a keratin layer attached to the skin's surface and make the skin softer. Catalase and superoxide dismutase can be used to inhibit oxidation reactions in cosmetics and slow down the aging process of products [6]. In food processing, amylase is used to break down starch, for example, in the making of syrup, bread and beer [7,8]. Protease can hydrolyze proteins to produce functional peptides [9], make cheese and tenderize meat [10]. Enzymes can also be used to increase the extraction efficiency of functional compounds in food. This Special Issue investigates enzymes for biomedical, cosmetic and food applications.

Shih's article (contribution 1) discusses three enzyme hydrolysates, Dur-A, Dur-B and Dur-C, derived from *Durvillaea antarctica* biomass using viscozyme, cellulase and α-amylase, respectively. Through ^{1}H-NMR analysis, these extracts were found to contain fucose-containing sulfated polysaccharides with distinct structural qualities. Notably, Dur-A, Dur-B and Dur-C demonstrated significant antioxidant activities, as confirmed by DPPH, ABTS and ferrous ion-chelating analyses. Moreover, these extracts displayed the ability to inhibit key enzymes associated with metabolic syndrome, including angiotensin I-converting enzyme (ACE), α-glucosidase, α-amylase and pancreatic lipase. In particular, Dur-B exhibited superior antioxidant and antimetabolic syndrome effects compared to the other extracts. These findings suggest that these enzyme hydrolysates, especially Dur-B, are promising natural antioxidants and antimetabolic syndrome agents for applications in various health-oriented products like food, cosmetics and nutraceuticals.

Weng's article (contribution 2) reports the immobilization of recombinant endoglucanase (CelA) on regenerated cellulose (RC) membrane modified using two different approaches, one to generate the immobilized metal ion affinity membranes RC-EPI-IDA-Co^{2+} (IMAMs) for coordination coupling and another to develop the aldehyde functional group membranes RC-EPI-DA-GA (AMs) for covalent bonding. A recombinant endoglucanase (CelA) originating from the cellulosome of *Clostridium thermocellum* was expressed in *Escherichia coli* and then immobilized on RC-EPI-IDA-Co^{2+} (IMAM) and RC-EPI-DA-GA (AM) membranes. The characteristics of the immobilized enzyme, along with its preliminary purification, were assessed in comparison to a free enzyme. Moreover, an appreciable enzyme activity till 5 cycles of reusability achieved in this study paves a way for developing an economical process.

Sharma's article (contribution 3) provided a detailed overview of the production of industrial enzymes from microbes using agro-industrial food waste. Enzymes are versatile biocatalysts with immense potential to transform the food industry and lignocellulosic biorefineries. Microbial enzymes offer cleaner and greener methods of producing fine chemicals and compounds. The review highlights novel strategies for food enzyme immobilization and their potential applications in the food industry. Moreover, deeper insights into the development of engineered enzymes for sustainably processing waste biomass into diverse bioproducts are highlighted. In short, this review discusses recent developments in the production of essential industrial enzymes from agro-industrial food waste and the application of inexpensive immobilization and enzyme engineering approaches for sustainable development.

Thirametoakkhara's article (contribution 4) explored the production and applications of two endoxylanases, Xyn45 and Xyn23, derived from the bacterium *Bacillus halodurans*. Xyn45 was obtained through recombinant *E. coli*, while a mixture of nonrecombinant Xyn45 and Xyn23 was acquired from *B. halophilus*. The study revealed that combining these enzymes led to enhanced catalytic activity compared to using Xyn45 alone. Furthermore, the researchers investigated the impact of xylooligosaccharides (XOS) derived from oil palm empty fruit bunches on the growth and metabolism of probiotic strains, *Bifidobacteria* and *Lactobacilli*. The findings demonstrated that XOS, particularly xylobiose, effectively induced the secretion of endoxylanases in *B. halophilus*, highlighting their potential for industrial applications. This research offers valuable insights into the synergistic effects of endoxylanases and the utilization of XOS to promote the growth of beneficial bacteria.

Cunha's article (contribution 5) prepared a new heterogeneous biocatalyst by applying the cross-linking enzyme aggregates (CLEAs) technique to non-commercial β-glucosidase from *Aspergillus niger* produced via solid-state fermentation. The effects of relevant factors on the immobilization process, such as the soy protein isolate and glutaraldehyde concentrations, were evaluated using a central composite rotatable design (CCRD). The influence of certain factors on the hydrolytic activity of the immobilized enzyme (pH, temperature and thermal stability) was evaluated, and it was compared with its soluble form in order to contribute to the advancement of enzyme immobilization technology and consolidate the results of utilizing β-glucosidase in bioprocesses.

Ma's article (contribution 6) presented a significant advancement in biocatalysis by establishing a bienzymatic, parallel cascade that combines aryl alcohol oxidases with peroxygenases for the selective oxidation of benzylic alcohols to corresponding aromatic aldehydes. Aromatic aldehydes are important aromatic compounds for the flavor and fragrance industries. A parallel cascade combining aryl alcohol oxidase from *Pleurotus eryngii* (PeAAOx) and unspecific peroxygenase from the basidiomycete *Agrocybe aegerita* (AaeUPO) was used to convert aromatic primary alcohols into high-value aromatic aldehydes. In a partially optimized system, up to an 84% conversion rate of 50 mM veratryl alcohol into veratryl aldehyde was achieved through a self-sufficient aerobic reaction. The findings offer insights into industrial applications in flavor compound synthesis.

Nabil-Adam's article (contribution 7) investigated the antidiabetic, anti-inflammatory and antioxidant competency of *Jania rubens* polyphenolic extract (JRPE) through its interactions with α-amylase, lipase and trypsin enzymes. An HPLC analysis revealed the dominance of twelve polyphenolic compounds. The antioxidant and antibacterial activities demonstrated by *Jania rubens* extract suggest its potential as a natural source for combating oxidative stress and microbial infections. Computational analyses further corroborated the ability of these polyphenolics to form complexes with digestive enzymes. The findings provide a foundation for understanding the anti-obesity and antidiabetes characteristics of *Jania rubens* polyphenolic compounds.

Xing's article (contribution 8) presented several important findings regarding the enzymatic modification of bovine lactoferrin (bLf) and its potential applications. Laccase-mediated pectin–ferulic acid conjugate (PF) and transglutaminase (TG) treatments were studied for their ability to enhance the diversity and abundance of bLf peptide fragments during in vitro simulated gastrointestinal digestion. The encapsulation of bLf by PF led to enhanced diversity and abundance of active peptide fragments, especially for long-chain species. The TG treatment on the lactoferrin–pectin–ferulic acid conjugate (LfPFTG) demonstrated an influence on the final gastrointestinal digest, highlighting the potential of TG-induced crosslinking between bLf chains. The identified peptides, with enhanced abundances in the LfPFG digest, hold promise for various applications in nutraceuticals, clinical therapy or as components in antibacterial vaccines.

Statkeviˇcius's article (contribution 9) identified 11 full-length fold type IV aminotransferases (ATs) that were successfully expressed and used for substrate profiling. Three of them (AT-872, AT-1132 and AT-4421) were active toward (R)-methylbenzylamine. The research showed the high thermostability of specific ATs, such as AT-872 and AT-1132, along with their broad substrate spectrum, particularly AT-872 and AT-4421. The purified proteins showed activity with L- and D-amino acids and various aromatic compounds, such as (R)-1-aminotetraline. The study demonstrated the close relationship between AT-1132 and branched-chain amino acid transaminases (BCATs) through the protein sequence, and its specificity for amino acceptors was similar to that for BCATs, which are most active with α-ketoglutarate. AT-872 and AT-4421 were homologous to the D-amino acid transaminases (DAATs), which prefer pyruvate as an amino acceptor. These properties make these enzymes potential biocatalysts for the production of valuable amino compounds. These findings offer valuable insights into the potential of these enzymes in the synthesis of chiral drugs and the environmentally friendly preparation of chiral amines.

Gupta's article (contribution 10) reported the significance of various natural compounds and extracts in the fields of photoprotection and skincare. It extensively covers studies on microalgae, peptides and antioxidants, highlighting their potential applications in inhibiting melanogenesis, reducing oxidative stress and protecting against skin cancer. The role of antioxidants, such as vitamins C and E, in enhancing the photoprotective properties of sunscreens was illustrated. It discusses the relevance of various topical options containing DNA repair enzymes, antioxidants and growth factors in preventing skin aging and cancer. Moreover, the potential of photolyases, which are enzymes that repair DNA damage caused by ultraviolet radiation, and their uses in skincare are discussed.

Tran's article (contribution 11) investigated the metabolic processes and pharmacological effects of the synthetic cannabinoid JWH-019 ((1-Hexyl-1H-indol-3-yl)-1-naphthalenyl-methanone). Through a series of experiments utilizing human liver microsomes (HLMs) and recombinant P450 enzymes, 6-OH JWH-019 was found to be the primary oxidative metabolite in HLMs, and CYP1A2 was the principal enzyme responsible for the mono-hydroxylation of JWH-019. The research not only enhances our understanding of the metabolic pathways of this synthetic cannabinoid but also provides crucial insights into its pharmacological activity.

de Carvalho-Silva's article (contribution 12) investigated the stability and performance of immobilized tannase, an enzyme derived from *Aspergillus ficuum*, using two different methods: their entrapment in calcium alginate beads and their covalent attachment to magnetic nanoparticles. The kinetic and thermodynamic behavior of the enzyme under various conditions was studied. The findings indicated that the magnetic nanoparticle immobilization method results in the lowest activation energy, suggesting its suitability for cost-effective industrial applications. Additionally, shelf-life tests revealed that the enzyme remained stable for a longer period of time when immobilized on magnetic nanoparticles.

Arnold's article (contribution 13) showed the two-step enzymatic synthesis of (S)-norlaudanosoline ((S)-NLS), a crucial precursor for opioids, using lyophilized whole-cell biocatalysts containing ω-transaminase (Tam) and norcoclaurine synthase (NCS) enzymes. The research revealed that the addition of a substantial quantity of cells containing the NCS enzyme resulted in a high yield of an optically pure chiral product. The cells were immobilized to enable their retention in fixed-bed reactors for batch and flow system comparisons. The flow system produced a substantial amount of (S)-NLS, allowing for its long-term practical application for opioid precursor synthesis.

Funding: This research received no external funding.

Conflicts of Interest: The authors declare no conflict of interest.

List of Contributions

1. Shih, M.-K.; Hou, C.-Y.; Dong, C.-D.; Patel, A.K.; Tsai, Y.-H.; Lin, M.-C.; Xu, Z.-Y.; Perumal, P.K.; Kuo, C.-H.; Huang, C.-Y. Production and characterization of *Durvillaea antarctica* enzyme extract for antioxidant and anti-metabolic syndrome effects. *Catalysts* **2022**, *12*, 1284.
2. Weng, Z.-H.; Nargotra, P.; Kuo, C.-H.; Liu, Y.-C. Immobilization of recombinant endoglucanase (CelA) from *Clostridium thermocellum* on modified regenerated cellulose membrane. *Catalysts* **2022**, *12*, 1356.
3. Sharma, V.; Tsai, M.-L.; Nargotra, P.; Chen, C.-W.; Kuo, C.-H.; Sun, P.-P.; Dong, C.-D. Agro-industrial food waste as a low-cost substrate for sustainable production of industrial enzymes: A critical review. *Catalysts* **2022**, *12*, 1373.
4. Thirametoakkhara, C.; Hong, Y.-C.; Lerkkasemsan, N.; Shih, J.-M.; Chen, C.-Y.; Lee, W.-C. Application of endoxylanases of Bacillus halodurans for producing xylooligosaccharides from empty fruit bunch. *Catalysts* **2022**, *13*, 39.
5. da Cunha, T.M.; Mendes, A.A.; Hirata, D.B.; Angelotti, J.A. Optimized conditions for preparing a heterogeneous biocatalyst via cross-linked enzyme aggregates (CLEAs) of β-glucosidase from *Aspergillus niger*. *Catalysts* **2022**, *13*, 62.
6. Ma, Y.; Li, Z.; Zhang, H.; Wong, V.K.W.; Hollmann, F.; Wang, Y. Bienzymatic cascade combining a peroxygenase with an oxidase for the synthesis of aromatic aldehydes from benzyl alcohols. *Catalysts* **2023**, *13*, 145.
7. Nabil-Adam, A.; Ashour, M.L.; Tamer, T.M.; Shreadah, M.A.; Hassan, M.A. Interaction of *Jania rubens* polyphenolic extract as an antidiabetic agent with α-amylase, lipase, and trypsin: In vitro evaluations and in silico studies. *Catalysts* **2023**, *13*, 443.
8. Xing, M.; Ji, Y.; Ai, L.; Xie, F.; Wu, Y.; Lai, P.F. Improving effects of laccase-mediated pectin-ferulic acid conjugate and transglutaminase on active peptide production in bovine lactoferrin digests. *Catalysts* **2023**, *13*, 521.
9. Statkevičius, R.; Vaitekūnas, J.; Stanislauskienė, R.; Meškys, R. Metagenomic type IV aminotransferases active toward (R)-methylbenzylamine. *Catalysts* **2023**, *13*, 587.
10. Gupta, A.; Singh, A.P.; Singh, V.K.; Singh, P.R.; Jaiswal, J.; Kumari, N.; Upadhye, V.; Singh, S.C.; Sinha, R.P. Natural sun-screening compounds and DNA-repair enzymes: Photoprotection and photoaging. *Catalysts* **2023**, *13*, 745.
11. Tran, N.; Fantegrossi, W.E.; McCain, K.R.; Wang, X.; Fujiwara, R. Identification of cytochrome P450 enzymes responsible for oxidative metabolism of synthetic cannabinoid (1-hexyl-1 H-indol-3-yl)-1-naphthalenyl-methanone (JWH-019). *Catalysts* **2023**, *13*, 1008.
12. de Carvalho-Silva, J.; da Silva, M.F.; de Lima, J.S.; Porto, T.S.; de Carvalho, L.B., Jr.; Converti, A. Thermodynamic and kinetic investigation on *Aspergillus ficuum* tannase immobilized in calcium alginate beads and magnetic nanoparticles. *Catalysts* **2023**, *13*, 1304.
13. Arnold, A.H.; Castiglione, K. Comparative evaluation of the asymmetric synthesis of (S)-norlaudanosoline in a two-step biocatalytic reaction with whole *Escherichia coli* Cells in batch and continuous Flow Catalysis. *Catalysts* **2023**, *13*, 1347.

References

1. Deckers, M.; Deforce, D.; Fraiture, M.-A.; Roosens, N.H. Genetically modified micro-organisms for industrial food enzyme production: An overview. *Foods* **2020**, *9*, 326. [CrossRef] [PubMed]
2. Shen, Y.; Xia, Y.; Chen, X. Research progress and application of enzymatic synthesis of glycosyl compounds. *Appl. Microbiol. Biotechnol.* **2023**, *107*, 5317–5328. [CrossRef] [PubMed]
3. Eletskaya, B.Z.; Berzina, M.Y.; Fateev, I.V.; Kayushin, A.L.; Dorofeeva, E.V.; Lutonina, O.I.; Zorina, E.A.; Antonov, K.V.; Paramonov, A.S.; Muzyka, I.S. Enzymatic Synthesis of 2-Chloropurine Arabinonucleosides with Chiral Amino Acid Amides at the C6 Position and an Evaluation of Antiproliferative Activity In Vitro. *Int. J. Mol. Sci.* **2023**, *24*, 6223. [CrossRef] [PubMed]
4. Kuo, C.-H.; Tsai, M.-L.; Wang, H.-M.D.; Liu, Y.-C.; Hsieh, C.; Tsai, Y.-H.; Dong, C.-D.; Huang, C.-Y.; Shieh, C.-J. Continuous production of DHA and EPA ethyl esters via lipase-catalyzed transesterification in an ultrasonic packed-bed bioreactor. *Catalysts* **2022**, *12*, 404. [CrossRef]
5. Baek, Y.; Lee, J.; Son, J.; Lee, T.; Sobhan, A.; Lee, J.; Koo, S.-M.; Shin, W.H.; Oh, J.-M.; Park, C. Enzymatic Synthesis of Formate Ester through Immobilized Lipase and Its Reuse. *Polymers* **2020**, *12*, 1802. [CrossRef] [PubMed]
6. Rinnerthaler, M.; Bischof, J.; Streubel, M.K.; Trost, A.; Richter, K. Oxidative stress in aging human skin. *Biomolecules* **2015**, *5*, 545–589. [CrossRef] [PubMed]
7. Ben Rejeb, I.; Charfi, I.; Baraketi, S.; Hached, H.; Gargouri, M. Bread Surplus: A Cumulative Waste or a Staple Material for High-Value Products? *Molecules* **2022**, *27*, 8410. [CrossRef] [PubMed]
8. Dabija, A.; Ciocan, M.E.; Chetrariu, A.; Codină, G.G. Maize and Sorghum as Raw Materials for Brewing, a Review. *Appl. Sci.* **2021**, *11*, 3139. [CrossRef]
9. Ucak, I.; Afreen, M.; Montesano, D.; Carrillo, C.; Tomasevic, I.; Simal-Gandara, J.; Barba, F.J. Functional and Bioactive Properties of Peptides Derived from Marine Side Streams. *Mar. Drugs* **2021**, *19*, 71. [CrossRef] [PubMed]
10. Mohd Azmi, S.I.; Kumar, P.; Sharma, N.; Sazili, A.Q.; Lee, S.-J.; Ismail-Fitry, M.R. Application of Plant Proteases in Meat Tenderization: Recent Trends and Future Prospects. *Foods* **2023**, *12*, 1336. [CrossRef] [PubMed]

 *catalysts*

Article

Production and Characterization of *Durvillaea antarctica* Enzyme Extract for Antioxidant and Anti-Metabolic Syndrome Effects

Ming-Kuei Shih [1,†], Chih-Yao Hou [2,†], Cheng-Di Dong [3,4,5], Anil Kumar Patel [5], Yung-Hsiang Tsai [2], Mei-Chun Lin [2], Zheng-Ying Xu [2], Pitchurajan Krishna Perumal [5], Chia-Hung Kuo [2,*] and Chun-Yung Huang [2,*]

[1] Graduate Institute of Food Culture and Innovation, National Kaohsiung University of Hospitality and Tourism, No. 1, Songhe Road, Xiaogang District, Kaohsiung City 812301, Taiwan
[2] Department of Seafood Science, National Kaohsiung University of Science and Technology, No. 142, Haijhuan Road, Nanzih District, Kaohsiung City 81157, Taiwan
[3] Department of Marine Environmental Engineering, National Kaohsiung University of Science and Technology, No. 142, Haijhuan Road, Nanzih District, Kaohsiung City 81157, Taiwan
[4] Sustainable Environment Research Center, Department of Marine Environmental Engineering, National Kaohsiung University of Science and Technology, No. 142, Haijhuan Road, Nanzih District, Kaohsiung City 81157, Taiwan
[5] Institute of Aquatic Science and Technology, National Kaohsiung University of Science and Technology, No. 142, Haijhuan Road, Nanzih District, Kaohsiung City 81157, Taiwan
[*] Correspondence: kuoch@nkust.edu.tw (C.-H.K.); cyhuang@nkust.edu.tw (C.-Y.H.); Tel.: +886-7-3617141 (ext. 23646) (C.-H.K.); +886-7-3617141 (ext. 23606) (C.-Y.H.)
[†] These authors contributed equally to this work.

Citation: Shih, M.-K.; Hou, C.-Y.; Dong, C.-D.; Patel, A.K.; Tsai, Y.-H.; Lin, M.-C.; Xu, Z.-Y.; Perumal, P.K.; Kuo, C.-H.; Huang, C.-Y. Production and Characterization of *Durvillaea antarctica* Enzyme Extract for Antioxidant and Anti-Metabolic Syndrome Effects. *Catalysts* **2022**, *12*, 1284. https://doi.org/10.3390/catal12101284

Academic Editor: Francesco Secundo

Received: 22 September 2022
Accepted: 19 October 2022
Published: 21 October 2022

Publisher's Note: MDPI stays neutral with regard to jurisdictional claims in published maps and institutional affiliations.

Abstract: In this study, three enzyme hydrolysate termed Dur-A, Dur-B, and Dur-C, were produced from *Durvillaea antarctica* biomass using viscozyme, cellulase, and α-amylase, respectively. Dur-A, Dur-B, and Dur-C, exhibited fucose-containing sulfated polysaccharide from chemical composition determination and characterization by FTIR analyses. In addition, Dur-A, Dur-B, and Dur-C, had high extraction yields and low molecular weights. All extracts determined to have antioxidant activities by DPPH (2,2-diphenyl-1-picrylhydrazyl), ABTS (2,20-azino-bis(3-ethylbenzothiazoline-6-sulfonic acid) diammonium salt), and ferrous ion-chelating methods. All extracts were also able to positively suppress the activities of key enzymes involved in metabolic syndrome: angiotensin I-converting enzyme (ACE), α-amylase, α-glucosidase, and pancreatic lipase. In general, Dur-B exhibited higher antioxidant and higher anti-metabolic syndrome effects as compared to the other two extracts. Based on the above health promoting properties, these extracts (especially Dur-B) can be used as potential natural antioxidants and natural anti-metabolic syndrome agents in a variety of food, cosmetic, and nutraceutical products for health applications.

Keywords: antioxidant; *Durvillaea antarctica*; enzymatic extraction; fucose-containing sulfated polysaccharide; metabolic syndrome

1. Introduction

Oxidative stress occurs when the net amount of reactive oxygen species (ROS) exceeds the antioxidant capacity. This may happen as a result of a general increase in ROS generation, a depression of antioxidant systems, or both [1]. ROS such as superoxide anion radical ($O2^{-\bullet}$), hydrogen peroxide (H_2O_2), hydroxyl radical ($^\bullet OH$), singlet oxygen ($^1O_2$), and nitric oxide ($NO^\bullet$), are metabolic products and may exist in the environment [2]. Excessive production of ROS may damage cellular DNA, proteins, and lipids, resulting in altered biochemical compounds, corroded cell membranes, and possibly even metabolic syndrome [1,3]. Thus, antioxidants are needed to delay or prevent the oxidation of cellular oxidizable substrates by promoting scavenging of ROS and preventing the production of ROS [4].

Diabetes mellitus is a chronic disease characterized by hyperglycemia, a condition in which blood glucose levels are elevated. There are two types of diabetes mellitus, type 1 and type 2, which differ in insulin production and utilization. Type 1 diabetes mellitus is caused by insufficient insulin production by the pancreas, as a result of autoimmune disease. The disease is managed by insulin injections. In type 2 diabetes, there is sufficient insulin but the body's cells are insulin resistant, a condition which is thought to be caused by multiple factors such as overweight, high cholesterol, and physical inactivity. The vast majority (95%) of diabetic patients have type 2, and strong associations with dietary intake and reduced physical activity have been reported [5]. Metabolic syndrome is a constellation of abnormalities, which include abdominal obesity, dyslipidemia, high blood glucose/impaired glucose tolerance, and high blood pressure. It has been well established that metabolic syndrome increases the risk of overt diabetes mellitus [6]. The prevalence of cardiovascular disease (CVD) is rising rapidly, regardless of age and gender, and the mortality rate of CVD is high. CVD kills more than 18 million people worldwide annually [7]. In fact, CVD is not a single disease; rather it is a group of disorders related to the heart and blood vessels, and includes coronary heart disease, rheumatic heart disease, and cerebrovascular disease, as well as other conditions. Approximately 80% of CVD-related deaths are due to cardiac arrest and strokes, which involve the deposition of fat in blood vessels. A number of other factors are also thought to play a role in CVDs, such as a poor diet, high intake of oils, alcohol consumption, hereditary factors, and low physical activity [8]. Obesity is a metabolic condition in which the body has gained excess weight, primarily as a consequence of the deposition of excess fat. It is considered a gateway for various other metabolic syndromes, such as insulin resistance, dyslipidemia, hypertension, hyperthyroidism, impaired glucose metabolism, and stroke [9], which have been shown to be associated with CVD. Increased oxidative stress has been identified as a key factor in metabolic syndrome and its related pathologies, and it has been postulated that oxidative stress is mechanistically involved in the progression of this disease. ROS play an important role in the development and progression of CVDs [10]. Furthermore, oxidative stress has been shown to play an important role in the mechanism of micro- and macrovascular complications in metabolic syndrome [11]. Hence, it is crucially important to identify and develop antioxidants (particularly those from naturally occurring sources) that are capable of eliminating ROS. Antioxidant-containing nutraceuticals could be useful in treating or preventing metabolic syndrome-related disorders.

In recent decades, there has been a growing awareness that traditional foods and naturally occurring compounds could be hugely beneficial to digestion and other aspects of human health. Indeed, the importance of consuming complete and nutritious food is increasingly being recognized [12,13]. Marine algae and seaweeds have attracted considerable interest due to their nutritional components and highly bioactive properties. Seaweeds are a part of the traditional diet of coastal-dwelling cultures, particularly in Asia, and are also used in folk medicines. It has been reported that Japan, North Korea, Taiwan, and China, are the largest consumers of seaweeds in Asia. Japanese people consume approximately 5.3 g seaweed per day^{-1}. Interestingly, Japan also has one of the longest lifespans worldwide, as well as low rates of CVDs [5,8,9]. At the cellular level, seaweed-derived polysaccharides exhibit high biological activity, such as antioxidant, anti-cancer, anti-inflammatory, and immunomodulatory properties, although these activities are limited by various characteristics, including molecular weight, glycosidic bonds, solubility, spatial configuration, and main-chain configuration [14]. Thus, these biological activities can be enhanced by applying various molecular modification techniques that can be classified into physical, chemical, and biological methods. The physical method involves reducing the molecular weight of polysaccharides by breaking the glucoside chain using ultrasonic disruption [15], radiation [16], or microwave denaturation [17]. These high energies result in denaturation of glucoside bonds, reducing them to oligomers and monomers. In the chemical method, functional groups such as sulfate, phosphate, selenium, iron, and alkyl groups, are introduced to the polysaccharides, which improves solubility and bioactivity [14]. The biological method also makes use of molecular modification, which is achieved by enzymatic hydrolysis of glycosidic bonds. In comparison with the two aforementioned methods, the enzymatic method is more accurate, more efficient, highly specific, has a higher yield-coefficient, releases novel bio-compounds, and has few side effects. Moreover, a recent study suggested that the enzyme-assisted method promotes the bioactivity of cell wall polymers during extraction [18].

In the present study, various hydrolysates of *D. antarctica* were produced using the enzyme-assisted extraction (EAE) method, and their antioxidant and inhibitory activities against key metabolic syndrome-related enzymes (ACE, α-amylase, α-glucosidase, and pancreatic lipase) were investigated. The EAE method has attracted a lot of attention in the past decade as it exhibits high selectivity and it allows a large amount of biological active compounds to be recovered [19]. Kulshreshtha et al. (2015) employed commercial protease and carbohydrases to extract bioactive materials from *Codium fragile* and *Chondrus crispus* and achieved an extraction yield of 40–70% (dry basis) [19]. Nguyen et al. (2020) applied combined commercial enzymes, cellulase (Cellic® CTec2) and alginate lyase, which were derived from *Sphingomonas* sp. in a single-step process, to extract fucoidan from brown seaweeds, *Fucus distichus* subsp. *evanescens* and *Saccharina latissimi* [20], and reported an extraction yield of 29–40%. Moreover, Olivares-Molina and Fernández (2016) used an enzymatic method (cellulase and α-amylase) to extract compounds from three brown seaweeds, *Lessonia nigrescens*, *D. antarctica*, and *Macrocystis pyrifera*, and found the extraction yields ranged from 9–38% [18]. Herein, we used viscozyme, cellulose, and α-amylase to extract bioactive substances from *D. antarctica*. Viscozyme® L (Novozyme Corp.) is a blend of β-glucanases, hemicellulases, pectinases, and xylanases, which is a cell wall-degrading enzyme complex produced by *Aspergillus* sp. Cellulase is a member of the glycoside hydrolase family, and is secreted by a number of cellulolytic microorganisms. Cellulase produced by *Aspergillus niger* catalyzes the hydrolysis of endo-1,4-β-D-glycosidic linkages in cellulose, cereal beta-D-glucans, lichenin, and the cellooligosaccharides cellotriose to cellohexaose. As cellulose molecules are strongly bound to one another, it is more difficult to break it down by cellulolysis compared with polysaccharides such as starch. Microbial amylases are exoenzymes that are utilized in various industrial applications, such as bread making, maltose syrups, and the fermentation of soya sauce, miso, and so on. The α-amylase derived from *Aspergillus oryzae* acts as a catalyst in the hydrolysis of α-1,4 glycosidic bonds in soluble starches and other related substrates. These substrates are then further broken down, releasing short oligosaccharides and α-limit dextrins.

Different enzymes exert different degrees of action on target polysaccharides, and the activities of these enzymes have been analyzed to determine the optimum conditions [14,21]. In the present study, *Durvillaea antarctica*, a type of brown seaweed, was investigated in an experimental model. Seaweed samples were hydrolyzed using three different enzymes, viscozyme, cellulase, and α-amylase, while maintaining a uniform condition in order to compare the physicochemical properties of the hydrolysates and their biological activities. As this study was specifically concerned with investigating antioxidants and their potential role in metabolic syndrome-related enzyme inhibition, various antioxidant assays such as DPPH (2,2-diphenyl-1-picrylhydrazyl), ABTS (2,20-azino-bis(3-ethylbenzothiazoline-6-sulfonic acid) diammonium salt), and ferrous ion-chelating ability, were employed. The physiochemical characterization of hydrolyzed products was performed using FTIR and NMR techniques to analyze functional groups. Chemical composition, monosaccharide composition, and molecular weight analyses, were performed by various assays and HPLC techniques. Finally, the metabolic syndrome-related enzyme inhibition studies were conducted by evaluating the inhibition of angiotensin I-converting enzyme (ACE), α-glucosidase, α-amylase, and pancreatic lipase enzymes. The overall aim of these experiments was to explore the potential of enzymatic hydrolysates from *D. antarctica* to serve as a natural antioxidant for use in the food, cosmetic, and nutraceutical industries and which may be beneficial in the prevention or treatment of metabolic syndrome.

2. Results

2.1. Preparation of Enzymatic Extracts (Dur-A, Dur-B, and Dur-C) from D. antarctica, and Physicochemical Characteristics of Dur-A, Dur-B, and Dur-C

In the current study, *D. antarctica* was utilized for hydrolysis using viscozyme, cellulase, or α-amylase in order to obtain hydrolysis products, namely Dur-A, Dur-B, and Dur-C. Table 1 shows the hydrolysis variables of Dur-A, Dur-B, and Dur-C, and the extraction yields of Dur-A, Dur-B, and Dur-C were 38.4% ± 1.1%, 43.0% ± 0.2%, and 41.4% ± 0.6%, respectively, suggesting that Dur-B and Dur-C have higher extraction yields than that of Dur-A. However, the difference was not remarkable. Table 2 shows the results of the physicochemical properties of Dur-A, Dur-B, and Dur-C. The size exclusion chromatographic method revealed the MW distribution in Dur-A, Dur-B, and Dur-C (Table 2). Among these three hydrolysis products, two peaks were observed. For Dur-A, the peak for higher MW was 171.4 kDa, with a peak area of approximately 99.5% and the peak for lower MW was 8.61 kDa, with a peak area of approximately 0.5%. For Dur-B, the peak for higher MW was 215.9 kDa with a peak area approximately 99.5%, and the peak for lower MW was 10.0 kDa, with a peak area of approximately 0.5%. For Dur-C, the peak for higher MW was 183.3 kDa, with a peak area of approximately 51.6%, and the peak for lower MW was 2.19 kDa, with a peak area approximately 48.4%. These data indicate that Dur-C had more oligosaccharides (the molecular weight was about 2.19 kDa) as compared to Dur-A and Dur-B. Total sugar, fucose, sulfate, uronic acid, alginic acid, polyphenols, and proteins contents, as well as the monosaccharide composition of Dur-A, Dur-B, and Dur-C, are presented in Table 2. The total sugar contents of Dur-A, Dur-B, and Dur-C were 44.4% ± 0.2%, 40.4% ± 0.3%, and 72.3% ± 0.6%, respectively. The fucose contents of Dur-A, Dur-B, and Dur-C were 13.5% ± 0.5%, 18.2% ± 0.3%, and 19.9% ± 0.3%, respectively. In general, Dur-C had the highest total sugar content and fucose content as compared to Dur-A and Dur-B. The sulfate contents of Dur-A, Dur-B, and Dur-C were 38.5% ± 0.0%, 41.2% ± 0.0%, and 24.7% ± 0.1%, respectively. The percentage contents of uronic acid in Dur-A, Dur-B, and Dur-C were 16.9% ± 0.3% 16.7% ± 0.6%, and 16.8% ± 0.2%, respectively. The alginic acid content percentages of Dur-A, Dur-B, and Dur-C were 6.99% ± 0.20%, 6.50% ± 0.20%, and 7.27% ± 0.54%, respectively. The percentage contents of polyphenols in Dur-A, Dur-B, and Dur-C were 0.46% ± 0.02%, 0.39% ± 0.02%, and 0.42% ± 0.04%, respectively. Moreover, the percentage contents of protein in Dur-A, Dur-B, and Dur-C were 1.08% ± 0.02%, 0.93% ± 0.03%, and 0.97% ± 0.03%, respectively. Generally, Dur-B had the highest sulfate content, and the uronic acid content, alginic acid content, polyphenols content, and proteins content in Dur-A, Dur-B, and Dur-C were similar. Table 2 also shows the monosaccharide compositions of Dur-A, Dur-B, and Dur-C. In brief, fucose, rhamnose, and glucuronic acid were the most prevalent sugar units in Dur-A. Fucose, rhamnose, and galacturonic acid were the most prevalent sugar units in Dur-B; fucose, rhamnose, galacturonic acid, glucose, and xylose were the most prevalent sugar units in Dur-C. Taken together, the monosaccharide compositions of Dur-A, Dur-B, and Dur-C varied. In summary, Dur-A, Dur-B, and Dur-C, had low molecular weights and showed the characteristic compositions of fucose-containing sulfated polysaccharides.

FTIR and NMR techniques were used to perform structural analyses of Dur-A, Dur-B, and Dur-C. Figure 1 shows the IR bands at 3401 cm^{-1} and 2940 cm^{-1} which correspond to OH and H_2O stretching vibration and C–H stretching of the pyranoid ring or the C-6 group of the fucose and galactose units [22,23]. Absorption bands were detected at 1621 cm^{-1} and 1421 cm^{-1} indicating the scissoring vibration of H_2O and vibrations of the in-plane ring CCH, COH, and OCH, which is a known absorption pattern of polysaccharides [22–24]. The peaks at 1230 cm^{-1} and 1055 cm^{-1} can be attributed to the presence of asymmetric stretching of S=O and C–O–C stretching vibrations in the ring or C–O–H of the glucosidal bond [22,23]. Bands at 900 cm^{-1} corresponded to C1–H bending vibration in β-anomeric units (probably galactose) [25]. The peak at 820 cm^{-1} corresponded to the bending vibrations of C-O-S of sulfate [26]. The bands at 620 cm^{-1} may correspond to symmetric O=S=O deformations [25]. Figure 1 provides evidence that Dur-A, Dur-B, and Dur-C exhibited the characteristic IR absorptions of fucose-containing sulfated polysaccharides, but the differences in their FTIR results are not obvious. Figure S1A depicts the ^{1}H-NMR spectra of Dur-A, Dur-B, and Dur-C. The signals at 4.57 ppm and 4.46 ppm indicate the presence of H-2 in a 2-sulfated fucopyranose residue [27], and the signal at 4.13 ppm (4[H]) is suggestive of the presence of 3-linked α-L-fucose [24]. Signals with a ppm of 4.07/3.95 (6[H]/6′[H]) denote the presence of a (1-6)-β-D-linked galacton [28]. The signals detected at 3.78 ppm and 3.72 ppm may be indicative of the existence of (3[H]) 4-linked β-D-galactose and (4[H]) 2,3-linked α-β-mannose, respectively [24]. In addition, the signal at 1.32 ppm indicates the presence of C6 methyl protons of L-fucopyranose [29]. Due to the dissimilarity of the ^{1}H-NMR spectra, it can thus be deduced that Dur-A, Dur-B, and Dur-C may have different structures. In addition, Figure S1B shows the ^{13}C-NMR spectra of Dur-A, Dur-B, and Dur-C. The signals at 75.86 ppm attributes to the presence of C-5 of M-block in a mannuronate homo-oligomer [30], indicating the mannuronic acid of alginic acid is probably detectable in Dur-A, Dur-B, and Dur-C.

Table 1. Hydrolysis variables and extraction yields of enzymatic hydrolysis products (Dur-A, Dur-B, and Dur-C) from *D. antarctica*.

Variables of Hydrolysis	Dur-A	Dur-B	Dur-C
Enzyme used	Viscozyme	Cellulase	α-Amylase
Hydrolysis conditions	pH 6.0, 40 °C, 17 h	pH 6.0, 40 °C, 17 h	pH 6.0, 40 °C, 17 h
Extraction Yield	**Dur-A**	**Dur-B**	**Dur-C**
Extraction yield (%)	38.4 ± 1.1 [1,a]	43.0 ± 0.2 [b]	41.4 ± 0.6 [b]

[1] Values are mean ± SD (n = 3); values in the same row with different letters (in [a] and [b]) are significantly different ($p < 0.05$).

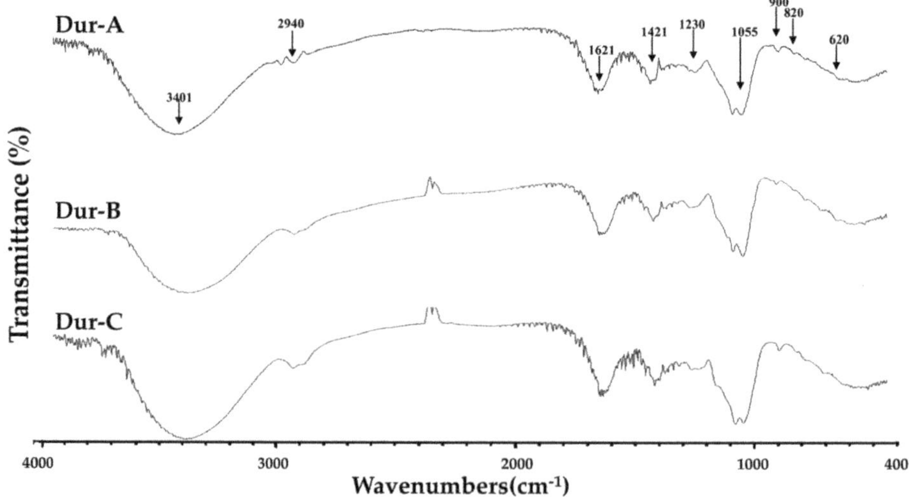

Figure 1. FTIR spectra for Dur-A, Dur-B, and Dur-C. Absorption bands at 3401, 2940, 1621, 1421, 1230, 1055, 900, 820, and 620 cm^{-1} are indicated.

Table 2. Physicochemical analyses for Dur-A, Dur-B, and Dur-C.

Molecular Weight	Dur-A	Dur-B	Dur-C
Peak 1 (MW (kDa)/Peak area (%))	171.4/99.5	215.9/99.5	183.3/51.6
Peak 2 (MW (kDa)/Peak area (%))	8.61/0.5	10.0/0.5	2.19/48.4
Chemical Composition	**Dur-A**	**Dur-B**	**Dur-C**
Total sugar (%) [1]	44.4 ± 0.2 [b]	40.4 ± 0.3 [a]	72.3 ± 0.6 [c]
Fucose (%) [1]	13.5 ± 0.5 [a]	18.2 ± 0.3 [b]	19.9 ± 0.3 [c]
Sulfate (%) [1]	38.5 ± 0.0 [b]	41.2 ± 0.0 [c]	24.7 ± 0.1 [a]
Uronic acid (%) [1]	16.9 ± 0.3 [a]	16.7 ± 0.6 [a]	16.8 ± 0.2 [a]
Alginic acid (%) [1]	6.99 ± 0.20 [a]	6.50 ± 0.20 [a]	7.27 ± 0.54 [a]
Polyphenols (%) [1]	0.46 ± 0.02 [a]	0.39 ± 0.02 [a]	0.42 ± 0.04 [a]
Proteins (%) [1]	1.08 ± 0.02 [b]	0.93 ± 0.03 [a]	0.97 ± 0.03 [a]
Monosaccharide Composition (Molar Ratio)	**Dur-A**	**Dur-B**	**Dur-C**
Fucose	1	1	1
Rhamnose	0.16	1.71	1.69
Glucuronic acid	0.12	0.02	0.05
Galacturonic acid	0.08	0.19	0.23
Glucose	0.05	0.05	0.79
Galactose	0.03	0.01	0.01
Xylose	0.02	0.05	0.16

[1] Total sugars (%), fucose (%), sulfate (%), uronic acid (%), alginic acid (%), polyphenols (%), and protein (%), = $(g/g_{sample, dry basis}) \times 100$; Values are mean ± SD (n = 3); values in the same row with different letters (in [a], [b], and [c]) are significantly different ($p < 0.05$).

2.2. Effect of Enzymatic Extracts (Dur-A, Dur-B, and Dur-C) on Antioxidant Activities

The antioxidant activities of Dur-A, Dur-B, and Dur-C, were examined by DPPH, ABTS, and ferrous ion-chelating analyses, and the data are presented in Figure 2. The scavenging effects of Dur-A, Dur-B, and Dur-C on DPPH free radicals are shown in Figure 2A. These extracts exhibited DPPH scavenging properties in a dose-dependent manner. The observed results also suggest that Dur-B has the most potent effect on DPPH scavenging ability, followed by Dur-A, and then Dur-C. The scavenging effects of Dur-A, Dur-B, and Dur-C on ABTS$^{\bullet+}$ free radicals are shown in Figure 2B. These extracts exhibited ABTS$^{\bullet+}$ scavenging properties in a dose-dependent manner. The observed results suggest that Dur-A has the most potent effect on ABTS$^{\bullet+}$ radical scavenging ability, followed by Dur-B, and then Dur-C. The scavenging effects of Dur-A, Dur-B, and Dur-C on ferrous ion-chelating are shown in Figure 2C. These extracts exhibited ferrous ion-chelating activities in a dose-dependent manner. The observed results suggest that Dur-B has the most potent effect on ferrous ion-chelating activity, followed by Dur-A, and then Dur-C. All of these extracts showed antioxidant activities and thus further investigations are warranted to explore their anti-metabolic syndrome capacities.

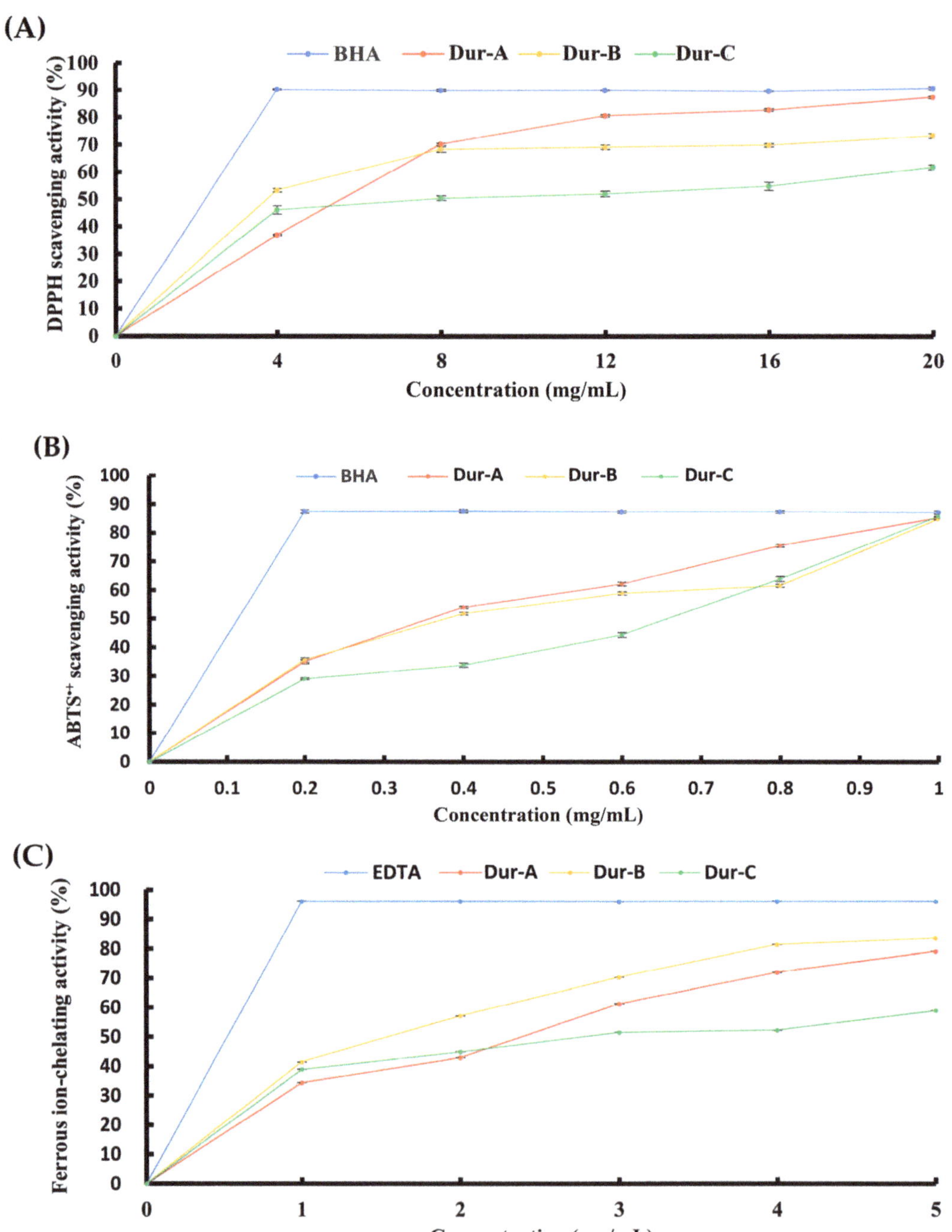

Figure 2. Antioxidant activities of Dur-A, Dur-B, and Dur-C. (**A**) DPPH radical-scavenging activity for Dur-A, Dur-B, Dur-C, and BHA. (**B**) ABTS radical-scavenging activity for Dur-A, Dur-B, Dur-C, and BHA. (**C**) Ferrous ion-chelating activity for Dur-A, Dur-B, Dur-C, and EDTA.

2.3. Effect of Enzymatic Extracts (Dur-A, Dur-B, and Dur-C) on the Inhibition against Key Metabolic Syndrome-Related Enzymes

The inhibitory activities against key metabolic syndrome-related enzymes of Dur-A, Dur-B, and Dur-C were investigated, and the results are shown in Table 3. For the inhibition of ACE, the concentration of Dur-A, Dur-B, and Dur-C used was 1 mg/mL, and the inhibitory rates for Dur-A, Dur-B, and Dur-C were 72.5% ± 1.4%, 80.7% ± 1.6%, and 62.9% ± 0.6%, respectively. The observed results suggest that Dur-B had the most ACE inhibitory activity, followed by Dur-A, and then Dur-C. For the inhibition of α-amylase, the concentration of Dur-A, Dur-B, and Dur-C used was 3 mg/mL, and the inhibitory rates for Dur-A, Dur-B, and Dur-C were 53.3% ± 0.4%, 69.8% ± 0.5%, and 61.6% ± 1.0%, respectively. The observed results suggest that Dur-B had the most α-amylase inhibitory activity, followed by Dur-C, and then Dur-A. For the inhibition of α-glucosidase, the concentration of Dur-A, Dur-B, and Dur-C used was 3 mg/mL, and the inhibitory rates for Dur-A, Dur-B, and Dur-C were 26.9% ± 0.6%, 30.1% ± 4.8%, and 60.6% ± 0.9%, respectively. The observed results suggest that Dur-C has the most α-glucosidase inhibitory activity, followed by Dur-A and Dur-B. For the inhibition of pancreatic lipase, the concentration of Dur-A, Dur-B, and Dur-C used was 0.3 mg/mL, and the inhibitory rates for Dur-A, Dur-B, and Dur-C were 38.1% ± 5.0%, 33.6% ± 6.0%, and 17.1% ± 3.0%, respectively. The observed results suggest that Dur-A and Dur-B have better pancreatic lipase inhibitory activity than Dur-C. In summary, all extracts showed inhibitory effects on metabolic syndrome-related enzymes.

Table 3. Inhibitory activities against key metabolic syndrome-related enzymes of Dur-A, Dur-B, and Dur-C.

Inhibitory Activities	Relative Inhibitors [1]	Dur-A	Dur-B	Dur-C
Angiotensin I-converting enzyme (ACE) [2]	93.6 ± 0.9 [2,3,d]	72.5 ± 1.4 [b]	80.7 ± 1.6 [c]	62.9 ± 0.6 [a]
α-Amylase [2]	81.7 ± 0.4 [2,d]	53.3 ± 0.4 [a]	69.8 ± 0.5 [c]	61.6 ± 1.0 [b]
α-Glucosidase [2]	97.7 ± 0.2 [2,c]	26.9 ± 0.6 [a]	30.1 ± 4.8 [a]	60.6 ± 0.9 [b]
Pancreatic lipase [2]	95.8 ± 2.5 [2,c]	38.1 ± 5.0 [b]	33.6 ± 6.0 [b]	17.1 ± 3.0 [a]

[1] Inhibitor for Angiotensin I-Converting Enzyme: captopril (1.10 μg/mL); inhibitor for α-amylase: acarbose (0.25 mg/mL); inhibitor for α-glucosidase: acarbose (0.60 mg/mL); inhibitor for lipase: orlistat (20 μg/mL). [2] For the inhibition of angiotensin I-converting enzyme, the concentration of Dur-A, Dur-B, and Dur-C used was 1 mg/mL; for the inhibition of α-amylase, the concentration of Dur-A, Dur-B, and Dur-C used was 3 mg/mL; for the inhibition of α-glucosidase, the concentration of Dur-A, Dur-B, and Dur-C used was 3 mg/mL; for the inhibition of pancreatic lipase, the concentration of Dur-A, Dur-B, and Dur-C used was 0.3 mg/mL. [3] Values are mean ± SD (n = 3); values in the same row with different letters (in [a], [b], [c], and [d]) are significantly different ($p < 0.05$).

3. Discussion

In this study, we investigated the antioxidant and inhibitory activities of enzymatic extract of *D. antarctica* biomass against key metabolic syndrome-related enzymes (ACE, α-amylase, α-glucosidase, and pancreatic lipase), which were extracted using the EAE method. The extraction yields of *D. antarctica* obtained with viscozyme (Dur-A), cellulase (Dur-B), and α-amylase (Dur-C), were 38.4%, 43.0%, and 41.4%, respectively (Table 1), indicating that Dur-B and Dur-C had higher extraction yields than that of Dur-A. However, the difference in yield was not significant. Hammed et al. (2017) investigated viscozyme, cellulase, and amyloglucosidase, as potential enzymes for the extraction of sulfated polysaccharides from the brown seaweed *Turbinaria turbinate* under different extraction conditions. This study obtained a maximum extraction yield of about 24–25% [21]. In short, the present study used an extraction model that produced remarkable extraction yields that were similar or superior to previously reported findings.

With respect to bioactive polysaccharides, the molecular structures, sulfate content, type of sugar, and molecular weight, play important roles in their biological functions.

Among these variables, the molecular weight of polysaccharides appears to be the most important factor that determines the biological characteristics [31]. High molecular weight polysaccharides do not exhibit good solubility and cannot be processed easily, thereby limiting their capacity to penetrate cells. However, polysaccharides that have a low molecular weight possess more biological functions, such as immunostimulation, anticancer, antioxidant activities, and anticoagulation [2,32,33]. The EAE method has been successfully used to obtain LMW polysaccharides with superior biological functions as compared to compounds obtained using conventional hot-water extraction [34]. In the current study, among the three extracts, higher molecular weight distribution was between 171.4 kDa and 215.9 kDa, and the lower molecular weight distribution was found between 2.19 kDa and 10.0 kDa (Table 2). It has been reported that the molecular weight of a fucose-containing sulfated polysaccharide extracted by hot water was approximately 690.8 kDa (higher molecular weight) and 327.1 kDa (lower molecular weight) [35], and therefore the EAE method applied herein facilitated the production of polysaccharides with lower molecular weights. Furthermore, Dur-C contained more short oligosaccharides (approx. 2.19 kDa) in comparison with those of Dur-A and Dur-B (Table 2), suggesting that α-amylase was more effective for releasing short oligosaccharides compared to the other two enzymes. Whether the oligosaccharides content of extracts is related to their biological functions remains to be elucidated.

Table 2 depicts the chemical compositions of Dur-A, Dur-B, and Dur-C. It can be seen that Dur-C has the highest total sugar content and fucose content, but has the lowest sulfate content, in comparison with Dur-A and Dur-B. Other compounds, such as uronic acid content, polyphenols content, and alginic acid content, showed similar amounts among Dur-A, Dur-B, and Dur-C. Regarding monosaccharide compounds (Table 2), the sugars in Dur-A were largely fucose, rhamnose, and glucuronic acid; the sugars in Dur-B were predominantly fucose, rhamnose, and galacturonic acid; in Dur-C, the predominant sugars were fucose, rhamnose, galacturonic acid, xylose, and glucose. The results imply that these extracts are largely fucose-containing sulfated polysaccharides. FTIR is a powerful tool that is used for the detection of a range of functional groups. It can also provide information on the chemical composition of compounds [36]. The data shown in Figure 1 indicate that Dur-A, Dur-B, and Dur-C, exhibit the characteristic IR peaks of fucose-containing sulfated polysaccharide. We then assessed the ^{1}H-NMR spectra of Dur-A, Dur-B, and Dur-C (Figure S1A), and the results showed notable differences among Dur-A, Dur-B, and Dur-C. As there were differences in total sugar content, fucose content, sulfate content, monosaccharide composition, and structure among Dur-A, Dur-B, and Dur-C, we were interested in further characterizing their biological activities, particularly with respect to their antioxidant properties and their potential to exert mitigating effects against metabolic syndrome.

We evaluated the antioxidant activities of Dur-A, Dur-B, and Dur-C, using DPPH, ABTS, and ferrous ion-chelating analyses. DPPH is a widely used technique for evaluating antioxidant activity within a relatively short duration [37]. Figure 2A shows the DPPH radical-scavenging activities of Dur-A, Dur-B, and Dur-C, and BHA (as a reference). The results reveal that all extracts exhibited DPPH radical-scavenging activity dose-dependently. The IC_{50} values (the concentration of a sample capable of scavenging 50% of DPPH radicals) of the extracted samples (Dur-A, Dur-B, and Dur-C) on DPPH radical-scavenging activity were 5.60, 3.78, and 7.52 mg/mL, respectively (Figure 2A). Hence, Dur-B showed the highest DPPH radical-scavenging activity, followed by Dur-A, and then Dur-C. The key mechanism of the ABTS radical decolorization assay involves the decolorization of $ABTS^{\bullet+}$ when it reacts with a hydrogen-donating antioxidant [38]. Figure 2B shows the $ABTS^{\bullet+}$ scavenging properties of Dur-A, Dur-B, and Dur-C, and BHA (as a reference). In all extracts, the $ABTS^{\bullet+}$ scavenging activity was found to have occurred dose-dependently. The IC_{50} values (the concentration of a sample capable of scavenging 50% of $ABTS^{\bullet+}$) of the extracted samples (Dur-A, Dur-B, and Dur-C) on $ABTS^{\bullet+}$ scavenging activity were found to be 0.36, 0.38, and 0.66 mg/mL, respectively (Figure 2B). Therefore, Dur-A showed the highest

ABTS$^{\bullet+}$ scavenging activity, followed by Dur-B, and then Dur-C. However, there was no notable difference between Dur-A and Dur-B. Ferrozine is capable of quantitatively forming complexes with Fe^{2+}, and absorbance is seen at 562 nm. In the presence of a chelating agent, the complexes are disrupted resulting in a decrease in the red color of the complexes. Hence, measuring the color reduction is an indicator of the ferrous ion-chelating effect [39]. Ferrous ions have been shown to stimulate lipid peroxidation and are known to be an effective pro-oxidant in food systems [40]. Figure 2C depicts the ferrous ion-chelating characteristics of Dur-A, Dur-B, and Dur-C, and EDTA (as a reference). The results reveal that all of the extracts demonstrated ferrous ion-chelating activity in a dose-dependent manner. The IC_{50} values (the concentration of a sample capable of chelating 50% of ferrous ion) of the extracted samples (Dur-A, Dur-B, and Dur-C) on ferrous ion-chelating activity were found to be 2.40, 1.55, and 2.78 mg/mL, respectively (Figure 2C). Thus, Dur-B showed the highest ferrous ion-chelating activity, followed by Dur-A, and then Dur-C. These results, taken together, demonstrate that all three of the tested extracts (Dur-A, Dur-B, and Dur-C) possessed antioxidant capabilities. Overall, the antioxidant activities of Dur-B were superior in comparison with the other extracts. The high antioxidant activities of Dur-B may be related to its high sulfate content (Table 2). Further studies on oversulfation or desulfation of polysaccharides are required to better understand the antioxidant properties of Dur-B.

In previous studies, it has been suggested that the predominant external sources of oxidative stress, i.e., cigarette smoke and air pollution, are associated with the development of metabolic syndrome and CAD. There is also evidence that the antioxidant characteristics of pharmacological agents, such as statins, metformin, angiotensin II type I receptor blockers (ARBs), and ACE inhibitors, are helpful in preventing and treating cardiovascular complications of metabolic syndrome [10]. As Dur-A, Dur-B, and Dur-C, demonstrated antioxidant capabilities, we conducted additional experiments to explore their potential to combat metabolic syndrome.

Metabolic syndrome is a pathological condition that is characterized by dysfunctional metabolism of fats, proteins, carbohydrates, and other substances in the body. It is thought that metabolic syndrome is a precursor to cardiovascular and cerebrovascular diseases and diabetes. The preventive and therapeutic strategies that are applied in the management of these life-threatening ailments typically involve the use of synthetic drugs, which often cause adverse side effects. Therefore, nutritional modalities have been explored as alternative approaches aimed at preventing or managing metabolic syndrome [41]. Fucose-containing sulfated polysaccharide has been reported to ameliorate metabolic syndrome-related disorders, such as hypertension, obesity, hyperglycemia, and hyperlipidemia, via various regulatory mechanisms [41]. Hence, we investigated the inhibitory effects of Dur-A, Dur-B, and Dur-C, on metabolic syndrome-related enzymes, including ACE, α-glucosidase, α-amylase, and pancreatic lipase. ACE plays a vital role in blood pressure regulation and normal cardiovascular function. It catalyzes the conversion of angiotensin I to angiotensin II, which raises blood pressure. Therefore, inhibiting ACE could be useful in the control of hypertension [42]. The findings displayed in Table 3 indicate that for captopril, a positive control, the ACE inhibitory activity was 93.6% at a concentration of 1.10 μg/mL. Previous studies suggested that the IC_{50} values of crude extracts obtained from *Sargassum siliquosum* and *Sargassum polycystum* biomass ranged between 0.03–1.53 mg/mL for ACE inhibitory activity [43]. Thus, a concentration of 1 mg/mL for Dur-A, Dur-B, and Dur-C was used for ACE inhibitory experiments. The ACE inhibitory activity of our extracted samples (Dur-A, Dur-B, and Dur-C) were 72.5%, 80.7%, and 62.9%, at a concentration of 1 mg/mL (Table 3). Among the studied samples, Dur-B displayed the most potent ACE inhibitory activity, which was consistent with its antioxidant activities (Figure 2) and sulfate content (Table 2), both of which were highest in Dur-B. The most widely applied therapeutic approach in the treatment of type 2 diabetes mellitus is to reduce postprandial hyperglycemia. The role of α-amylase is to break up large polysaccharides into sugars. Naturally occurring α-amylase inhibitors, therefore, could be a useful therapeutic tool for treating postprandial hyperglycemia, as they could decrease the release of glucose from starch [44]. α-glucosidase is an enzyme that hydrolyses starch and disaccharides to glucose. The inhibition of α-glucosidase has been applied in the management of type 2 diabetes [45]. α-glucosidase inhibitors could blunt the rapid hydrolysis of dietary carbohydrates, thereby suppressing postprandial hyperglycemia. The inhibitory effects of Dur-A, Dur-B, Dur-C, and acarbose (a commercial inhibitor of α-amylase and α-glucosidase) on digestive enzymes (α-amylase and α-glucosidase), are displayed in Table 3. Regarding the inhibition of α-amylase by acarbose, at a concentration of 0.25 mg/mL, the inhibitory activity was 81.7%. Previous studies suggested that the IC_{50} values of fucoidan obtained from *Ascophyllum nodosum* ranged between 0.12–4.64 mg/mL [46]. Thus, a concentration of 3 mg/mL for Dur-A, Dur-B, and Dur-C, was used for α-amylase inhibitory experiments. The α-amylase inhibitory activity of the extracted samples in this study (Dur-A, Dur-B, and Dur-C) were 53.3%, 69.8%, and 61.6%, at a concentration of 3 mg/mL (Table 3). Among the tested samples, Dur-B showed the strongest α-amylase inhibitory activity, which was in the line with its antioxidant activities (Figure 2) and sulfate content (Table 2), both of which were highest for Dur-B. For the inhibition of α-glucosidase by acarbose, at a concentration of 0.60 mg/mL, the inhibitory activity was 97.7%. For the ease of comparison with α-amylase, 3 mg/mL concentration of Dur-A, Dur-B, and Dur-C, was applied for α-glucosidase inhibitory experiments. The α-glucosidase inhibitory activity of the extracted samples (Dur-A, Dur-B, and Dur-C) were 26.9%, 30.1%, and 60.6%, at a concentration of 3 mg/mL (Table 3). Among the tested samples, Dur-C displayed the most potent α-glucosidase inhibitory activity, followed by Dur-B and Dur-A. Lipase, which is predominantly produced in the pancreas, hydrolyses lipids to form fatty acids, which are then absorbed in the digestive tract [45]. In order for fat to be absorbed it must first be digested, and therefore, the inhibition of pancreatic lipase, the most important enzyme in the digestion of dietary triglycerides, is one of the various methods that can be employed to reduce the uptake of fat. This inhibitory approach can be utilized to manage weight and potentially reduce the severity of obesity [47]. Fucose-containing sulfated polysaccharide has been shown to have an excellent attenuating effect on lipid accumulation in 3T3-L1 adipocytes [48]. As shown in Table 3, orlistat, which was used as a positive control, at a concentration of 20 μg/mL, had a pancreatic lipase inhibitory activity of 95.8%. Previous studies suggested that the IC_{50} values of extracts obtained from *Ascophyllum nodosum, Fucus vesiculosus, and Pelvetia canaliculata* ranged between 0.119–0.969 mg/mL for lipase inhibitory activity [49]. Thus, a concentration of 0.3 mg/mL for Dur-A, Dur-B, and Dur-C, was used for lipase inhibitory experiments. The pancreatic lipase inhibitory activity of the extracted samples (Dur-A, Dur-B, and Dur-C) were 38.1%, 33.6%, and 17.1%, at a concentration of 0.3 mg/mL (Table 3). Among the tested samples, Dur-A and Dur-B had the strongest pancreatic lipase inhibitory activity, followed by Dur-C. Thus, these data suggest that Dur-A, Dur-B, and Dur-C, can suppress the activities of some of the key enzymes involved in metabolic syndrome, i.e., ACE, α-glucosidase, α-amylase, and pancreatic lipase.

4. Materials and Methods

4.1. Materials

A sample of *D. antarctica* was purchased from a local grocery market in Kaohsiung City, Taiwan. It was oven-dried, and then kept in plastic bags at 4 °C until use. L-fucose, L-rhamnose, D-glucose, D-glucuronic acid, D-galactose, D-xylose, D-galacturonic acid, gallic acid, sodium carbonate, potassium sulfate, 2,2-diphenyl-1-picrylhydrazyl (DPPH), 2-azino-bis(3-ethylbenzothiazoline-6-sulphonic acid) diammonium salt (ABTS), 3-(2-pyridyl)-5,6-bis (4-phenylsulfonic acid)-1,2,4-triazine (ferrozine), 2,4,6-Tris(2-pyridyl)-s-triazine (TPTZ), sodium acetate trihydrate, thioglycolic acid solution, bovine serum albumin (BSA), potassium bromide (KBr), potassium persulfate, sodium sulphite, ferrous chloride, hippuryl-histidyl-leucine (HHL), hyppuric acid (HA), and purified angiotensin I-converting enzyme (ACE) (EC 3.4.15.1) from rabbit lung, 2,2,2-trifluoroacetic acid (TFA), captopril, α-amylase, acarbose, α-glucosidase, lipase, orlistat, and Bradford reagent, were obtained from Sigma-

Aldrich (St. Louis, MO, USA). All other reagents were of analytical grade or the best grade available.

4.2. Seaweed Extraction by Enzymes

The enzymatic extraction was conducted using a hydrolytic reaction of viscozyme ($\geq$100 FBGU/g), cellulase (~0.8 units/mg solid), or α-amylase ($\geq$5 units/mg solid) with the seaweeds according to previously published methods [18,50], with minor modifications. One hundred mL of the ddH$_2$O (adjust to pH 6.0) was added to 1 g of the dried algal sample, and 100 μL of viscozyme or 100 mg of cellulase or α-amylase were mixed in. The conditions for the enzymatic reactions were 40 °C for 17 h with continuous shaking (250 rpm). The samples were centrifuged at 8000 rpm at 4 °C for 30 min and then vacuum-filtered through 0.45-μm PVDF filters to remove the unhydrolyzed residues. The extracts were frozen and freeze dried, and the resulting powder was stored at -20 °C for subsequent analysis. The extraction yield was calculated using the following equation:

$$\text{Extraction yield (\%)} = (g_A/g_B) \times 100 \tag{1}$$

where g_A represents the weight of the extracted solid on a dry basis, and g_B is the weight of the sample on a dry basis.

4.3. Chemical Methods

The phenol-sulfuric acid colorimetric method was used to determine the total sugar content, and galactose was used as the standard. The fucose content was determined by the previous method [51], and L-fucose was used as the standard. Protein in the extract was quantified by the Bradford method using BSA as the standard. Uronic acids were estimated by the colorimetric method using D-galacturonic acid as the standard [37]. Alginate content was measured according to the previous method [38]. Polyphenols were analyzed by the Folin-Ciocalteu method and gallic acid was used as the standard. Sulfate content was determined by first hydrolyzing the sample with 1 N HCl solution for 5 h at 105 °C. The hydrolysate was then quantified based on the percentage of sulfate composition using Dionex ICS-1500 Ion Chromatography (Sunnyvale, CA, USA) with an IonPac AS9-HC column (4 $\times$ 250 mm) at a flow rate of 1 mL/min at 30 °C with conductometric detection. The eluent was 9 mM Na$_2$CO$_3$, and K$_2$SO$_4$ was utilized as the standard.

4.4. Analysis of Monosaccharide Composition

The monosaccharide composition was analyzed according to our previously described method [52], using L-fucose, D-xylose, D-galactose, D-glucose, D-glucuronic acid, L-rhamnose, and D-galacturonic acid, as the standards.

4.5. NMR Spectroscopy

A 20 mg sample was dissolved in 550 μL of 99.9% deuterium oxide (D$_2$O) in a NMR tube and ^{1}H NMR spectrum was recorded at 60 °C on a Varian 400-MR (400 MHz) spectrometer (Agilent Technologies, Santa Clara, CA, USA) for proton detection. The proton chemical shift was expressed in ppm.

4.6. Molecular Weight Analysis

The molecular weight analysis of the polysaccharides was conducted according to the method of Yang [37]. The standards used to calibrate the column were various dextrans with different molecular weights (1, 12, 50, 150, and 670 kDa), which were obtained from Sigma-Aldrich (Sigma-Aldrich, St. Louis, MO, USA).

4.7. FTIR Spectroscopy

The FTIR analysis was performed according to the method of Huang [53]. In brief, sample and KBr (*w/w*, 1:50) were mixed and ground evenly until particles measured less than 2.5 μm in size. The transparent KBr pellets were made at 500 kg/cm^2 under vacuum. The FTIR spectra were obtained using a FT-730 spectrometer (Horiba, Kyoto, Japan), and the absorbance was read between 400 and 4000 cm^{-1}. Pellet of KBr alone was used as a background.

4.8. DPPH Radical Scavenging Activity

The DPPH radical scavenging activity was determined according to a method described elsewhere [54]. Briefly, 50 μL of sample was added to 150 μL 0.1 mM freshly prepared DPPH solution (in methanol). The mixture was shaken vigorously for 1 min, left to stand for 30 min in the dark at room temperature, and the absorbance of all sample solutions was measured at 517 nm using a microplate reader (SPECTROstar Nano; BMG Labtech, Ortenberg, Germany). The radical-scavenging activity was calculated using the following equation:

$$\text{DPPH}_{\text{radical-scavenging}}\ (\%) = \left[1 - \frac{A_{\text{sample}}}{A_{\text{control}}}\right] \times 100 \tag{2}$$

where A_{sample} is the absorbance of the methanol solution of DPPH with tested samples, and A_{control} represents the absorbance of the methanol solution of DPPH without the sample.

4.9. ABTS Radical Cation Scavenging Activity

The scavenging activity of the samples against ABTS radical cation was measured according to a method described elsewhere [35]. In brief, the ABTS$^{\bullet+}$ solution was prepared by reacting 5 mL of ABTS solution (7 mM) with 88 μL of potassium persulfate (140 mM), and the mixture was kept in the dark at room temperature for 16 h. The solution was diluted with 95% ethanol to obtain an absorbance of 0.70 ± 0.05 at 734 nm. To start the assay, 100 μL diluted ABTS$^{\bullet+}$ solution was mixed with 100 μL of various sample solutions. The mixture was allowed to react at room temperature for 6 min, and the absorbance of all sample solutions at 734 nm was measured using a microplate reader (SPECTROstar Nano; BMG Labtech, Germany). The blank was prepared in the same manner, except that distilled water was used instead of the sample. The activity of scavenging ABTS$^{\bullet+}$ was calculated according to the following equation:

$$\text{ABTS}_{\text{cation radical-scavenging}}\ (\%) = \left[1 - \frac{A_{\text{sample}}}{A_{\text{control}}}\right] \times 100 \tag{3}$$

where A_{sample} is the absorbance of ABTS with tested samples, and A_{control} represents the absorbance of ABTS without the sample.

4.10. Ferrous Ion-Chelating Activity

The ferrous ion-chelating activity of polysaccharides was measured using the previous method [35]. Briefly, 200 μL of sample, 740 μL of methanol, and 20 μL of FeCl$_2$ solution (2 mM) were mixed. The mixture was incubated for 30 s followed by the addition of 5 mM ferrozine (40 μL). After allowing the reaction to continue for 10 min at room temperature, the absorbance of the mixture was measured at 562 nm using a microplate reader (SPECTROstar Nano; BMG Labtech, Germany). The chelating activity of ferrous ion was calculated as follows:

$$\text{Chelating activity (\%)} = (1 - A_{\text{sample}}/A_{\text{control}}) \times 100$$

where A_{control} represents absorbance without the sample, and A_{sample} is absorbance with tested samples.

4.11. ACE-Inhibiting Activity

The ACE-inhibiting activities of samples were assayed by measuring the concentration of hippuric acid liberated from HHL according to a previous method [42], with some modifications. For each assay, 50 μL of sample solution with 50 μL of ACE solution (a purified enzyme from rabbit lung) (60 mU in sodium borate buffer, pH 8.3) were preincubated at 37 °C for 20 min and then incubated with 200 μL of substrate (0.67 mM HHL in 0.05 M sodium borate buffer at pH 8.3) at 37 °C for 60 min. The reaction was then stopped by adding 250 μL of 1 N HCl. Hippuric acid concentration was determined using a Waters HPLC system (Waters Corp., Milford, MA, USA) with an Inspire C18 column (4.6 mm × 250 mm, 5 μm) (Dikma Technologies Inc., Lake Forest, CA, USA). The column was operated at 25 °C. The mobile phase consisted of 0.1% (v/v) TFA in 50% methanol. The spectra were monitored at 228 nm and performed at a flow rate of 0.8 mL/min.

4.12. α-Amylase Inhibitory Activity

A volume of 40 μL of sample, positive control (acarbose), or negative control (distilled water) were added to 20 μL α-amylase solution (2 U/mL in 0.02 M sodium phosphate buffer pH 6.9). Test tubes were incubated at room temperature for 1 h. Later, 20 μL of 1% potato soluble starch solution (previously dissolved in 0.02 M sodium phosphate buffer pH 6.9 and boiled for 15 min) was added to each tube and incubated at 37 °C for 10 min. Finally, 40 μL of dinitrosalicylic acid solution was added, and the tubes were placed in a 100 °C water bath for 8 min. A volume of 780 μL of distilled water was added to the mixture, after centrifugation at 12,000× g for 5 min, and the absorbance was read at 540 nm in a microplate reader (SPECTROstar Nano; BMG Labtech, Germany). The percentage inhibition was calculated relative to the negative control with 100% enzyme activity.

4.13. α-Glucosidase Inhibition

The α-glucosidase inhibition assay was performed following a previous method [55]. A quantity of 72 μL of samples, positive control (acarbose), or negative control (distilled water) were added to 72 μL of rat intestine α-glucosidase (1 U/mL in 0.1 M maleate buffer pH 6.9). Test tubes were incubated at 37 °C for 30 min. After pre-incubation, 144 μL of substrate (2 mM maltose or 20 mM sucrose) was added to each tube. The reaction mixtures were incubated at 37 °C for 20 min. Finally, reactions were stopped by adding 576 μL of 1 M Na_2CO_3, after centrifugation at 12,000× g for 5 min, and the absorbance was read at 405 nm in a microplate reader (SPECTROstar Nano; BMG Labtech, Germany). The percentage inhibition was calculated relative to the negative control with 100% enzyme activity.

4.14. Lipase Inhibition Assay

The inhibition of pancreatic lipase activity was performed using the modified method with p-nitrophenyl palmitate (p-NPP) as the substrate and porcine pancreatic lipase [56]. A quantity of 330 μL of samples, positive control (orlistat), or negative control (distilled water) were added to 70 μL of pancreatic lipase (0.4 U). Test tubes were incubated at 37 °C for 30 min. After pre-incubation, 300 μL of substrate (0.8 mmol L^{-1} p-NPP) was added to each tube. The reaction mixtures were incubated at 37 °C for 30 min. After centrifugation at 12,000× g for 5 min, the absorbance was read at 405 nm in a microplate reader (SPECTROstar Nano; BMG Labtech, Germany). The percentage inhibition was calculated relative to the negative control with 100% enzyme activity.

4.15. Statistical Analysis

All experiments were performed in triplicate, and the results were the average of three independent experiments. Measurements were presented as means ± standard deviation. Statistical analysis was performed using the Statistical Package for the Social Sciences (SPSS). The results obtained were analyzed using one-way analysis of variance (ANOVA), followed by Duncan's Multiple Range tests. A probability value of $p < 0.05$ was considered statistically significant.

5. Conclusions

All the extracted samples in this study had high extraction yields, low molecular weights, and the characteristics of fucose-containing sulfated polysaccharides. These extracts appeared to have different structural qualities as demonstrated by ^{1}H-NMR. Dur-A, Dur-B, and Dur-C, showed antioxidant activities as determined by DPPH, ABTS, and ferrous ion-chelating analyses, and were capable of suppressing the activities of some of the key enzymes involved in metabolic syndrome, i.e., ACE, α-glucosidase, α-amylase, and pancreatic lipase. Therefore, the three extracts evaluated in this study (particularly Dur-B) show promise as naturally occurring antioxidants and anti-metabolic syndrome agents that could be used in various food, cosmetic, and nutraceutical products. Further studies are thus warranted using an in vivo model to confirm the abilities of Dur-A, Dur-B, and Dur-C, to prevent metabolic syndrome-related health ailments.

Supplementary Materials: The following supporting information can be downloaded at: https://www.mdpi.com/article/10.3390/catal12101284/s1, Figure S1: NMR analyses of Dur-A, Dur-B, and Dur-C. (A) ^{1}H-NMR spectra of Dur-A, Dur-B, and Dur-C. (B) ^{13}C-NMR spectra of Dur-A, Dur-B, and Dur-C. The characteristic peaks are indicated in each graph.

Author Contributions: Conceptualization, M.-K.S. and C.-H.K.; methodology, Y.-H.T.; software, C.-Y.H. (Chih-Yao Hou); validation, M.-C.L., C.-D.D., and C.-Y.H. (Chun-Yung Huang); formal analysis, P.K.P. and M.-C.L.; investigation, Y.-H.T., C.-H.K., and C.-D.D.; resources, C.-H.K. and C.-Y.H. (Chih-Yao Hou); data curation, Z.-Y.X., M.-C.L., and M.-K.S.; writing—original draft preparation, M.-K.S. and C.-H.K.; writing—review and editing, A.K.P., C.-H.K., and C.-Y.H. (Chun-Yung Huang); supervision, M.-K.S. and C.-Y.H. (Chih-Yao Hou); project administration, C.-Y.H. (Chih-Yao Hou) and C.-Y.H. (Chun-Yung Huang); funding acquisition, C.-Y.H. (Chih-Yao Hou), M.-K.S., and C.-Y.H. (Chun-Yung Huang). All authors have read and agreed to the published version of the manuscript.

Funding: This research was funded by grants provided by the National Science and Technology Council, Taiwan, grant numbers MOST 111-2221-E-992-024, awarded to Chun-Yung Huang; 110-2320-B-992-001-MY3 and 111-2622-E-992-002, awarded to Chih-Yao Hou; and 111-2221-E-328-001-MY3, awarded to Ming-Kuei Shih, respectively.

Data Availability Statement: Data is contained within the article.

Conflicts of Interest: The authors declare no conflict of interest.

References

1. Roberts, C.K.; Sindhu, K.K. Oxidative stress and metabolic syndrome. *Life Sci.* **2009**, *84*, 705–712. [CrossRef]
2. Wang, J.; Zhang, Q.; Zhang, Z.; Song, H.; Li, P. Potential antioxidant and anticoagulant capacity of low molecular weight fucoidan fractions extracted from *Laminaria japonica*. *Int. J. Biol. Macromol.* **2010**, *46*, 6–12. [CrossRef]
3. Sallmyr, A.; Fan, J.; Rassool, F.V. Genomic instability in myeloid malignancies: Increased reactive oxygen species (ROS), DNA double strand breaks (DSBs) and error-prone repair. *Cancer Lett.* **2008**, *270*, 1–9. [CrossRef]
4. Halliwell, B. Antioxidants in human health and disease. *Annu. Rev. Nutr.* **1996**, *16*, 33–50. [CrossRef] [PubMed]
5. Sharifuddin, Y.; Chin, Y.X.; Lim, P.E.; Phang, S.M. Potential bioactive compounds from seaweed for diabetes management. *Mar. Drugs* **2015**, *13*, 5447–5491. [CrossRef] [PubMed]
6. Haffner, S.M. The metabolic syndrome: Inflammation, diabetes mellitus, and cardiovascular disease. *Am. J. Cardiol.* **2006**, *97*, 3–11. [CrossRef]
7. Frieden, T.R.; Jaffe, M.G. Saving 100 million lives by improving global treatment of hypertension and reducing cardiovascular disease risk factors. *J. Clin. Hypertens.* **2018**, *20*, 208. [CrossRef] [PubMed]
8. Cardoso, S.M.; Pereira, O.R.; Seca, A.M.; Pinto, D.C.; Silva, A.M. Seaweeds as preventive agents for cardiovascular diseases: From nutrients to functional foods. *Mar. Drugs* **2015**, *13*, 6838–6865. [CrossRef]
9. Seca, A.M.; Pinto, D.C. Overview on the antihypertensive and anti-obesity effects of secondary metabolites from seaweeds. *Mar. Drugs* **2018**, *16*, 237. [CrossRef] [PubMed]
10. Hutcheson, R.; Rocic, P. The metabolic syndrome, oxidative stress, environment, and cardiovascular disease: The great exploration. *Exp. Diabetes Res.* **2012**, *2012*, 271028. [CrossRef]
11. Folli, F.; Corradi, D.; Fanti, P.; Davalli, A.; Paez, A.; Giaccari, A.; Perego, C.; Muscogiuri, G. The role of oxidative stress in the pathogenesis of type 2 diabetes mellitus micro-and macrovascular complications: Avenues for a mechanistic-based therapeutic approach. *Curr. Diabetes Rev.* **2011**, *7*, 313–324. [CrossRef] [PubMed]

12. Zaky, A.A.; Simal-Gandara, J.; Eun, J.B.; Shim, J.H.; Abd El-Aty, A. Bioactivities, applications, safety, and health benefits of bioactive peptides from food and by-products: A review. *Front. Nutr.* **2021**, *8*, 815640. [CrossRef] [PubMed]

13. Lafarga, T.; Acién-Fernández, F.G.; Garcia-Vaquero, M. Bioactive peptides and carbohydrates from seaweed for food applications: Natural occurrence, isolation, purification, and identification. *Algal Res.* **2020**, *48*, 101909. [CrossRef]

14. Li, S.; Xiong, Q.; Lai, X.; Li, X.; Wan, M.; Zhang, J.; Yan, Y.; Cao, M.; Lu, L.; Guan, J. Molecular modification of polysaccharides and resulting bioactivities. *Compr. Rev. Food Sci. Food Saf.* **2016**, *15*, 237–250. [CrossRef] [PubMed]

15. Zhong, K.; Zhang, Q.; Tong, L.; Liu, L.; Zhou, X.; Zhou, S. Molecular weight degradation and rheological properties of schizophyllan under ultrasonic treatment. *Ultrason. Sonochem.* **2015**, *23*, 75–80. [CrossRef] [PubMed]

16. Choi, W.S.; Ahn, K.J.; Lee, D.W.; Byun, M.W.; Park, H.J. Preparation of chitosan oligomers by irradiation. *Polym. Degrad. Stab.* **2002**, *78*, 533–538. [CrossRef]

17. Tao, Y.; Xu, W. Microwave-assisted solubilization and solution properties of hyperbranched polysaccharide. *Carbohydr. Res.* **2008**, *343*, 3071–3078. [CrossRef] [PubMed]

18. Olivares-Molina, A.; Fernández, K. Comparison of different extraction techniques for obtaining extracts from brown seaweeds and their potential effects as angiotensin I-converting enzyme (ACE) inhibitors. *J. Appl. Phycol.* **2016**, *28*, 1295–1302. [CrossRef]

19. Kulshreshtha, G.; Burlot, A.S.; Marty, C.; Critchley, A.; Hafting, J.; Bedoux, G.; Bourgougnon, N.; Prithiviraj, B. Enzyme-assisted extraction of bioactive material from *Chondrus crispus* and *Codium fragile* and its effect on herpes simplex virus (HSV-1). *Mar. Drugs* **2015**, *13*, 558–580. [CrossRef]

20. Nguyen, T.T.; Mikkelsen, M.D.; Tran, V.H.N.; Trang, V.T.D.; Rhein-Knudsen, N.; Holck, J.; Rasin, A.B.; Cao, H.T.T.; Van, T.T.T.; Meyer, A.S. Enzyme-assisted fucoidan extraction from brown macroalgae *Fucus distichus* subsp. *evanescens* and *Saccharina latissima*. *Mar. Drugs* **2020**, *18*, 296. [CrossRef]

21. Hammed, A.; Jaswir, I.; Simsek, S.; Alam, Z.; Amid, A. Enzyme aided extraction of sulfated polysaccharides from Turbinaria turbinata brown seaweed. *Int. Food Res. J.* **2017**, *24*, 1660–1666.

22. Movasaghi, Z.; Rehman, S.; ur Rehman, D.I. Fourier transform infrared (FTIR) spectroscopy of biological tissues. *Appl. Spectrosc. Rev.* **2008**, *43*, 134–179. [CrossRef]

23. Shao, P.; Pei, Y.P.; Fang, Z.X.; Sun, P.L. Effects of partial desulfation on antioxidant and inhibition of DLD cancer cell of *Ulva fasciata* polysaccharide. *Int. J. Biol. Macromol.* **2014**, *65*, 307–313. [CrossRef]

24. Palanisamy, S.; Vinosha, M.; Marudhupandi, T.; Rajasekar, P.; Prabhu, N.M. Isolation of fucoidan from *Sargassum polycystum* brown algae: Structural characterization, in vitro antioxidant and anticancer activity. *Int. J. Biol. Macromol.* **2017**, *102*, 405–412. [CrossRef] [PubMed]

25. Synytsya, A.; Bleha, R.; Synytsya, A.; Pohl, R.; Hayashi, K.; Yoshinaga, K.; Nakano, T.; Hayashi, T. Mekabu fucoidan: Structural complexity and defensive effects against avian influenza A viruses. *Carbohydr. Polym.* **2014**, *111*, 633–644. [CrossRef] [PubMed]

26. Vijayabaskar, P.; Vaseela, N.; Thirumaran, G. Potential antibacterial and antioxidant properties of a sulfated polysaccharide from the brown marine algae *Sargassum swartzii*. *Chin. J. Nat. Med.* **2012**, *10*, 421–428. [CrossRef]

27. Tako, M.; Nakada, T.; Hongou, F. Chemical characterization of fucoidan from commercially cultured *Nemacystus decipiens* (Itomozuku). *Biosci. Biotechnol. Biochem.* **1999**, *63*, 1813–1815. [CrossRef]

28. Immanuel, G.; Sivagnanavelmurugan, M.; Marudhupandi, T.; Radhakrishnan, S.; Palavesam, A. The effect of fucoidan from brown seaweed *Sargassum wightii* on WSSV resistance and immune activity in shrimp *Penaeus monodon* (Fab). *Fish Shellfish Immunol.* **2012**, *32*, 551–564. [CrossRef]

29. Synytsya, A.; Kim, W.J.; Kim, S.M.; Pohl, R.; Synytsya, A.; Kvasnička, F.; Čopíková, J.; Park, Y.I. Structure and antitumour activity of fucoidan isolated from sporophyll of Korean brown seaweed *Undaria pinnatifida*. *Carbohydr. Polym.* **2010**, *81*, 41–48. [CrossRef]

30. Zhang, Z.; Yu, G.; Zhao, X.; Liu, H.; Guan, H.; Lawson, A.M.; Chai, W. Sequence analysis of alginate-derived oligosaccharides by negative-ion electrospray tandem mass spectrometry. *J. Am. Soc. Mass Spectrom.* **2006**, *17*, 621–630. [CrossRef]

31. García-Vaquero, M.; Rajauria, G.; O'Doherty, J.V.; Sweeney, T. Polysaccharides from macroalgae: Recent advances, innovative technologies and challenges in extraction and purification. *Food Res. Int.* **2017**, *99*, 1011–1020.

32. Hsiao, W.C.; Hong, Y.H.; Tsai, Y.H.; Lee, Y.C.; Patel, A.K.; Guo, H.R.; Kuo, C.H.; Huang, C.Y. Extraction, biochemical characterization, and health effects of native and degraded fucoidans from *Sargassum crispifolium*. *Polymers* **2022**, *14*, 1812. [CrossRef] [PubMed]

33. Choi, J.i.; Kim, H.J. Preparation of low molecular weight fucoidan by gamma-irradiation and its anticancer activity. *Carbohydr. Polym.* **2013**, *97*, 358–362. [CrossRef] [PubMed]

34. Song, Y.R.; Sung, S.K.; Jang, M.; Lim, T.G.; Cho, C.W.; Han, C.J.; Hong, H.D. Enzyme-assisted extraction, chemical characteristics, and immunostimulatory activity of polysaccharides from Korean ginseng (*Panax ginseng* Meyer). *Int. J. Biol. Macromol.* **2018**, *116*, 1089–1097. [CrossRef]

35. Huang, C.Y.; Wu, S.J.; Yang, W.N.; Kuan, A.W.; Chen, C.Y. Antioxidant activities of crude extracts of fucoidan extracted from *Sargassum glaucescens* by a compressional-puffing-hydrothermal extraction process. *Food Chem.* **2016**, *197*, 1121–1129. [CrossRef]

36. Amir, R.M.; Anjum, F.M.; Khan, M.I.; Khan, M.R.; Pasha, I.; Nadeem, M. Application of Fourier transform infrared (FTIR) spectroscopy for the identification of wheat varieties. *J. Food Sci. Technol.* **2013**, *50*, 1018–1023. [CrossRef]

37. Yang, W.N.; Chen, P.W.; Huang, C.Y. Compositional characteristics and in vitro evaluations of antioxidant and neuroprotective properties of crude extracts of fucoidan prepared from compressional puffing-pretreated *Sargassum crassifolium*. *Mar. Drugs* **2017**, *15*, 183. [CrossRef]

38. Huang, C.Y.; Kuo, C.H.; Chen, P.W. Compressional-puffing pretreatment enhances neuroprotective effects of fucoidans from the brown seaweed *Sargassum hemiphyllum* on 6-hydroxydopamine-induced apoptosis in SH-SY5Y cells. *Molecules* **2018**, *23*, 78. [CrossRef] [PubMed]
39. Wang, C.Y.; Wu, T.C.; Hsieh, S.L.; Tsai, Y.H.; Yeh, C.W.; Huang, C.Y. Antioxidant activity and growth inhibition of human colon cancer cells by crude and purified fucoidan preparations extracted from *Sargassum cristaefolium*. *J. Food Drug Anal.* **2015**, *23*, 766–777. [CrossRef]
40. Wang, C.Y.; Wu, T.C.; Wu, C.H.; Tsai, Y.H.; Chung, S.M.; Hong, Y.H.; Huang, C.Y. Antioxidant, anti-inflammatory, and HEP G2 cell growth-inhibitory effects of aqueous-ethanol extracts obtained from non-puffed and compressional-puffed *Sargassum crassifolium*. *J. Mar. Sci. Technol.* **2020**, *28*, 5.
41. Devalaraja, S.; Jain, S.; Yadav, H. Exotic fruits as therapeutic complements for diabetes, obesity and metabolic syndrome. *Food Res. Int.* **2011**, *44*, 1856–1865. [CrossRef] [PubMed]
42. Huang, C.Y.; Tsai, Y.H.; Hong, Y.H.; Hsieh, S.L.; Huang, R.H. Characterization and antioxidant and angiotensin I-converting enzyme (ACE)-inhibitory activities of gelatin hydrolysates prepared from extrusion-pretreated milkfish (*Chanos chanos*) scale. *Mar. Drugs* **2018**, *16*, 346. [CrossRef]
43. Nagappan, H.; Pee, P.P.; Kee, S.H.Y.; Ow, J.T.; Yan, S.W.; Chew, L.Y.; Kong, K.W. Malaysian brown seaweeds *Sargassum siliquosum* and *Sargassum polycystum*: Low density lipoprotein (LDL) oxidation, angiotensin converting enzyme (ACE), α-amylase, and α-glucosidase inhibition activities. *Food Res. Int.* **2017**, *99*, 950–958. [CrossRef]
44. Apostolidis, E.; Kwon, Y.I.; Shetty, K.; Apostolidis, E.; Kwon, Y.I. Potential of cranberry-based herbal synergies for diabetes and hypertension management. *Asia Pac. J. Clin. Nutr.* **2006**, *15*, 433–441.
45. Sakulnarmrat, K.; Konczak, I. Composition of native Australian herbs polyphenolic-rich fractions and in vitro inhibitory activities against key enzymes relevant to metabolic syndrome. *Food Chem.* **2012**, *134*, 1011–1019. [CrossRef] [PubMed]
46. Kim, K.T.; Rioux, L.E.; Turgeon, S.L. Alpha-amylase and alpha-glucosidase inhibition is differentially modulated by fucoidan obtained from *Fucus vesiculosus* and *Ascophyllum nodosum*. *Phytochemistry* **2014**, *98*, 27–33. [CrossRef]
47. Kaisoon, O.; Konczak, I.; Siriamornpun, S. Potential health enhancing properties of edible flowers from Thailand. *Food Res. Int.* **2012**, *46*, 563–571. [CrossRef]
48. Huang, C.Y.; Kuo, C.H.; Lee, C.H. Antibacterial and antioxidant capacities and attenuation of lipid accumulation in 3T3-L1 adipocytes by low-molecular-weight fucoidans prepared from compressional-puffing-pretreated *Sargassum crassifolium*. *Mar. Drugs* **2018**, *16*, 24. [CrossRef]
49. Chater, P.I.; Wilcox, M.; Cherry, P.; Herford, A.; Mustar, S.; Wheater, H.; Brownlee, I.; Seal, C.; Pearson, J. Inhibitory activity of extracts of Hebridean brown seaweeds on lipase activity. *J. Appl. Phycol.* **2016**, *28*, 1303–1313. [CrossRef]
50. Heo, S.J.; Park, E.J.; Lee, K.W.; Jeon, Y.J. Antioxidant activities of enzymatic extracts from brown seaweeds. *Bioresour. Technol.* **2005**, *96*, 1613–1623. [CrossRef]
51. Wu, T.C.; Hong, Y.H.; Tsai, Y.H.; Hsieh, S.L.; Huang, R.H.; Kuo, C.H.; Huang, C.Y. Degradation of *Sargassum crassifolium* fucoidan by ascorbic acid and hydrogen peroxide, and compositional, structural, and in vitro anti-Lung cancer analyses of the degradation products. *Mar. Drugs* **2020**, *18*, 334. [CrossRef] [PubMed]
52. Hsiao, H.H.; Wu, T.C.; Tsai, Y.H.; Kuo, C.H.; Huang, R.H.; Hong, Y.H.; Huang, C.Y. Effect of oversulfation on the composition, structure, and in vitro anti-lung cancer activity of fucoidans extracted from *Sargassum aquifolium*. *Mar. Drugs* **2021**, *19*, 215. [CrossRef] [PubMed]
53. Huang, C.Y.; Kuo, J.M.; Wu, S.J.; Tsai, H.T. Isolation and characterization of fish scale collagen from tilapia (*Oreochromis* sp.) by a novel extrusion-hydro-extraction process. *Food Chem.* **2016**, *190*, 997–1006. [CrossRef] [PubMed]
54. Shiao, W.C.; Wu, T.C.; Kuo, C.H.; Tsai, Y.H.; Tsai, M.L.; Hong, Y.H.; Huang, C.Y. Physicochemical and antioxidant properties of gelatin and gelatin hydrolysates obtained from extrusion-pretreated fish (*Oreochromis* sp.) scales. *Mar. Drugs* **2021**, *19*, 275. [CrossRef]
55. Eruygur, N.; Dincel, B.; Dincel, N.G.K.; Ucar, E. Comparative study of in vitro antioxidant, acetylcholinesterase and butyrylcholinesterase activity of alfalfa (*Medicago sativa* L.) collected during different growth stages. *Open Chem.* **2018**, *16*, 963–967. [CrossRef]
56. Tiss, M.; Souiy, Z.; ben Abdeljelil, N.; Njima, M.; Achour, L.; Hamden, K. Fermented soy milk prepared using kefir grains prevents and ameliorates obesity, type 2 diabetes, hyperlipidemia and Liver-Kidney toxicities in HFFD-rats. *J. Funct. Foods* **2020**, *67*, 103869. [CrossRef]

 *catalysts*

Article

Immobilization of Recombinant Endoglucanase (CelA) from *Clostridium thermocellum* on Modified Regenerated Cellulose Membrane

Zi-Han Weng [1,†], Parushi Nargotra [2,†], Chia-Hung Kuo [2,*] and Yung-Chuan Liu [1,*]

[1] Department of Chemical Engineering, National Chung Hsing University, Taichung 402, Taiwan
[2] Department of Seafood Science, National Kaohsiung University of Science and Technology, Kaohsiung 811, Taiwan
* Correspondence: kuoch@nkust.edu.tw (C.-H.K.); ycliu@dragon.nchu.edu.tw (Y.-C.L.);
   Tel.: +886-7-3617141 (ext. 23646) (C.-H.K.); +886-4-22853769 (Y.-C.L.)
† These authors contributed equally to this work.

Abstract: Cellulases are being widely employed in lignocellulosic biorefineries for the sustainable production of value-added bioproducts. However, the high production cost, sensitivity, and non-reusability of free cellulase enzymes impede their commercial applications. Enzyme immobilization seems to be a potential approach to address the aforesaid complications. The current study aims at the production of recombinant endoglucanase (CelA) originated from the cellulosome of *Clostridium thermocellum* in *Escherichia coli* (*E. coli*), followed by immobilization using modified regenerated cellulose (RC) membranes. The surface modification of RC membranes was performed in two different ways: one to generate the immobilized metal ion affinity membranes RC-EPI-IDA-Co^{2+} (IMAMs) for coordination coupling and another to develop aldehyde functional group membranes RC-EPI-DA-GA (AMs) for covalent bonding. For the preparation of IMAMs, cobalt ions expressed the highest affinity effect compared to other metal ions. Both enzyme-immobilized membranes exhibited better thermal stability and maintained an improved relative activity at higher temperatures (50–90 °C). In the storage analysis, 80% relative activity was retained after 15 days at 4 °C. Furthermore, the IMAM- and AM-immobilized CelA retained 63% and 53% relative activity, respectively, after being reused five times. As to the purification effect during immobilization, IMAMs showed a better purification fold of 3.19 than AMs. The IMAMs also displayed better kinetic parameters, with a higher Vmax of 15.57 U mg^{-1} and a lower Km of 36.14 mg mL^{-1}, than those of AMs. The IMAMs were regenerated via treatment with stripping buffer and reloaded with enzymes and displayed almost 100% activity, the same as free enzymes, up to 5 cycles of regeneration.

Keywords: enzyme immobilization; recombinant endoglucanase; regenerated cellulose membrane; IMAM; immobilized metal ion affinity membrane

Citation: Weng, Z.-H.; Nargotra, P.; Kuo, C.-H.; Liu, Y.-C. Immobilization of Recombinant Endoglucanase (CelA) from *Clostridium thermocellum* on Modified Regenerated Cellulose Membrane. *Catalysts* **2022**, *12*, 1356. https://doi.org/10.3390/catal12111356

Academic Editor: Maria H. L. Ribeiro

Received: 8 October 2022
Accepted: 1 November 2022
Published: 3 November 2022

Publisher's Note: MDPI stays neutral with regard to jurisdictional claims in published maps and institutional affiliations.

1. Introduction

The ever-increasing reliance and over-exploitation of non-renewable fossil fuels, owing to accelerated urbanization and globalization, have led to their acute shortage, besides raising grave environmental concerns due to greenhouse gas emissions such as methane and carbon dioxide [1]. A large increase in methane emission (774 tera grams/year) due to tremendous anthropogenic activities has been reported. Additionally, the adverse effects of methane as a greenhouse gas are roughly 25 times greater than those of carbon dioxide [2]. Therefore, to meet the world's energy demand and mitigate climate change issues, the development of an alternative renewable and clean resource of energy is a prerequisite [3]. Plant biomass in the form of lignocellulosic biomass serves as a promising inexpensive feedstock for the production of biofuels such as bioethanol, biohydrogen, biomethanol, etc. to fuel the transportation sector [4–6]. Among these, presently, bioethanol is the most common biofuel, owing to the transportation/storage and economical production constraints associated with biohydrogen and biomethanol, respectively [6]. Lignocellulosic biomass, such as forest materials and agro-industrial and municipal wastes, consist of an intricate network of biopolymers, such as cellulose, hemicellulose, and lignin [7]. The cellulose and hemicellulose fractions from the biomass can be hydrolyzed using cellulases and hemicellulases, respectively, into monomeric sugars, which are subsequently fermented to produce bioethanol [8]. The intense bonding between cellulose, hemicellulose, and lignin leads to biomass recalcitrance, which can be reduced by employing various pretreatment strategies, such as chemical, enzymatic, or microbial strategies [1]. Whole-cell-based cultures are also useful and economical for biomass treatment to release sugars [9]. While whole-cell-based cultures require nutrients and energy for growth and enzyme production, the enzymatic hydrolysis of cellulose is faster and yields more product than microbial action because it is a specific catalytic reaction that only needs ideal physical conditions (temperature, pH, and time) for efficient hydrolysis. Additionally, the product yield can also be enhanced by using purified enzymes, whereas whole-cell-based cultures may lead to the production of various unwanted enzymes/products, which may further interfere with the desired enzymatic hydrolysis reaction. The cellulase enzymes for industrial applications are naturally produced from various bacterial and fungal species on a large scale. However, the low concentrations, unfavorable conditions, specific substrate requirements, and high production cost, among others, are the few limitations associated with natural enzyme production [10]. Therefore, the focus has been shifted to the production of enzymes through recombinant technology using *Escherichia coli* as a host. Since *E. coli* can grow profusely on any growth medium, this approach offers the advantage of enabling the scaling up of the production of recombinant cellulase without the use of any particular substrates [10].

The production of cellulolytic enzymes for enzymatic hydrolysis in lignocellulosic biorefineries is a major bottleneck due to the high production cost of enzymes, which significantly impacts the cost-economics of biomass-to-biofuel conversion [5]. Furthermore, the high sensitivity of free enzymes, difficulty in separation from the reaction medium, non-reusability, and poor stability in extreme environmental conditions (temperature and pH) are several other hurdles that reduce the overall viability of the process. Consequently, the whole catalytic process is substantially hindered by their low operational stability, economic constraints, and short lifetime [11,12]. To overcome the aforementioned challenges and increase the enzyme's catalytic stability for the enzymatic saccharification of complex biopolymers such as cellulose and hemicellulose, enzyme immobilization appears to be a promising method [5]. Immobilization not only enhances enzyme stability in robust environments but also enables the recovery and reuse of enzymes for multiple cycles. This improves the overall process economy and efficient output in terms of high product yield [13]. Enzyme immobilization involves the binding or localization of enzymes onto a support surface or within a specific matrix. Immobilized enzymes replicate the natural mode of action and are more resistant to harsh milieu when they are bound to a support surface [14,15]. Enzyme immobilization can be achieved through physical interactions such as encapsulation and entrapment and chemical interactions such as covalent bonding, electrostatic interactions, hydrogen bonds, hydrophobic interactions, van der Waals interactions, and magnetic nanoparticles [5,12,16]. Enzymes immobilized via the covalent bond formation method can be used in reaction catalysis more frequently as it provides a more stable connection between the enzymes and support [12]. In this method, the surface of the carrier is first activated to produce functional groups (-COOH, $-NH_2$, and -SH), which are then covalently linked with the enzyme using a cross-linker. Glutaraldehyde (GA) is used as a cross-linking reagent in covalent enzyme immobilization, which usually reacts with the amino group of the enzyme [17,18]. Moreover, it also prevents the leaching of the enzyme during recycling and maintains the robustness of the immobilized system [19]. Efficient enzyme immobilization is influenced by various factors, including the type of immobilization method and enzyme, immobilization conditions, enzyme loading, protein concentrations, etc. [20].

Enzyme purification is one of the crucial steps towards its biochemical and kinetic characterization for diverse industrial applications [21]. Moreover, the production of recombinant cellulases in *E. coli* also produces various unwanted proteins, making the purification of the protein a prerequisite [22]. Immobilized metal ion affinity chromatography (IMAC) has emerged as a well-accepted technology for protein purification from a bench scale to an industrial scale. Conventionally, it was based on the affinity of transitions metal ions, such as Cu(II), Co(II), Fe(III), Zn(II), and Ni(II), towards histidine, tryptophan, and cysteine in aqueous solution, which was further extended to include the concept of using metal ions immobilized on a substrate to separate and purify proteins in liquids [23]. The interaction between an immobilized metal ion and electron donor groups on the surface of proteins forms chelated complexes that provide the basis for protein adsorption. The availability of histidine residues controls the actual protein binding in IMAC. The high specificity of IMAC is based on the difference between the strength of the various complexes formed [23]. The targeted protein can be eluted by using a low-pH buffer, a competitive displacement, or chelating agents.

The use of granular solid carriers in IMAC results in a pressure drop and often hinders the reaction. This shortcoming of IMAC can be overcome by using immobilized metal affinity membranes (IMAMs) [22]. This is the most widely used technique for the purification of proteins with mainly surface-exposed amino acids, such as histidine, tryptophan, cysteine, and tyrosine, or polyhistidine-tagged biomolecules [24,25]. Numerous membranes, viz., cellulose nitrate, regenerated cellulose (RC), and cellulose acetate, are used to prepare IMAMs; however, regenerated cellulose membranes (RCMs) offer various advantages, such as high hydrophilicity, chemical stability, biocompatibility, strong mechanical properties, and low non-specific adsorption [22]. Previously, our laboratory has purified penicillin G acylase using a modified RC-based IMAM technique. The membrane was successfully reused more than eight times after activation by regeneration [26].

Consequently, in the present study, the expression plasmid of endoglucanase (CelA) from the cellulosome of *Clostridium thermocellum* was constructed and expressed in *E. coli* and immobilized on a modified RC membrane. Two different membranes, i.e., IMAM (RC-EPI-IDA-Co^{2+}; EPI, epichlorohydrin; IDA, iminodiacetic acid) for coordination coupling and the aldehyde functional group membrane (AM) RC-EPI-DA-GA (DA, 1,4-Diaminobutane) for covalent bonding, were prepared by subjecting the RC membranes to two different surface modifications. Moreover, the characterization of free and immobilized enzymes has been executed, and the difference in their kinetics is also discussed.

2. Results and Discussion

2.1. Cloning of Cellulase Gene in Plasmid

The cellulase gene CelA was inserted in plasmid pET21b to form the construct pET21b-CelA-his, which was further transformed into *E. coli* strains, namely, *E. coli* ER2566 and *E. coli* BL21. The plasmid was extracted and subjected to PCR and gene shearing. A clear band was visible after PCR, which is indicative of the presence of the target gene CelA fragment, with a molecular weight of 1.6 kb (Figure S1). Similar to the present study, the cellulase gene from *Bacillus subtilis* and *B. licheniformis* ATCC 14580 was cloned in *E. coli*, where the band appeared near 1.5 and 1.6 kb, respectively [10,27].

2.2. Recombinant Cellulase Expression under Different Conditions

2.2.1. Effect of Different Hosts and Temperatures

After the pET21b-CelA-his plasmid was successfully transformed into *E. coli* ER2566 and *E. coli* BL21, protein expression analysis was carried out. The amount of protein expression is positively correlated with temperature, as the higher the temperature, the shorter the incubation time and the faster the protein expression. However, an induction at higher temperatures may result in the formation of the insoluble proteinaceous matter known as inclusion bodies, which blocks protein translocation and cannot be used [28]. In the current study, the induction was set at 15 °C for 24 h and at 37 °C for 6 h to observe the performance of the target protein at different induction temperatures and times. The protein expression was confirmed by the presence of a bright band after sodium dodecyl sulfate polyacrylamide gel electrophoresis (SDS-PAGE) of the target protein of a molecular weight of about 60 kDa, irrespective of the host and culture conditions used (Figure 1a). The integrated color intensity of each band was calculated by ImageJ software to quantify the target protein (Figure 1b). The results indicated that the maximum amount of protein was obtained when *E. coli* ER2566 was used as the expression host under the culture condition of 37 °C for 6 h. Therefore, *E. coli* ER2566 was used as the expression host in subsequent experiments and was induced and cultured at 37 °C for 6 h.

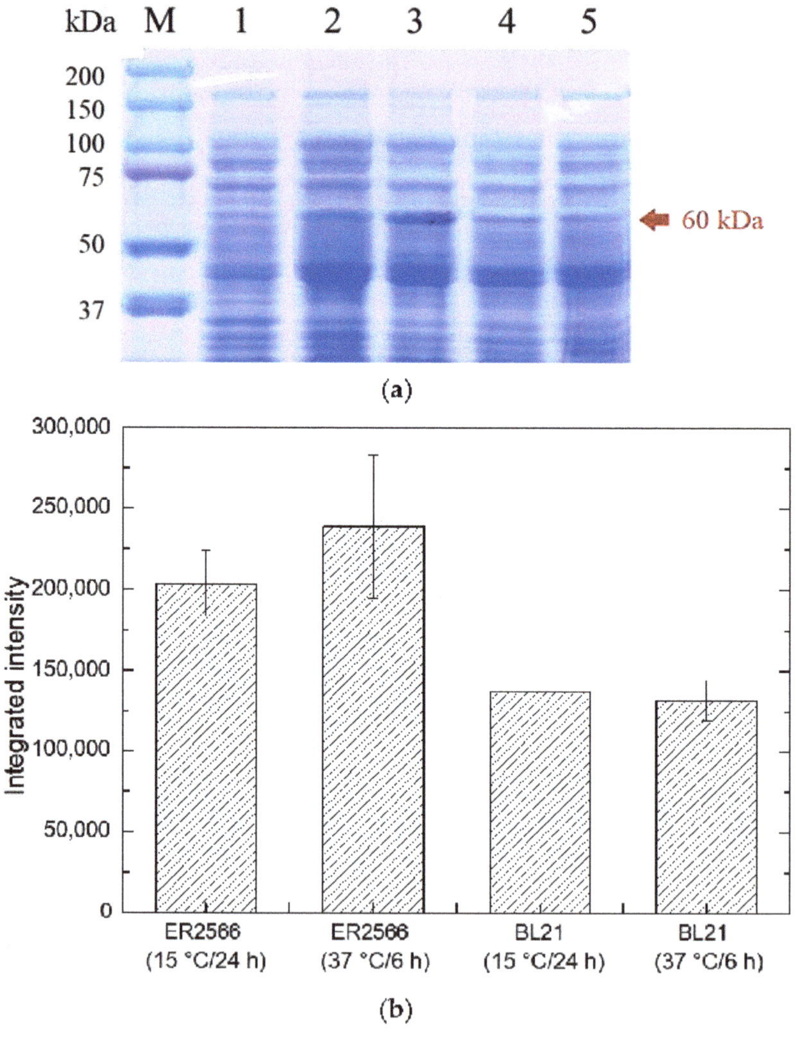

Figure 1. (a) SDS-PAGE electropherogram of pET21b-CelA-his with different hosts and temperatures. Lane M: marker (kDa); Lane 1: pET21b; Lane 2: pET21b-CelA-his/ER2566 (15 °C/24 h); Lane 3: pET21b-CelA-his/ER2566 (37 °C/6 h); Lane 4: pET21b-CelA-his/BL21 (15 °C/24 h); Lane 5: pET21b-CelA-his/BL21 (37 °C/6 h). (b) ImageJ analysis of protein content of pET21b-CelA-his in different hosts and temperatures.

2.2.2. Effect of Different Inducer Concentrations

The production of recombinant proteins in *E. coli* has been extensively carried out using a T7-promoter-based expression system, with isopropyl β-D-1-thiogalactopyranoside (IPTG) as an inducer [28]. Recombinant protein expression is greatly influenced by the concentration and timing of the addition of the inducer IPTG. Higher concentrations of IPTG, apart from impacting the cost-economics of the process, may cause cell toxicity, whereas lower concentrations result in inadequate protein induction, thus, reducing the efficiency. Therefore, the regulation of IPTG concentration is vital for successful protein expression [29]. In this study, different IPTG concentrations, viz., 0.1, 0.3, 0.5, and 0.7 mM were used. A sharp band of the target protein of molecular weight 60 kDa was observed after induction, which indicated that a higher amount of target protein was produced after induction compared to without induction (Figure 2a). At the IPTG concentration of 0.1 mM, the sharpest and brightest band was observed. The amount of target protein was estimated by determining the integrated color intensity of each band by using ImageJ software (Figure 2b). Therefore, IPTG was used at the 0.1 mM concentration as the inducer in subsequent experiments. In another research work, Mühlmann et al. demonstrated the relationship between inducer concentration and induction temperature and time on the metabolic state of *E. coli.* and the formation of the product. Their results exhibited a decrease in the optimal inducer concentration to 0.05 mM IPTG from 0.1 mM IPTG at 28 °C and at elevated temperatures (34 and 37 °C) [30]. These findings indicate that lower IPTG concentrations are favorable at higher temperatures.

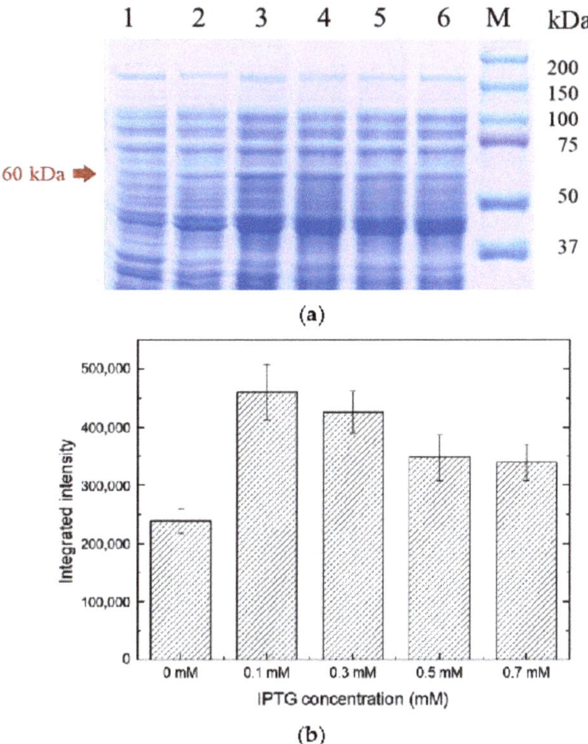

Figure 2. (a) SDS-PAGE electrophoresis of recombinant protein induced by different IPTG concentrations. Lane M: marker (kDa); Lane 1: pET21b; Lane 2: not induced; Lane 3: 0.1 mM; Lane 4: 0.3 mM; Lane 5: 0.5 mM; Lane 6: 0.7 mM. (b) ImageJ analysis of the protein amount induced by different IPTG concentrations.

2.3. Characterization of Modified RC Membranes

The RC membrane used in this study is reinforced with nonwoven cellulose, which has been tested not to be digested by CelA prior to immobilization experiments. The preparation steps of RC-EPI-IDA and RC-EPI-DA-GA membranes for CelA immobilization are shown in Schemes 1 and 2. The Fourier-transformed infrared spectroscopy (FTIR) analysis was performed after the surface modification of RC membranes to understand the functional group changes that occurred in membranes post-modification in contrast to RC membranes without modifications. Figure 3a shows the FTIR spectra of the RC membrane and the RC-EPI-IDA membrane after surface modification with EPI and IDA (before metal chelation). The intensity of absorption peak at 901 cm^{-1} for the OH group in the unmodified RC membrane was reduced in the modified RC membrane, indicating the activation of the membrane by EPI [31]. The characteristic absorption peak of the epoxy group in the range of 950 to 815 cm^{-1} in the spectrum of the EPI-activated RC membrane was observed [31]. After adding IDA, the absorption peaks at 1774 and 1196 cm^{-1} for the carboxylate group were visible [31]. The peaks for the C-N and C=O groups of IDA were also observed at 1368 and 1680 cm^{-1}, respectively, confirming the successful modification of the surface of the RC membrane by EPI and IDA.

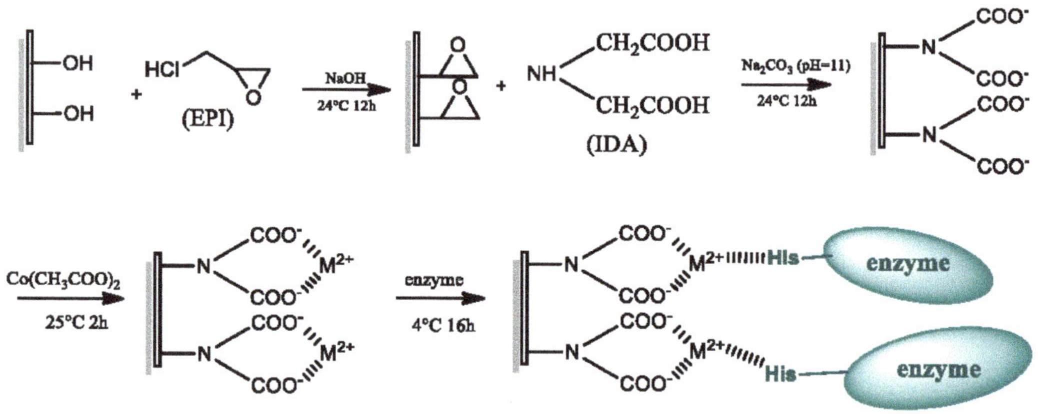

Scheme 1. RC-EPI-IDA membrane preparation.

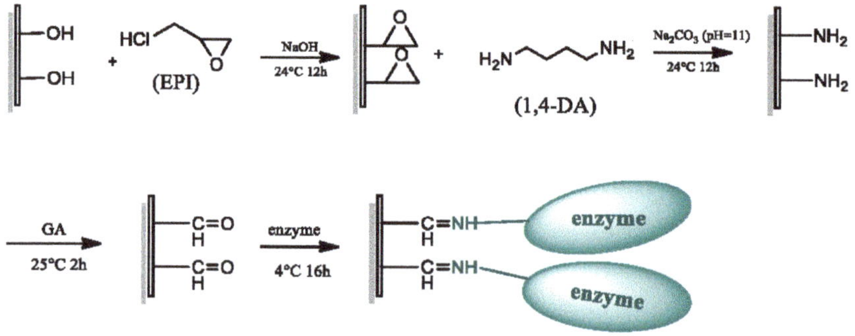

Scheme 2. RC-EPI-DA-GA membrane preparation.

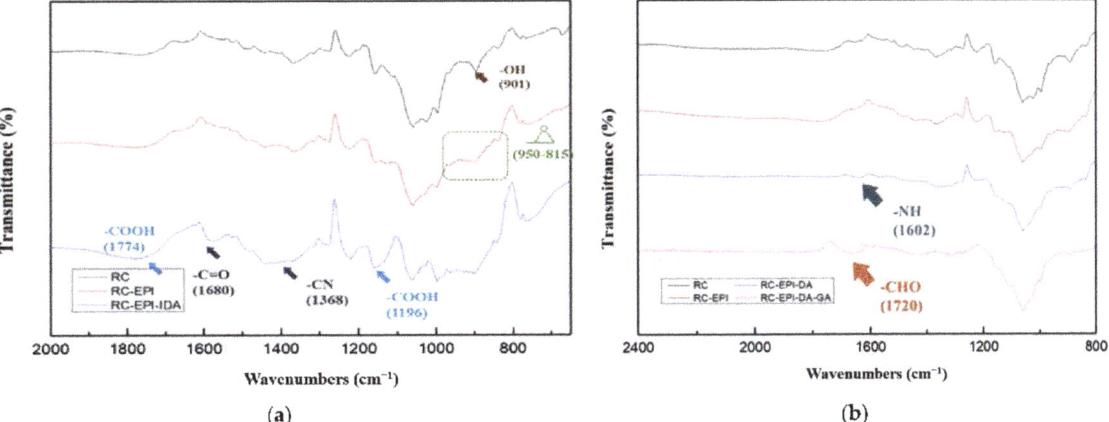

Figure 3. FTIR characterization of (**a**) RC-EPI-IDA membrane and (**b**) RC-EPI-DA-GA membrane.

Figure 3b shows the FTIR spectra of the RC membrane and the RC-EPI-DA-GA membrane after surface modification with 1,4-DA and GA. Modification by 1,4-DA was confirmed by the presence of an absorption band at 1602 cm^{-1}, corresponding to N-H bending in 1,4-DA [32]. A peak at 1720 cm^{-1}, representing the free aldehyde group of GA, confirmed the successful membrane activation by GA [33].

2.4. Metal Ion Selection for IMAM Chelation and Effect on Enzyme Activity

In order to couple metal ions with the modified membrane RC-EPI-IDA, first, the effect of metal ions on the activity of free enzyme CelA must be determined. Therefore, the effect of various metal ions, viz., Zn^{2+}, Al^{3+}, Cu^{2+}, Ni^{2+} Fe^{3+}, Mn^{2+}, and Co^{2+}, on the activity of free CelA enzymes was studied. The initial activity of the free enzymes was 1.25 U/mL, which was taken as control (100%). The Zn^{2+}, Al^{3+}, Cu^{2+}, and Ni^{2+} metal ion solutions exhibited an inhibitory effect on the endoglucanase activity. However, the CelA endoglucanase activity was improved in the Fe^{3+}, Mn^{2+}, and Co^{2+} metal ion solutions (Figure S2). Metal ions facilitate the binding of enzymes to the substrate to enhance enzymatic action by activating electrophiles or nucleophiles via electron transfer reactions. Metal ions may also help in maintaining three-dimensional active confirmation by bringing the enzymes and substrate into close proximity via a coordinate [21]. However, the enzyme–substrate complex may sometimes be broken up by metal ions action, changing the proper interactions between the substrate and the enzyme and resulting in a loss of activity [21]. The metal ions Mn^{2+} and Co^{2+} have been reported to improve the hydrolysis activity of endoglucanase in the literature. CTendo7's endoglucanase activity was increased in the presence of Mn^{2+} and Co^{2+}, which are heterologously expressed in *Pichia pastoris* [34].

The amount and activity of the adsorbed protein were measured to calculate the specific activity of the enzyme after the chelation of different metal ions on IMAM and enzyme immobilization. The results indicated that IMAMs with Cu^{2+} on chelation were able to adsorb the most amount of protein, but the specific activity (0.80 U mg^{-1}) was lower than in other metal ions, which might be due to inhibition of CelA endoglucanase by Cu^{2+} (Figure 4). Cu^{2+} is known for its inhibitory effect on many glycoside hydrolases due to the auto-oxidation of proximal cysteine residues, which results in the formation of inter- and intra-molecular disulfide bridges [34]. The amount of protein adsorbed by other metal ions was almost similar; however, Co^{2+} and Mn^{2+} significantly increased the endoglucanase activity, which further resulted in high specific activity (2.07 and 2.00 U mg^{-1}) in the case of these two metal ions (Figure 4). Therefore, the IMAMs coupled with Co^{2+} and Mn^{2+} were further analyzed to check their reuse ability. It was observed that the activity of the IMAM-Mn^{2+} immobilized enzyme decreased very quickly after several repeated uses

(Figure 5). Co^{2+}, being an intermediate metal ion, tends to form a covalent bond with oxygen, sulfur, nitrogen, and histidine and, thus, may improve the catalytic efficiency of the enzyme [35]. Therefore, Co^{2+} was used for the chelation of IMAMs in the subsequent experiments. The chelation was executed with 50 mM cobalt ion solution for 3 h, and the immobilized enzyme was referred to as RC-EPI-IDA-Co^{2+}-CelA. IMAMs chelated with Co^{2+} have been used for the simultaneous purification and immobilization of the xylanase enzyme, which is heterologously expressed in *E. coli*. The presence of Co^{2+} metal ions increased the activity of both free and immobilized enzymes [22].

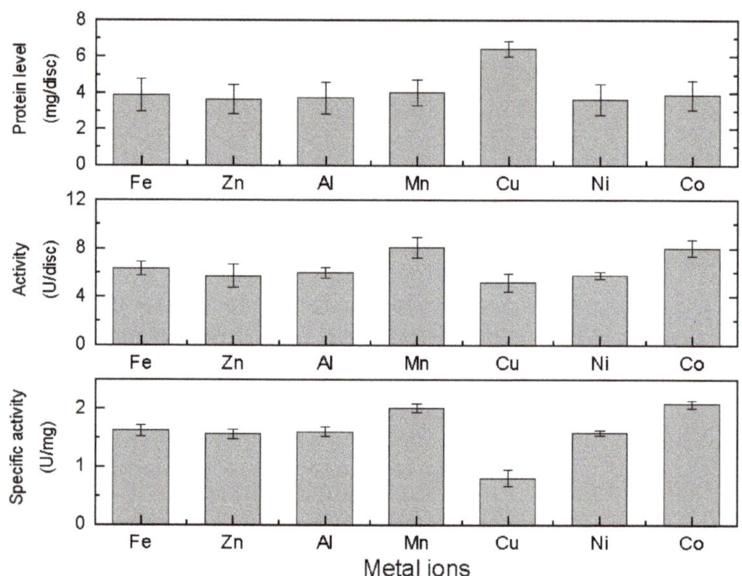

Figure 4. Protein adsorption capacity, activity, and specific activity of IMAMs chelated with different metal ions on CelA.

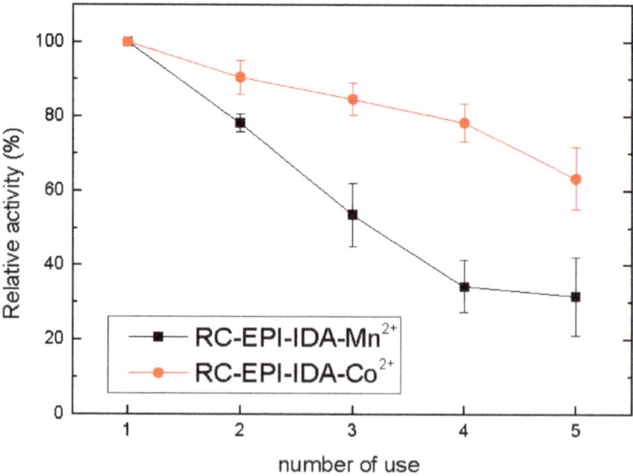

Figure 5. Reusability of immobilized enzymes using IMAM-Co^{2+} and IMAM-Mn^{2+}.

2.5. Effect of GA Concentration and Treatment Time

The enzyme immobilization was also executed using an RC-EPI-DA-GA membrane, wherein GA was used as an activator, and the immobilized enzyme was referred to as RC-EPI-DA-GA-CelA. GA, a bi-functional chemical, condenses aldehyde on the surface of the RC membrane to produce active α, β-unsaturated aldehyde, which subsequently binds to an amino group of protein, resulting in the formation of a complex and stable product via covalent bonding [18]. The immobilization of the enzyme is greatly influenced by GA concentration and the duration of its treatment. Therefore, in the current study, GA concentration (0.05, 0.1, 0.2, 0.5, and 1%, *v/v*) and its treatment time (1–5 h) were optimized to evaluate the effect on enzyme activity. It was observed that at the GA concentration of 0.5% (*v/v*), the maximum amount of protein immobilization occurred. However, the maximum CelA endoglucanase activity of 8.1 U was obtained at 0.1% (*v/v*) GA concentration, which was reduced at 0.2% (*v/v*) (Figure 6). The decrease in CelA endoglucanase activity at higher GA concentrations is attributed to the fact that higher GA concentrations alter the enzyme conformation since it is a protein denaturant and also increase the rigidity of the enzyme because of too much cross-linking, which reduces enzyme activity [18,36]. Moreover, insufficient endoglucanase-GA cross-linking may also cause the activity to decline at lower GA concentrations [36].

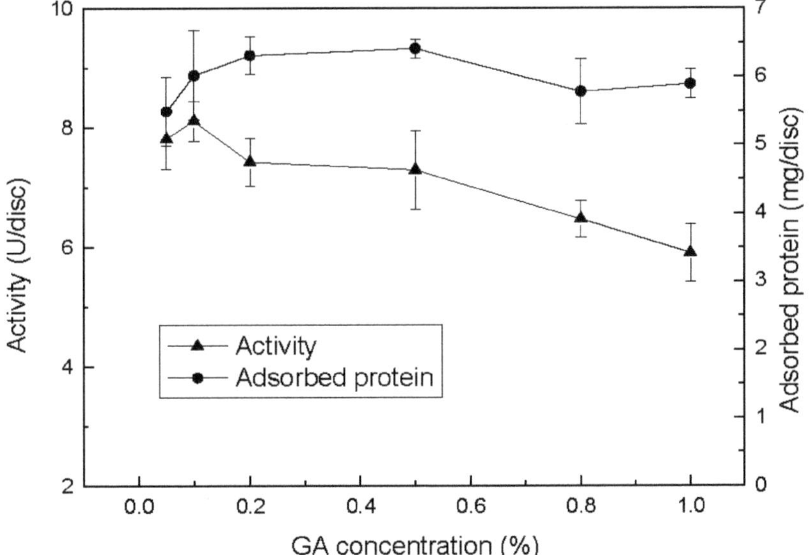

Figure 6. Effect of different concentrations of GA on CelA endoglucanase activity, and protein adsorption.

In the case of GA treatment time with RC-EPI-DA, the maximum CelA endoglucanase activity was found at 2 h of treatment time. A reduction in CelA endoglucanase activity was seen when the treatment time was 4 h (Figure S3). The optimization of GA treatment time is crucial for efficient enzyme activity as the longer duration of treatment may result in interactions at multiple sites on the carrier, thereby reducing the number of sites for enzyme immobilization, whereas shorter treatment time may not be sufficient to completely activate the functional groups on the matrix for enzyme immobilization, leading to reduced activity [18]. Therefore, in the subsequent experiments, the GA treatment was carried out for 2 h at a concentration of 0.1% (*v/v*). There are many reported studies wherein GA has been used as a cross-linking agent, and emphasis has been laid on the impact of GA concentration and treatment time. In a study, cross-linked cellulase aggregates (C-CLEAs) were synthesized, and GA was used as a cross-linker. The concentrations of GA and cross-linking time were optimized, and the maximum cellulase activity was achieved at 2% (*v/v*) concentration with 3.5 h of cross-linking time [37]. Similarly, the maximum activity of the inulinase enzyme was observed at 1.59% concentration and the treatment time of 4 h [18].

2.6. Purification Analysis

For a thorough examination of the catalytic, kinetic, structural, and functional properties of enzymes, enzyme purification is considered a crucial step [21]. Therefore, the purification efficiency achieved from the immobilization of enzymes on two different membranes was estimated. A 3.192 and 1.536 purification fold was achieved with RC-EPI-IDA-Co^{2+}-CelA and RC-EPI-DA-GA-CelA, respectively (Table 1). A single band of purified CelA was obtained on SDS-PAGE (Figure S4). The CelA endoglucanase enzyme contained His-tag, which could bind more strongly to the IMAM membrane because of an affinity for Co^{2+} metal ions and, therefore, result in the high purification of the enzyme [23]. IMAMs have been used for the purification of His-tagged recombinant serine hydroxymethyl transferase using various metal ions, and 5-fold purification has been achieved with Co^{2+} ions [37].

Table 1. Activity and purification effect of free and immobilized enzymes.

Sample	Activity (U/mL or U/disc)	Protein (mg/mL)	Specific Activity (U/mg)	Purification Fold
Free CelA [a]	1.250	1.322	0.946	1.00
RC-EPI-IDA-Co^{2+}-CelA [b]	9.42	3.12	3.02	3.19
RC-EPI-DA-GA-CelA [b]	7.24	4.98	1.45	1.54

[a] activity expressed in U/mL; [b] activity expressed in U/disc.

2.7. Optimal pH and Enzyme Stability

Any industrially significant enzyme must be able to maintain its catalytic activity and stability in extreme pH environments. In acidic or alkaline pH, the ionic group of the enzyme shows varying degrees of protonation, which result in a change in enzyme configuration and, ultimately, affect enzyme activity [22]. In our present study, the effect of different pH levels and the pH stability of free enzymes and immobilized CelA endoglucanase was studied. The maximum activity (100%) of free enzymes was obtained at optimal pH 5, which increased to pH 6 (100%) for RC-EPI-IDA-Co^{2+}-CelA and RC-EPI-DA-GA-CelA enzymes (Figure 7a). The activity of free enzymes decreased at pH 6; however, immobilized enzymes demonstrated high relative activity (80–60%) over a broad pH range (pH 4–9). The high activity retention of immobilized enzymes in a wide pH range may be attributed to the stable rigidity and confirmation of the enzyme provided by multipoint attachment due to interaction between the charged groups of the enzyme and carrier [38,39]. The results were aligned with previously reported studies. The optimal pH value of the endoglucanase of *Arachniotus citrinus* was increased from 4.9 to 5.6 after immobilization on polyacrylamide gel [40]. Similarly, cellulase immobilization on sodium alginate-polyethylene glycol showed maximum activity at the optimal pH value of 5, whereas for free cellulase, the optimal pH was 4 [41]. The current and previous findings indicate that cellulase shows proficient activity in the pH range of 4–6.

The stability of free enzymes and immobilized CelA endoglucanase at varying pH ranges was also analyzed, as the enzyme activity is highly influenced by its storage environment. The enzyme was exposed to various pH values for 2 h at 4 °C. The initial enzyme activity was taken as 100%. It was observed that the stability trend for free and immobilized enzymes over a pH range of 4–9 was similar, and enzymes exhibited approximately 80% relative activity in both free and immobilized forms (Figure 7b). It was evident from the results that immobilization did not lead to any change in the confirmation of CelA endoglucanase, and it was able to retain active three-dimensional orientation, indicated by efficient catalytic activity. The stability of free or immobilized enzymes at a wide pH range also exhibited its potential for industrial applications.

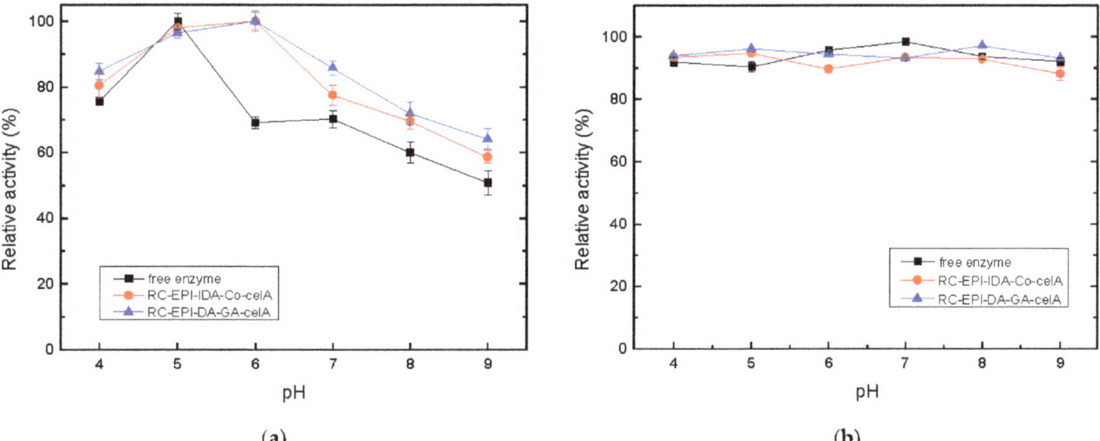

Figure 7. (**a**) Optimum pH and (**b**) pH stability of free and immobilized CelA endoglucanase.

2.8. Optimal Temperature and Thermo-Stability

Enzyme stability at extreme temperatures is also very crucial for industrial potential. High temperatures increase the rate of reaction by increasing the rate of effective collision between molecules. However, in an enzymatic reaction at a high temperature, the catalytic efficiency of the enzyme is reduced as the elevated temperature leads to changes in the active structure of the enzyme. Therefore, the optimal temperature and thermo-stability of free and immobilized enzymes were investigated. It was observed that the relative activity of the immobilized enzyme was higher compared to the free enzyme over a temperature range of 50–80 °C, with a temperature optima of 60 °C for free and RC-EPI-IDA-Co^{2+}-CelA enzymes and 70 °C for the RC-EPI-DA-GA-CelA enzyme (Figure 8a). The increase in optimal temperature for the RC-EPI-DA-GA-CelA enzyme could be explained by a decrease in the flexibility of the enzymes on a solid matrix, resulting in improved tolerance to unfolding and denaturation owing to conformational changes [39]. The results were inconsistent with previous studies, where the optimum temperature for free and immobilized cellulase is reported to be between 50–70 °C [39,41]. The temperature optima of cellulase enzyme from *Aspergillus fumigatus* immobilized on magnetic nanoparticles was 60 and 50 °C for free enzymes [39]. Likewise, the immobilization of cellulase on sodium alginate-polyethylene glycol resulted in an increased optimal temperature (70 °C) compared to free enzymes (50 °C) [41]. Therefore, the findings from this study and the literature substantiate that the optimal temperature falls within the range of 50–70 °C.

The thermal stability of enzymes is a crucial characteristic since enzyme activity can be affected by the extreme temperatures employed in various industrial processes. Therefore, the thermo-stability of free and immobilized CelA endoglucanases was also evaluated. The free and immobilized CelA endoglucanases were incubated at different temperatures (50–90 °C) for 1 h, and then, activity was measured. Figure 8b reveals that both the free enzyme and the immobilized enzyme exhibit 85% relative activity within a temperature range of 50–70 °C. The enzyme activity of the free enzymes drastically declined at 80 and at 90 °C, and the free enzyme relative activity decreased to 20%. However, in the case of immobilized enzymes, even though the reduction in activity was seen after 80 °C, the RC-EPI-IDA-Co^{2+}-CelA-immobilized enzymes retained 54% of the initial activity, and RC-EPI-DA-GA-CelA also retained 65% of the initial activity. The higher thermo-stability of the immobilized enzyme might be due to the more stable spatial configuration of the enzyme due to immobilization [41]. Thus, it was evident that the two immobilization methods made the enzyme more resistant to high temperatures and increased thermal stability over a wider temperature range.

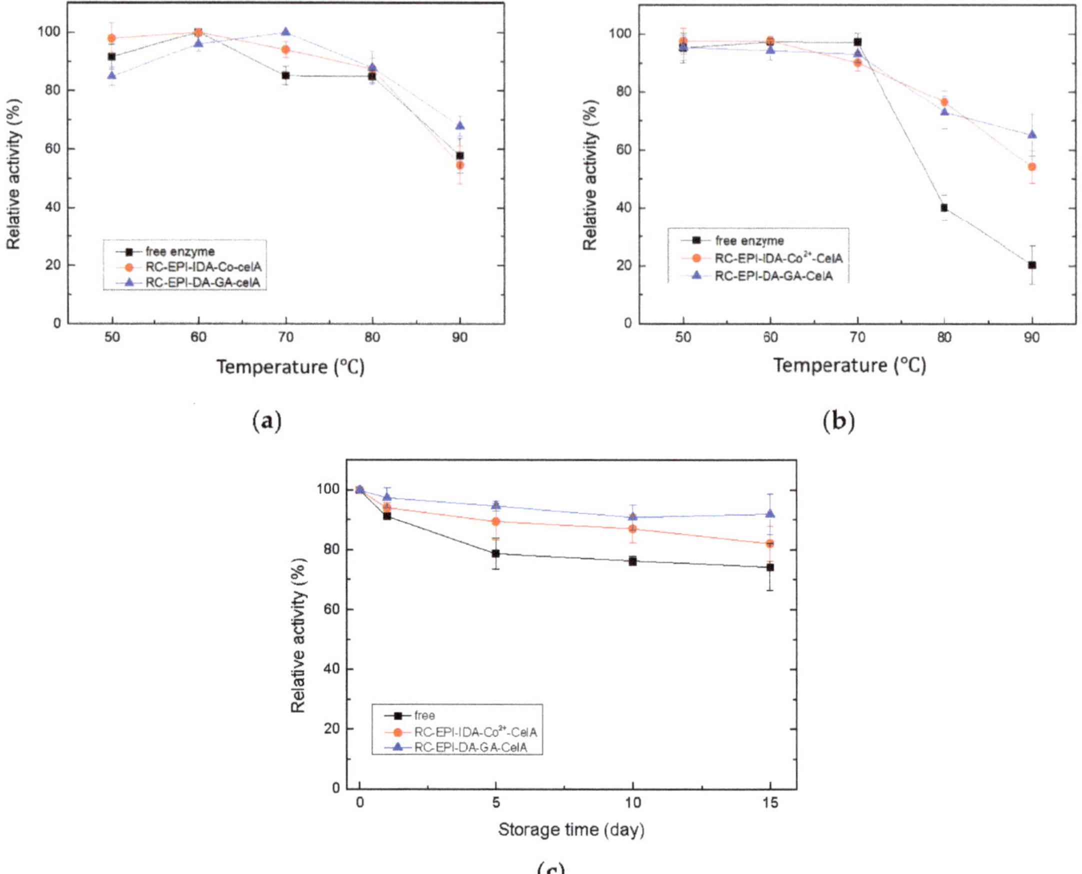

Figure 8. (**a**) Optimum temperature, (**b**) thermo-stability, and (**c**) storage stability of free and immobilized CelA endoglucanase.

The free and immobilized enzymes were analyzed for their storage stability at 4 °C for 5, 10, and 15 days. The activity on the 0th day was considered as the initial activity. Figure 8c reveals that RC-EPI-DA-GA-CelA maintained the best relative activity within 15 days, followed by RC-EPI-IDA-Co^{2+}-CelA. Both immobilization methods showed better preservation compared with free enzymes, retaining 91% and 82% of their relative activities after 15 days, respectively (Figure 8c). The results indicated that immobilization increased the shelf life of an enzyme [42]. The increased stability at 4 °C could be attributed to the rigidity and decreased flexibility of enzymes due to immobilization.

2.9. Enzyme Reusability Analysis

The most important advantage of enzyme immobilization is the ability to reuse the enzymes. The reusability of enzymes is a decisive factor that significantly governs the techno-economically viability of any enzyme-driven process. Therefore, under the optimized conditions discussed in the above experiments, the activity of both membranes was repeatedly investigated (five times) to determine the change in relative activity after multiple uses. The results revealed that 63% and 53% of the relative activity was retained for RC-EPI-IDA-Co^{2+}-CelA and RC-EPI-DA-GA-CelA, respectively, after five times of reuse (Figure 9). The loss of relative activity could be attributed to inadequate binding between the enzyme and the support [43]. In previously reported studies, chitosan–cellulase nanomixtures immobilized in alginate beads retained only 18% activity after five cycles of reuse [43], whereas 59.42% enzyme relative activity was observed after five cycles when the cellulase was immobilized on sodium alginate-polyethylene glycol-chitosan [44]. Therefore, it can be concluded that the immobilization of enzymes on RC-EPI-IDA-Co and RC-EPI-DA-GA is economically beneficial for the process as the membranes can be reused multiple times and retain sufficient enzyme activity.

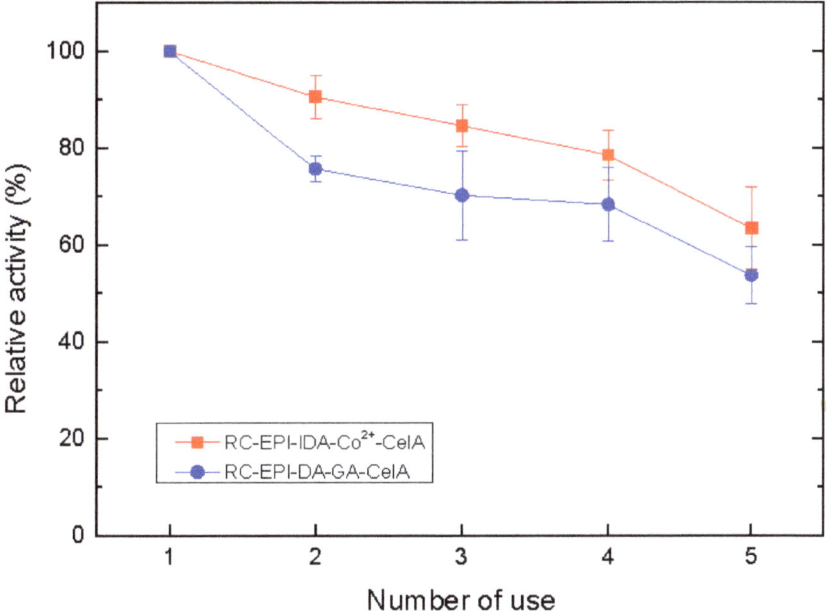

Figure 9. Reusability of RC-EPI-IDA-Co^{2+}-CelA and RC-EPI-DA-GA-CelA enzymes.

2.10. Kinetics Study of Free and Immobilized Enzymes

A kinetic parameter analysis of free and immobilized enzymes was carried out to study the catalytic rate of enzyme-catalyzed reactions. Free and immobilized enzyme activity was assayed at different substrate concentrations to determine K_m (Michaelis–Menton constant) and V_{max} (maximal velocity). The behavior of enzymes is best described by enzyme kinetic constants, i.e., K_m and V_{max}, which are dependent on varying concentrations of the substrate [45]. K_m defines the affinity of an enzyme towards a substrate. A lower K_m value indicates the higher affinity of an enzyme to its substrate and vice versa. V_{max} denotes a maximum rate of enzyme-catalyzed reaction when a substrate is converted into a product under specific conditions [46]. The Lineweaver Burk plot, depicted in Figure 10, shows a straight line between substrate concentrations (1/[S]) and reaction velocity (1/[V]). The K_m and V_{max} of free and immobilized enzymes were determined by analyzing the intercepts on the X- and Y-axes of the double reciprocal plot. The results revealed that immobilized enzymes have better V_{max} and K_m compared to free enzymes (Table 2). Among all enzymes, the RC-EPI-IDA-Co^{2+}-CelA enzyme had the lowest K_m value, which indicated its maximum affinity for the substrate, followed by RC-EPI-DA-GA-CelA, and then the free enzyme. Similar trends of results were found in the case of the V_{max} value. The decrease in the K_m value and the increase in the V_{max} value of immobilized enzymes compared to free enzymes could be attributed to the structural changes that occurred in the enzyme due to immobilization [47]. Moreover, it is speculated that the immobilized enzyme has the effect of preliminary purification due to IMAM immobilization, which eliminates protein impurities, lowers hindrances between the enzyme and the matrix, and increases affinity. Additionally, as described in Section 2.4, the cobalt ions can boost the activity of CelA endoglucanase, thereby increasing the maximal reaction rate and V_{max} and making the immobilization approach of IMAMs more effective. The results are consistent with other studies, wherein the decrease in K_m and increase in V_{max} values have been reported for different enzymes immobilized using IMAM/IMAC [47,48]. In our previous study, the K_m value of xylanase immobilized on IMAMs decreased compared to free enzymes (8.445 mg/mL), with a minimum K_m of 1.513 mg/mL for IMAM-Co^{2+}-CipA-XynCt. Similarly, a maximum V_{max} of 3.831 U/mg was achieved for IMAM-Co^{2+}-CipA-XynCt compared to free enzymes (2.235 U/mg) [22].

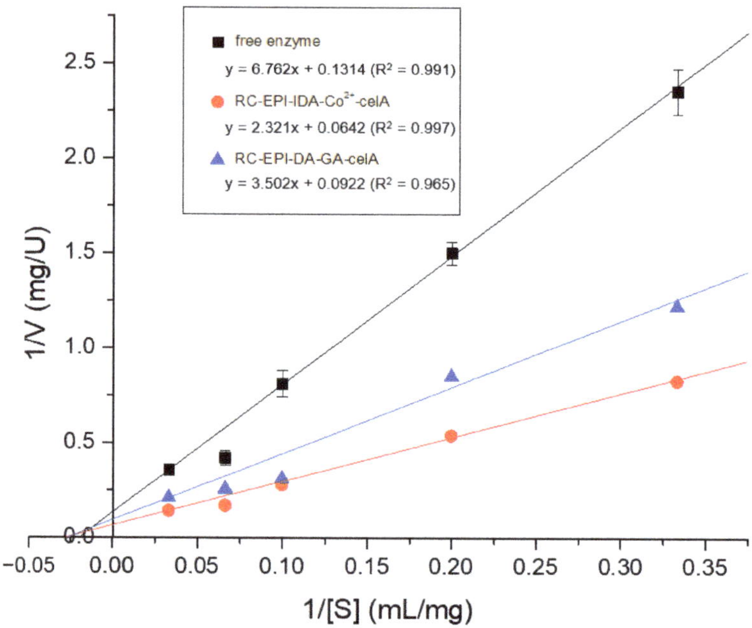

Figure 10. Lineweaver Burk plot mapping of free and immobilized CelA endoglucanase.

Table 2. Kinetic parameters of free and immobilized CelA endoglucanase.

Enzymes	V_{max} (U/mg)	K_m (mg/mL)
Free enzyme	7.61	51.45
RC-EPI-IDA-Co^{2+}-CelA	15.57	36.14
RC-EPI-DA-GA-CelA	10.85	37.99

2.11. Regeneration of IMAM

The IMAM prepared in this study was a reversible immobilized membrane, and the catalytic activity of the enzyme immobilized on the membrane decreased with multiple uses over a longer period of time. Thus, the regeneration of enzyme-immobilized IMAMs is important, wherein the enzyme and metal ions are desorbed and re-adsorbed again. There are two proposed approaches commonly applied to regenerate membranes. The use of imidazole is the most common way to desorb the His-tag target protein in many purification processes, while the use of ethylenediaminetetraacetic acid (EDTA) can desorb not only protein but also the chelated metal ion [49]. It can be seen in Figure 11 that the use of 250 mM imidazole resulted in a drop in relative activity to 38.4% of the original after five times of regeneration, while with the use of 100 mM EDTA to desorb metal ion, a relative activity of

51.2% was obtained after five times of regeneration. It is proposed that when the metal ion membrane is regenerated using the imidazole of EDTA, a thin layer of precipitate is formed on the IMAM membrane surface. These precipitates are mainly the cell debris and non-specific protein present in the crude enzyme extract, which adsorbs to the membrane surface. Although these precipitates do not interfere with metal ion adsorption, they, however, affect the adsorption capacity of the enzyme due to steric hindrance, resulting in lower enzyme activity [49]. Therefore, in the regeneration process, various other stripping solutions, alone or in combination, including EDTA, hydrodrochloric acid (HCl), sodium hydroxide (NaOH), and binding buffer, were investigated for the removal of precipitated contaminants.

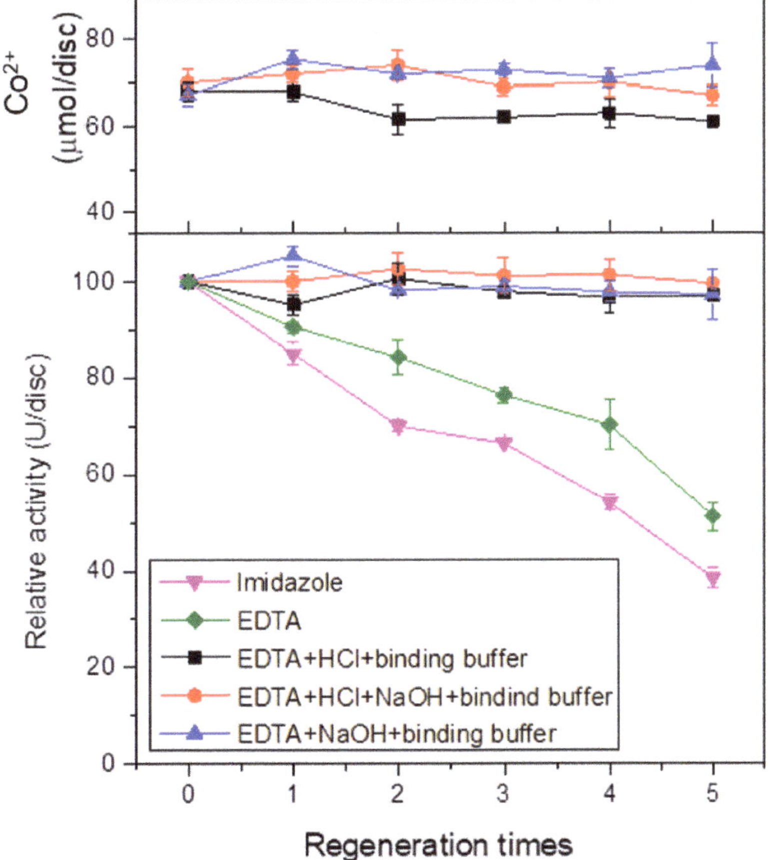

Figure 11. Impact of various stripping solutions on IMAM regeneration.

The combination of HCl or/and NaOH with EDTA sequentially in binding buffer as a stripping solution resulted in almost the same levels of metal ion and protein adsorption in the regenerated membrane (Figure 11). In addition, the adsorbed CelA endoglucanase activity after each regeneration process was almost similar to the original activity. This indicated that the proposed regeneration methods, by the addition of either HCl or NaOH, can effectively remove the impurity contaminants on the surface of IMAMs to regenerate a new IMAM for effective CelA coupling. This is the first report on developing and using reversible IMAMs as a matrix for recombinant endoglucanase immobilization.

3. Materials and Methods

3.1. Chemicals, E. coli Strains, and Plasmid Sources

The chemicals used in the study were purchased from Showa (Gunma, Japan), BD Biosciences (San Jose, CA, USA), Amresco (Solon, OH, USA), Fisher Scientific (San Jose, CA, USA), Sartorius (Gottingen, Germany), Sigma (St. Louis, MO, USA), BIO-RAD (Hercules, CA, USA), and Alfa Aesar (Karlsruhe, Germany) and were of analytical grade. An RC disc film, procured from Sartorius Stedim Biotech, was used as the solidified carrier, with a diameter of 4.7 cm and a pore size of 0.45 μm. The recombinant endocellulase (Endoglucanase, CelA) and anchoring region protein (Dockerin, docT) genes used were provided by the Department of Life Sciences, National Chung Hsing University (Taichung, Taiwan), and the Biodiversity Research Center, Academia Sinica (BRCAS, Taipei, Taiwan), respectively. The pET21b-CelA-docT-his plasmid was constructed by gene recombination technology in our laboratory (Figure S5). Moreover, *E. coli* DH5α was used as host cells for gene transformation, and *E. coli* ER2566 and *E. coli* BL21 (DE3) were used as host cells for recombinant protein expression. The source of plasmid and *E. coli* strains are mentioned in Table S1.

3.2. Gene Cloning and E. coli Culture

In this study, the endoglucanase (CelA) gene from the cellulosome of *Clostridium thermocellum* was expressed in *E. coli*. The competent cells to be used for transformation were prepared by using the heat shock method [50]. The procedure was performed under sterile conditions and at low temperatures. The transformation was achieved by adding plasmid DNA (pET21b-celA-docT-his) to 100 μL of previously prepared competent cells. The contents were placed on ice for 30 min, and plasmid was allowed into the channel on the membrane by calcium ions. After this, the contents were placed in a dry bath at 42 °C for 2 min and immediately placed on ice for 3–5 min. The presence of the targeted gene in the plasmid DNA was detected by polymerase chain reaction (PCR) using gene-specific primers. Plasmid DNA from the recombinant cells was extracted using the QIAprep® Spin Miniprep Kit (QIAGEN, Hilden, Germany) and stored at −20 °C. The PCR-amplified product was subjected to agarose gel electrophoresis for the confirmation of gene fragment size.

3.3. Gene Expression and Enzyme Extraction

After transformation, the bacterial cells were cultured on Luria–Bertani (LB) agar medium at 37 °C. A single colony was inoculated in 10 mL LB broth containing ampicillin (50 μg/mL) and incubated overnight at 37 °C and 200 rpm. For expression, 1% of the inoculum (O.D$_{600}$ = 0.6–0.8) to 100 mL was added to the LB broth medium and induced with IPTG (0.1–0.7 mM). The culture was incubated at 37 °C and 200 rpm for 6 h or 15 °C and 200 rpm for 24 h. Protein expression was confirmed with 10% SDS-PAGE.

The harvested broth was subjected to centrifugation at 4 °C/8500 rpm for 10 min to separate the supernatant. The obtained pellet was dissolved in lysis buffer (pH 8.0, 20 mM, Tris-buffer) solution. The bacterial cells were broken by using an ultrasonic sterilizer at 45% amplitude with a pulse of 30 s on and 30 s off for 10 min. The suspension was then subjected to centrifugation at 4 °C/8500 rpm for 20 min and the supernatant collected was considered crude CelA enzymes.

3.4. Preparation of IMAMs and Enzyme Immobilization

IMAMs were prepared according to the method of Liu et al. [26]. A piece of the RC membrane was placed in a glass jar, followed by the addition of 20 mL of 1.4 M NaOH and 5 mL of EPI (activator). The solution was reacted at 24 °C/150 rpm for 12 h. The membrane was washed three times with deionized water, and 20 mL of 1 M IDA (pH 11) was added thereafter. The reaction was carried out at 24 °C/100 rpm for 12 h. The modified membrane was washed with deionized water 3 times. Before the chelation of metal ions to the IMAM membrane, the impact of various metal ions on the enzyme activity of recombinant free cellulase was determined. Different metal ions, viz., Zn^{2+}, Al^{3+}, Cu^{2+},

Ni^{2+} Fe^{3+}, Mn^{2+}, and Co^{2+}, were used to study their effect on enzyme activity. The metal ion solution of concentrations of 1, 5, and 10 mM was used. The metal ion with a significant positive impact on enzyme activity was used to chelate the membrane by soaking the membrane in 20 mL of metal ion solutions at 25 °C for 2 h. The membrane was washed with deionized water 3 times to obtain IMAMs (Scheme 1). For enzyme immobilization, 20 mL (1.32 mg/mL) crude CelA with an activity of 1.25 U/mL was loaded onto one piece of IMAM and kept at 4 °C for 16 h. The scanning electron microscopy (SEM) images of the RC membrane before and after immobilization are shown in Figure S6.

3.5. Preparation of RC-EPI-DA-GA Membranes and Enzyme Immobilization

According to the method of Chen et al., a piece of the RC membrane was placed in a glass jar, 1.4 M NaOH (20 mL) and 5 mL EPI were added, and the reaction was carried out at 24 °C/150 rpm for 12 h [32]. After the reaction, the membrane was washed three times with deionized water, followed by the addition of 10 mL of 1 M DA (pH 11). The reaction was carried out at 24 °C for 12 h, and the membrane was washed with deionized water thrice. GA (10 mL of 0.1%, v/v) was used for activating the membrane. The membrane was incubated with GA at 25 °C for 2 h to obtain the RC-EPI-DA-GA membrane (Scheme 2). The impact of GA concentration and treatment time on enzyme activity was also determined. The enzyme was immobilized in the same manner as explained above.

3.6. Characterization of Thin Films by FTIR Analysis

The RC membranes before and after modification with EPI and IDA were subjected to FTIR characterization to directly measure the change in the functional group. FTIR analysis was carried out by using a Horiba FT-720 spectrometer (Horiba Ltd., Kyoto, Japan). The spectra (4000–400 cm^{-1}) were recorded with a resolution of 4 cm^{-1}, and 64 scans were performed per sample.

3.7. Assays

The cellulolytic activity of the enzyme was determined by using carboxymethyl cellulose sodium salt (CMC) as a substrate. The amount of reducing sugars liberated was estimated by the 3,5-dinitrosalicylic acid (DNS) method [51]. Precisely, free enzymes (0.2 mL) and one immobilized membrane were separately mixed with 0.8 mL of 1% CMC solution and 10 mL of 1% CMC solution, respectively, and incubated in a water bath at 60 °C/100 rpm for 10 min. After incubation, the reaction mixture of free enzymes was mixed with 1 mL DNS, whereas for the immobilized membrane, 1 mL of the hydrolytic product was taken after incubation and mixed with 1 mL DNS. The reaction mixture was kept at 100 °C for 10 min. The mixture was cooled down after incubation by adding 8 mL of deionized water, and sugars were quantified by measuring absorbance at 540 nm. The amount of enzyme required to release one micromole of glucose per min was defined as one unit (IU) of enzyme activity. The protein estimation was done using the Bradford assay, with bovine serum albumin (BSA) as the standard [52].

3.8. Effect of pH and Temperature

The optimum pH of free and immobilized enzymes was estimated. The enzymes were incubated with CMC substrate prepared at varying pH values, i.e., pH 4–9, using Britton–Robinson buffer. The reaction was executed at 60 °C/100 rpm for 10 min, and then, enzyme activity was evaluated. The pH stability of the enzyme was also studied. Free and immobilized enzymes were pre-incubated at pH 4–9 for 2 h at 4 °C, followed by cellulase activity analysis. Similarly, optimum temperatures of free and immobilized enzymes were also assessed. The enzyme was incubated with CMC substrate (pH 6), and the reaction was executed at different temperatures (50–90 °C) and 100 rpm for 10 min, followed by enzyme activity analysis. For the thermo-stability analysis, free and immobilized enzymes were pre-incubated at temperatures of 50–90 °C for 1 h at pH 6, followed by cellulase activity analysis. Enzyme storage stability was also evaluated. Free and immobilized enzymes were stored at 4 °C for 5, 10, and 15 days and an enzyme activity assay was performed. The ratio of the enzyme's activity to the maximum activity was used to determine the relative activity.

3.9. Kinetic Analysis

The Michaelis–Menten kinetic constant, i.e., K_m and the maximal velocity rate V_{max}, was calculated for the free and immobilized enzymes using the Lineweaver Burk double reciprocal plot. The free and immobilized enzymes were allowed to react with varying CMC substrate concentrations. The activities obtained at different substrate concentrations were used to plot the Michaelis–Menten equation double reciprocal graph (Equation (1)). The linear relationship on this plot was used to calculate the maximum reaction rate (V_{max}) and the Michaelis–Menten constant (K_m).

$$\frac{1}{V_0} = \left(\frac{K_m}{V_{max}}\right)\left(\frac{1}{S_0}\right) + \frac{1}{V_{max}} \tag{1}$$

where S_0 is the initial concentration of the substrate; V_0 is the initial rate of the reaction; K_m is the Michaelis–Menten constant; V_{max} is the maximum rate of reaction.

3.10. Reusability Study for Immobilized CelA

To test the reusability of the immobilized CelA, a disc of the prepared CelA membrane was used. Prior to each reusability cycle, the membrane was taken out of the reaction mixture and washed twice with PBS buffer. The reaction conditions and the residual CelA activity on the membrane were determined according to the methods listed in Section 3.7. The relative activity was calculated as the ratio of the residual CelA activity after each use to that of the original activity.

3.11. IMAM Regeneration

The removal of immobilized enzymes and coupled metal ions is imperative for the regeneration of the IMAM membrane. Various stripping solutions, viz., EDTA, HCl, NaOH, and binding buffer, were used, and the feasibility of stripping using different agents was investigated. The membrane regeneration with imidazole was carried out by washing the membrane thrice with deionized water and soaking it in 10 mL of 250 mM imidazole solution for 30 min, followed by rinsing twice with phosphate buffer (PB, 25 mM, pH 8). For stripping with EDTA alone, the membrane was subjected to washing with deionized water three times and soaked in 10 mL, 100 mM EDTA for 1 h. It was then rinsed twice with phosphate buffer (25 mM, pH 8). Membrane regeneration using EDTA sequentially combined with HCL and/or NaOH was also executed. For EDTA + HCl, treatment with 100 mM EDTA was carried out, as explained above, followed by immersing the membrane in 10 mL 0.5 M HCl for 10 min and then washing twice with deionized water. Similarly, for EDTA + NaOH, after soaking the membrane in 100 mM EDTA and washing, it was dipped in 10 mL 0.5 M NaOH for 10 min. For regeneration using combined EDTA + HCl + NaOH, the membrane was sequentially treated with each chemical, as explained above, and washed twice after every treatment. Finally, to complete the regeneration process for each treatment, the membrane was soaked in 10 mL binding buffer (300 mM NaCl, 25 mM PB buffer, pH 8) for 20 min [49]. The regeneration efficiency was evaluated by coupling metal ions and immobilizing the enzyme again to the membrane, as described in Section 3.4, after regenerating the membrane with different solutions, alone or in combination.

4. Conclusions

The current study is the first to report the immobilization of recombinant endoglucanase (CelA) on RC membranes modified using two different approaches, i.e., RC-EPI-IDA-Co^{2+} (IMAM) and RC-EPI-DA-GA. FTIR analysis revealed the successful modification of the RC membrane using EPI-IDA and DA-GA. The immobilization improved the temperature (50–90 °C) stability of the enzyme compared to free enzymes, whereas the pH (4–9) stability of free and immobilized enzymes was similar as both displayed approximately 80% relative activity. Enzyme immobilization aided in enzyme reusability, which is a crucial factor in maintaining the economic sustainability of the process. The immobilized enzyme was effectively used for five cycles, with 63% and 53% of relative activity for RC-EPI-IDA-Co^{2+}-CelA and RC-EPI-DA-GA-CelA, respectively. Moreover, the coupling of Co^{2+} ions on the IMAM increased the enzyme activity and preliminary purification due to the RC-EPI-IDA-Co^{2+}-CelA immobilization, leading to a 3.19-fold purification, followed by RC-EPI-DA-GA-CelA (1.54-fold purification). In comparison with free enzymes, the enzyme immobilized on the membrane had better kinetic characteristics (K_m and V_{max}). An effective five-times regeneration of RC-EPI-IDA-Co^{2+} was achieved with almost 100% activity. Although both approaches of immobilization improved enzyme characteristics and efficacy, RC-EPI-IDA-Co^{2+} showed a better performance than the RC-EPI-DA-GA-modified membrane for CelA immobilized with respect to purification and reusability. Therefore, IMAM immobilization emerges as a potential approach for increasing enzyme efficacy and reusability.

Supplementary Materials: The following supporting information can be downloaded at: https://www.mdpi.com/article/10.3390/catal12111356/s1, Figure S1: DNA electropherogram of pET21b-CelA-his plasmid in E. coli BL21 (a) and E. coli ER2566 (b), confirmed by gene cleavage. Lane M: 1 Kb marker; Lane 1–3: CelA digested with Nde I and Xho I, Figure S2: Effects of different metal ions on the activity of CelA endoglucanase enzyme, Figure S3: Effect of GA treatment time on CelA endoglucanase activity, Figure S4: SDS-PAGE electrophoresis of the His-tag-purified CelA. Lane M: marker (kDa); Lane 2: crude CelA; Lane 3: purified CelA, Figure S5: pET21b-CelA-docT gene construction; Figure S6: SEM micrographs of IMAM RC membrane (a) before (b) after immobilization; Table S1: Escherichia coli strains and plasmids used in the study.

Author Contributions: Conceptualization, Y.-C.L.; methodology, Z.-H.W.; validation, P.N.; formal analysis, Z.-H.W.; investigation, Z.-H.W.; writing—original draft preparation, P.N., C.-H.K. and Y.-C.L.; writing—review and editing, P.N., C.-H.K. and Y.-C.L.; supervision, Y.-C.L. All authors have read and agreed to the published version of the manuscript.

Funding: This study was supported by the Ministry of Science and Technology of Taiwan (grant no. MOST 108-2221-E-005-041-MY2).

Data Availability Statement: Data are contained within the article.

Conflicts of Interest: The authors declare no conflict of interest.

References

1. Sharma, V.; Tsai, M.-L.; Chen, C.-W.; Sun, P.-P.; Patel, A.K.; Singhania, R.R.; Nargotra, P.; Dong, C.-D. Deep eutectic solvents as promising pretreatment agents for sustainable lignocellulosic biorefineries: A review. *Bioresour. Technol.* **2022**, *360*, 127631. [CrossRef] [PubMed]
2. Patel, S.K.; Kalia, V.C.; Joo, J.B.; Kang, Y.C.; Lee, J.K. Biotransformation of methane into methanol by methanotrophs immobilized on coconut coir. *Bioresour. Technol.* **2020**, *297*, 122433. [CrossRef] [PubMed]
3. Nargotra, P.; Sharma, V.; Bajaj, B.K. Consolidated bioprocessing of surfactant-assisted ionic liquid-pretreated *Parthenium hysterophorus* L. biomass for bioethanol production. *Bioresour. Technol.* **2019**, *289*, 121611. [CrossRef] [PubMed]
4. Amjith, L.; Bavanish, B. A review on biomass and wind as renewable energy for sustainable environment. *Chemosphere* **2022**, *293*, 133579. [CrossRef] [PubMed]
5. Sharma, S.; Nargotra, P.; Sharma, V.; Bangotra, R.; Kaur, M.; Kapoor, N.; Paul, S.; Bajaj, B.K. Nanobiocatalysts for efficacious bioconversion of ionic liquid pretreated sugarcane tops biomass to biofuel. *Bioresour. Technol.* **2021**, *333*, 125191. [CrossRef] [PubMed]
6. Kumari, D.; Singh, R. Pretreatment of lignocellulosic wastes for biofuel production: A critical review. *Renew. Sustain. Energy Rev.* **2018**, *90*, 877–891. [CrossRef]
7. Vasić, K.; Knez, Ž.; Leitgeb, M. Bioethanol production by enzymatic hydrolysis from different lignocellulosic sources. *Molecules* **2021**, *26*, 753. [CrossRef]
8. Kuo, C.-H.; Lee, C.-K. Enhanced enzymatic hydrolysis of sugarcane bagasse by N-methylmorpholine-N-oxide pretreatment. *Bioresour. Technol.* **2009**, *100*, 866–871. [CrossRef]
9. Patel, S.K.; Singh, M.; Kumar, P.; Purohit, H.J.; Kalia, V.C. Exploitation of defined bacterial cultures for production of hydrogen and polyhydroxybutyrate from pea-shells. *Biomass Bioenergy* **2012**, *36*, 218–225. [CrossRef]

10. Vadala, B.S.; Deshpande, S.; Apte-Deshpande, A. Soluble expression of recombinant active cellulase in *E. coli* using *B. subtilis* (natto strain) cellulase gene. *J. Genet. Eng. Biotechnol.* **2021**, *19*, 1–7. [CrossRef]
11. Hosseini, S.H.; Hosseini, S.A.; Zohreh, N.; Yaghoubi, M.; Pourjavadi, A. Covalent immobilization of cellulase using magnetic poly (ionic liquid) support: Improvement of the enzyme activity and stability. *J. Agric. Food Chem.* **2018**, *66*, 789–798. [CrossRef]
12. Gennari, A.; Führ, A.J.; Volpato, G.; de Souza, C.F.V. Magnetic cellulose: Versatile support for enzyme immobilization—A review. *Carbohydr. Polym.* **2020**, *246*, 116646. [CrossRef] [PubMed]
13. Kumar, V.; Patel, S.K.; Gupta, R.K.; Otari, S.V.; Gao, H.; Lee, J.K.; Zhang, L. Enhanced saccharification and fermentation of rice straw by reducing the concentration of phenolic compounds using an immobilized enzyme cocktail. *Biotechnol. J.* **2019**, *14*, 1800468. [CrossRef]
14. Jia, J.; Zhang, W.; Yang, Z.; Yang, X.; Wang, N.; Yu, X. Novel magnetic cross-linked cellulase aggregates with a potential application in lignocellulosic biomass bioconversion. *Molecules* **2017**, *22*, 269. [CrossRef] [PubMed]
15. Kuo, C.-H.; Liu, Y.-C.; Chang, C.-M.J.; Chen, J.-H.; Chang, C.; Shieh, C.-J. Optimum conditions for lipase immobilization on chitosan-coated Fe_3O_4 nanoparticles. *Carbohydr. Polym.* **2012**, *87*, 2538–2545. [CrossRef]
16. Patel, S.K.; Anwar, M.Z.; Kumar, A.; Otari, S.V.; Pagolu, R.T.; Kim, S.Y.; Kim, I.W.; Lee, J.K. Fe_2O_3 yolk-shell particle-based laccase biosensor for efficient detection of 2,6-dimethoxyphenol. *Biochem. Eng. J.* **2018**, *132*, 1–8. [CrossRef]
17. Kuo, C.-H.; Chen, G.-J.; Twu, Y.-K.; Liu, Y.-C.; Shieh, C.-J. Optimum lipase immobilized on diamine-grafted PVDF membrane and its characterization. *Ind. Eng. Chem. Res.* **2012**, *51*, 5141–5147. [CrossRef]
18. Singh, R.S.; Chauhan, K.; Kaur, N.; Kumar, N. Inulinase immobilization onto glutaraldehyde activated duolite XAD for the production of fructose from inulin. *Biocatal. Agric. Biotechnol.* **2020**, *27*, 101699. [CrossRef]
19. Patel, S.K.; Choi, S.H.; Kang, Y.C.; Lee, J.K. Large-scale aerosol-assisted synthesis of biofriendly Fe_2O_3 yolk–shell particles: A promising support for enzyme immobilization. *Nanoscale* **2016**, *8*, 6728–6738. [CrossRef]
20. Patel, S.K.; Choi, H.; Lee, J.K. Multimetal-based inorganic–protein hybrid system for enzyme immobilization. *ACS Sustain. Chem. Eng.* **2019**, *7*, 13633–13638. [CrossRef]
21. Nargotra, P.; Sharma, V.; Sharma, S.; Bangotra, R.; Bajaj, B.K. Purification of an ionic liquid stable cellulase from *Aspergillus aculeatus* PN14 with potential for biomass refining. *Environ. Sustain.* **2022**, *5*, 313–323. [CrossRef]
22. Wong, H.-L.; Hu, N.-J.; Juang, T.-Y.; Liu, Y.-C. Co-immobilization of xylanase and scaffolding protein onto an immobilized metal ion affinity membrane. *Catalysts* **2020**, *10*, 1408. [CrossRef]
23. Cheung, R.C.F.; Wong, J.H.; Ng, T.B. Immobilized metal ion affinity chromatography: A review on its applications. *Appl. Microbiol. Biotechnol.* **2012**, *96*, 1411–1420. [CrossRef] [PubMed]
24. Xi, C.R.; Di Fazio, A.; Nadvi, N.A.; Patel, K.; Xiang, M.S.W.; Zhang, H.E.; Deshpande, C.; Low, J.K.; Wang, X.T.; Chen, Y. A novel purification procedure for active recombinant human DPP4 and the inability of DPP4 to bind SARS-CoV-2. *Molecules* **2020**, *25*, 5392. [CrossRef] [PubMed]
25. Wu, C.-Y.; Suen, S.-Y.; Chen, S.-C.; Tzeng, J.-H. Analysis of protein adsorption on regenerated cellulose-based immobilized copper ion affinity membranes. *J. Chromatogr. A* **2003**, *996*, 53–70. [CrossRef]
26. Liu, Y.-C.; ChangChien, C.-C.; Suen, S.-Y. Purification of penicillin G acylase using immobilized metal affinity membranes. *J. Chromatogr. B* **2003**, *794*, 67–76. [CrossRef]
27. Kim, D.; Ku, S. Bacillus cellulase molecular cloning, expression, and surface display on the outer membrane of *Escherichia coli*. *Molecules* **2018**, *23*, 503. [CrossRef]
28. Zhou, Y.; Lu, Z.; Wang, X.; Selvaraj, J.N.; Zhang, G. Genetic engineering modification and fermentation optimization for extracellular production of recombinant proteins using *Escherichia coli*. *Appl. Microbiol. Biotechnol.* **2018**, *102*, 1545–1556. [CrossRef]
29. Huleani, S.; Roberts, M.R.; Beales, L.; Papaioannou, E.H. *Escherichia coli* as an antibody expression host for the production of diagnostic proteins: Significance and expression. *Crit. Rev. Biotechnol.* **2022**, *42*, 756–773. [CrossRef]
30. Mühlmann, M.; Forsten, E.; Noack, S.; Büchs, J. Optimizing recombinant protein expression via automated induction profiling in microtiter plates at different temperatures. *Microb. Cell Factories* **2017**, *16*, 1–12. [CrossRef]
31. Lin, T.-N.; Lin, S.-C. Metal chelate-epoxy bifunctional membranes for selective adsorption and covalent immobilization of a His-tagged protein. *J. Biosci. Bioeng.* **2022**, *133*, 258–264. [CrossRef] [PubMed]
32. Chen, G.-J.; Kuo, C.-H.; Chen, C.-I.; Yu, C.-C.; Shieh, C.-J.; Liu, Y.-C. Effect of membranes with various hydrophobic/hydrophilic properties on lipase immobilized activity and stability. *J. Biosci. Bioeng.* **2012**, *113*, 166–172. [CrossRef] [PubMed]
33. Modenez, I.A.; Sastre, D.E.; Moraes, F.C.; Marques Netto, C.G. Influence of glutaraldehyde cross-linking modes on the recyclability of immobilized lipase B from *Candida antarctica* for transesterification of soy bean oil. *Molecules* **2018**, *23*, 2230. [CrossRef] [PubMed]
34. Hua, C.; Li, W.; Han, W.; Wang, Q.; Bi, P.; Han, C.; Zhu, L. Characterization of a novel thermostable GH7 endoglucanase from *Chaetomium thermophilum* capable of xylan hydrolysis. *Int. J. Biol. Macromol.* **2018**, *117*, 342–349. [CrossRef] [PubMed]
35. Chaga, G.S. Twenty-five years of immobilized metal ion affinity chromatography: Past, present and future. *J. Biochem. Biophys. Methods* **2001**, *49*, 313–334. [CrossRef]
36. Li, T.; Gong, X.; Yang, G.; Li, Q.; Huang, J.; Zhou, N.; Jia, X. Cross-linked enzyme aggregates (CLEAs) of cellulase with improved catalytic activity, adaptability and reusability. *Bioprocess Biosyst Eng.* **2022**, *45*, 865–875. [CrossRef]
37. Zeng, R.; Jin, B.K.; Yang, Z.H.; Guan, R.; Quan, C. Preparation of a modified crosslinked chitosan/polyvinyl alcohol blended affinity membrane for purification of his-tagged protein. *J. Appl. Polym. Sci.* **2019**, *136*, 47347. [CrossRef]

38. Ahmad, R.; Khare, S.K. Immobilization of *Aspergillus niger* cellulase on multiwall carbon nanotubes for cellulose hydrolysis. *Bioresour. Technol.* **2018**, *252*, 72–75. [CrossRef]
39. Kaur, P.; Taggar, M.S.; Kalia, A. Characterization of magnetic nanoparticle–immobilized cellulases for enzymatic saccharification of rice straw. *Biomass Convers. Biorefin.* **2021**, *11*, 955–969. [CrossRef]
40. Saleem, M.; Rashid, M.; Jabbar, A.; Perveen, R.; Khalid, A.; Rajoka, M. Kinetic and thermodynamic properties of an immobilized endoglucanase from *Arachniotus citrinus*. *Process Biochem.* **2005**, *40*, 849–855. [CrossRef]
41. Guo, R.; Zheng, X.; Wang, Y.; Yang, Y.; Ma, Y.; Zou, D.; Liu, Y. Optimization of cellulase immobilization with sodium alginate-polyethylene for enhancement of enzymatic hydrolysis of microcrystalline cellulose using response surface methodology. *Appl. Biochem. Biotechnol.* **2021**, *193*, 2043–2060. [CrossRef] [PubMed]
42. Abbaszadeh, M.; Hejazi, P. Metal affinity immobilization of cellulase on Fe_3O_4 nanoparticles with copper as ligand for biocatalytic applications. *Food Chem.* **2019**, *290*, 47–55. [CrossRef]
43. Saha, K.; Verma, P.; Sikder, J.; Chakraborty, S.; Curcio, S. Synthesis of chitosan-cellulase nanohybrid and immobilization on alginate beads for hydrolysis of ionic liquid pretreated sugarcane bagasse. *Renew. Energy* **2019**, *133*, 66–76. [CrossRef]
44. Wang, Y.; Feng, C.; Guo, R.; Ma, Y.; Yuan, Y.; Liu, Y. Cellulase immobilized by sodium alginate-polyethylene glycol-chitosan for hydrolysis enhancement of microcrystalline cellulose. *Process Biochem.* **2021**, *107*, 38–47. [CrossRef]
45. Singh, S.; Gupta, P.; Bajaj, B.K. Characterization of a robust serine protease from *Bacillus subtilis* K-1. *J. Basic Microbiol.* **2018**, *58*, 88–98. [CrossRef]
46. Chien, H.-I.; Tsai, Y.-H.; Wang, H.-M.D.; Dong, C.-D.; Huang, C.-Y.; Kuo, C.-H. Extrusion puffing pretreated cereals for rapid production of high-maltose syrup. *Food Chem. X* **2022**, *15*, 100445. [CrossRef]
47. Tüzmen, N.; Kalburcu, T.; Denizli, A. Immobilization of catalase via adsorption onto metal-chelated affinity cryogels. *Process Biochem.* **2012**, *47*, 26–33. [CrossRef]
48. Zhu, H.; Zhang, Y.; Yang, T.; Zheng, D.; Liu, X.; Zhang, J.; Zheng, M. Preparation of immobilized Alcalase based on metal affinity for efficient production of bioactive peptides. *LWT* **2022**, *162*, 113505. [CrossRef]
49. Chen, C.-I.; Ko, Y.-M.; Lien, W.-L.; Lin, Y.-H.; Li, I.-T.; Chen, C.-H.; Shieh, C.-J.; Liu, Y.-C. Development of the reversible PGA immobilization by using the immobilized metal ion affinity membrane. *J. Membr. Sci.* **2012**, *401*, 33–39. [CrossRef]
50. Morrison, D. Transformation and preservation of competent bacterial cells by freezing. In *Methods in Enzymology*; Elsevier: Amsterdam, The Netherlands, 1979; Volume 68, pp. 326–331.
51. Miller, G.L. Use of dinitrosalicylic acid reagent for determination of reducing sugar. *Anal. Chem.* **1959**, *31*, 426–428. [CrossRef]
52. Bradford, M.M. A rapid and sensitive method for the quantitation of microgram quantities of protein utilizing the principle of protein-dye binding. *Anal. Biochem.* **1976**, *72*, 248–254. [CrossRef]

Article

Application of Endoxylanases of *Bacillus halodurans* for Producing Xylooligosaccharides from Empty Fruit Bunch

Chanakan Thirametoakkhara [1], Yi-Cheng Hong [1], Nuttapol Lerkkasemsan [2], Jian-Mao Shih [1], Chien-Yen Chen [3] and Wen-Chien Lee [1,*]

1   Department of Chemical Engineering, National Chung Cheng University, 168 University Road, Min-Hsiung, Chiayi 62102, Taiwan
2   Department of Chemical Engineering, School of Engineering, King Mongkut's Institute of Technology Ladkrabang, Ladkrabang, Bangkok 10520, Thailand
3   Department of Earth and Environmental Sciences, National Chung Cheng University, 168 University Road, Min-Hsiung, Chiayi 62102, Taiwan
*   Correspondence: chmwcl@ccu.edu.tw

Abstract: Endo-1,4-β-xylanase catalyzes the random hydrolysis of β-1,4-D-xylosidic bonds in xylan, resulting in the formation of oligomers of xylose. This study aims to demonstrate the promise of endoxylanases from alkaliphilic *Bacillus halodurans* for the production of xylooligosaccharides (XOS) from oil palm empty fruit bunch (EFB) at high pH. Two enzyme preparations were employed: recombinant endoxylanase Xyn45 (GH10 xylanase) and nonrecombinant endoxylanases, a mixture of two extracellular endo-1,4-β-xylanases Xyn45 and Xyn23 (GH11 xylanase) produced by *B. halodurans*. EFB was first treated with an alkaline solution. Then, the dissolved xylan-containing fraction was retained, and a prepared enzyme was added to react at pH 8 to convert xylan into XOS. Compared with the use of only Xyn45, the combined use of Xyn45 and Xyn23 resulted in a higher yield of XOS, suggesting the synergistic effect of the two endoxylanases. The yield of XOS obtained from EFB was as high as 46.77% ± 1.64% (w/w), with the xylobiose-to-xylotriose ratio being 6:5. However, when the enzyme activity dose was low, the product contained more xylotriose than xylobiose. Four probiotic lactobacilli and bifidobacteria grew well on a medium containing XOS from EFB. The presence of XOS increased cell mass and reduced pH, suggesting that XOS promoted the growth of probiotics.

Keywords: endo-1,4-β-xylanases; xylooligosaccharides (XOS); empty fruit bunch (EFB); *Bacillus halodurans*; xylan

Citation: Thirametoakkhara, C.; Hong, Y.-C.; Lerkkasemsan, N.; Shih, J.-M.; Chen, C.-Y.; Lee, W.-C. Application of Endoxylanases of *Bacillus halodurans* for Producing Xylooligosaccharides from Empty Fruit Bunch. *Catalysts* **2023**, *13*, 39. https://doi.org/10.3390/catal13010039

Academic Editors: Chia-Hung Kuo, Chwen-Jen Shieh and Hui-Min David Wang

Received: 23 November 2022
Revised: 16 December 2022
Accepted: 21 December 2022
Published: 25 December 2022

1. Introduction

The alkaliphilic *Bacillus halodurans* can secrete enzymes to break down insoluble xylan, the main component of hemicellulose, into a carbon source for cell growth. According to the whole-genome sequencing data of *B. halodurans* C-125 [1], this alkaliphilic bacterium can synthesize endo-β-1,4-xylanase, acetylxylan esterase, α-L-arabinofuranosidase, α-glucuronidase, β-xylosidase, xylan β-1,4-xylosidase, and a reducing-end xylose-releasing exo-oligoxylanase to completely utilize xylan. During its growth on xylan-containing substrates, *B. halodurans* induces the production of some xylanolytic enzymes and their subsequent secretion into the environment; however, some enzymes are bound by cells [2,3]. Among xylanolytic enzymes, two isozymes of endo-β-1,4-xylanase are considered extracellular because they contain signal peptides. Our previous study indicated that *B. halodurans* BCRC 910501 can synthesize extracellular endo-1,4-β-D-xylanases, which are highly active and could effectively hydrolyze xylan extracted from various types of agricultural waste to produce xylo-oligosaccharides (XOS). These extracellular enzymes have potential for industrial application and are suitable for use in free and immobilized forms [4].

Endo-1,4-β-xylanase (EC 3.2.1.8) is an enzyme belonging to the class of glycoside hydrolases and catalyzes the random hydrolysis of β-1,4-D-xylosidic bonds within xylan [5].

Two endo-1,4-β-xylanases purified from *B. halodurans* BCRC 910501 had molecular weights of 45 and 23 kDa, respectively, and at 37 °C, they exhibited enzymatic activity over the pH range 5.0–11.0 [6]. This bacterium was previously named *Bacillus firmus* [6,7], but later reidentified as *B. halodurans* on the basis of the findings of 16S rDNA gene sequencing. The two cellulase-free endoxylanases are abbreviated Xyn45 and Xyn23, respectively, due to their molecular weights. Xyn45 belongs to glycoside hydrolase family 10 (GH10), which is characterized by high molecular weight and low pI. Xyn23 belongs to family 11 (GH11), which is characterized by low molecular weight and high pI [7]. These enzymes cause the cleavage of glycosidic bonds inside the xylan backbone, reducing the degree of polymerization of substrates and releasing xylose, with the main product being XOS [8–10]. The breakdown of xylan by endoxylanases mainly leads to the production of xylobiose and xylotriose [9]. Because of its relatively large molecular size, the endoxylanase of GH10 is more capable of attacking the nonreducing end of the glycosidic bond than that of GH11. The amino acid sequences of Xyn45 and Xyn23 are identical to those encoded by two endoxylanase genes in *B. halodurans* C-125: BH2120 and BH0899, respectively [11]. The gene (*BH0899*) encoding a GH11 *xylanase* from *B. halodurans* strain C-125 was overexpressed in *Kluyveromyces lactis* [12]. A cellulase-free GH11 endo-1,4-β-xylanase with a molecular mass of 24 kDa, as determined through sodium dodecyl sulfate–polyacrylamide gel electrophoresis (SDS–PAGE), was identified in *B. halodurans* PPKS-2 [13]. However, this endoxylanase exhibited maximal activity at pH 11 and 70 °C, which are slightly different from the conditions for maximal activity of the enzyme of *B. halodurans* BCRC 910501 [7].

Endoxylanases are key enzymes in the production of prebiotic XOS from biomass. XOS have varying degrees of polymerization (DP). They are oligomers composed of 2–10 xylose units through β-1,4-xylosidic linkages, and they are called xylobiose (DP2), xylotriose (DP3), xylotetraose (DP4), and so on depending on the number of xylose residues they contain. XOS can be produced from lignocellulosic biomass through alkaline extraction of xylan and subsequent enzymatic hydrolysis. Breaking the internal β-1,4-xylosidic bonds of xylan by enzymatic and/or chemical hydrolysis leads to XOS formation. Although some chemical methods are available for producing XOS [14], such as nonisothermal autohydrolysis treatment at elevated temperature, the enzymatic method is preferred in the food industry because it produces no undesirable side reactions or products [15]. Oil palm empty fruit bunch (EFB) is a type of lignocellulosic biomass produced in a large quantity during the production of palm oil; it accounts for approximately 20% of the byproducts produced during palm oil production [16]. In addition to cellulose and lignin, hemicellulose with xylan is the main component of EFB. The proportion of dry biomass in EFB varies widely, ranging from 14.62% to 33.6% (Table 1) [17–21].

Table 1. Chemical composition of EFB.

Cellulose (%)	Hemicellulose (%)	Lignin (%)	Ash (%)	Extractive (%)	References
37.26	14.62	31.64	6.69	1.34	[17]
35.8	19.9 *	32.1	-	-	[18]
59.7	22.1	18.1	-	-	[19]
47.6	28.1	13.1	-	-	[20]
28.3	36.6	35.1	-	-	[21]

*, counted xylan only; -, not determined.

Endoxylanases from *Thermomyces lanuginosus* have been used to produce XOS from EFB [22,23]. Before enzymatic hydrolysis by an immobilized enzyme in a packed bed column reactor for XOS production, xylan was extracted from EFB in one study by using a complex process. Alkaline extraction (12% NaOH) of EFB was performed in combination with steam explosion for prolonged treatment at elevated temperature (120 °C, 2 h), followed by further hydrolysis at 25 °C for 16 h. Finally, xylan was precipitated from the neutralized crude hydrolysate by using isopropanol [22]. In another process, a two-step chemical treatment was employed to recover the hemicellulose-rich fraction of EFB. EFB fibers were sequentially treated with peracetic acid and alkaline peroxide. Peracetic acid treatment removed approximately 50% of its lignin. The subsequent alkaline peroxide treatment recovered hemicellulose with xylan for enzymatic hydrolysis [23]. The present study explored the use of *B. halodurans* endo-1,4-β-xylanases to produce XOS from EFB, which was enzymatically reacted at high pH only after being subjected to alkali treatment.

This study determined whether XOS obtained from EFB can promote probiotic growth. Prebiotics are nondigestible food ingredients that can modulate microbiota to provide health benefits to the host. Moreover, prebiotics are considerably effective and essential for many medical applications. They do not contribute to the body's nutrition and are not absorbed but exert a profound effect on human gut flora. Prebiotics, such as XOS, can promote probiotic growth and enhance the function of probiotics to ensure a healthy balance [24]. Prebiotics could stimulate the growth of *Bifidobacterium* and *Lactobacillus* in the human intestine [25]. The main characteristics of XOS are heat resistance, acid and alkali resistance, sweet taste, and low calories. Moreover, XOS are not decomposed by bacteria in the oral cavity to produce acidic substances and cause tooth decay [26]. Recently, XOS have been recognized as novel non-digestible oligosaccharides widely used as a functional food ingredient or supplement. The global market for prebiotic XOS is growing rapidly due to their wide application in animal feed, human food additives, and medicine [27]. Given the growing global demand for prebiotics, a cost-effective method for producing XOS from various xylan-rich biomasses, including EFB, should be developed. The objective of this study was to confirm a reliable enzyme source for the efficient production of XOS from EFB. Since alkaline solutions are commonly used to dissolve xylan in EFB as a substrate, it is expected that enzymes can be applied preferentially at higher pH values. Xyn45 and Xyn23 from alkaliphilic *B. halodurans* are thus advantageous for XOS production.

2. Results

2.1. Production of Endo-1,4-β-Xylanases

B. halodurans BCRC 910501 was grown in Emerson medium containing both xylan and XOS, and endo-1,4-β-xylanases were synthesized and secreted into the medium. SDS–PAGE of extracellular proteins revealed bands corresponding to the endoxylanases Xyn45 and Xyn23, which have molecular masses of 45 and 23 kDa, respectively (Figure 1). Xylan can be extracted from agricultural waste, such as pineapple peel. Sodium hydroxide treatment destroyed the structure of lignocellulose and dissolved lignin and hemicellulose. Then, xylan was precipitated through soaking the alkali-soluble part in alcohol. XOS (including xylobiose and xylotriose) were obtained after enzymatic hydrolysis of xylan by endoxylanase, and both xylan and XOS could be used as substrates on which *B. halodurans* could synthesize endoxylanases. The xylanase activity per unit volume induced by pineapple peel xylan (0.5%, *w/w*) was 50.8 U/mL.

XOS containing mainly xylobiose and xylotriose induced the production of endoxylanases, and the yield of the enzyme could be improved, irrespective of the increase in the amount of xylobiose or xylotriose. To prepare the induction agent, the product generated through alkali treatment, alcohol precipitation and the enzymatic reaction of pineapple peel was filtered through a hollow fiber membrane to obtain a solution containing xylobiose (DP2) and xylotriose (DP3). The xylanase activity per unit volume increased to 135.6 U/mL after 5 days of incubation with 0.5% xylan plus 0.57% XOS (0.5% DP2 and 0.07% DP3; the xylan-DP2-DP3 ratio of 1:1:0.14 in Table 2). The specific activity was 106.1 U/mg protein. However, the production of enzymes with more XOS above this level led to reduced activity. Although the amount of protein decreased only slightly, the activity declined to 121.2 U/mL when the proportion of XOS was increased from 0.57% to 0.63% (0.5% DP2 and 0.13% DP3; the xylan-DP2-DP3 ratio of 1:1:0.26). Thus, an appropriate amount of XOS should be used as an inducer because bacteria can take up xylobiose and xylotriose and hydrolyze them into xylose. Once the intracellular xylose concentration has increased, bacteria no longer secrete enzymes.

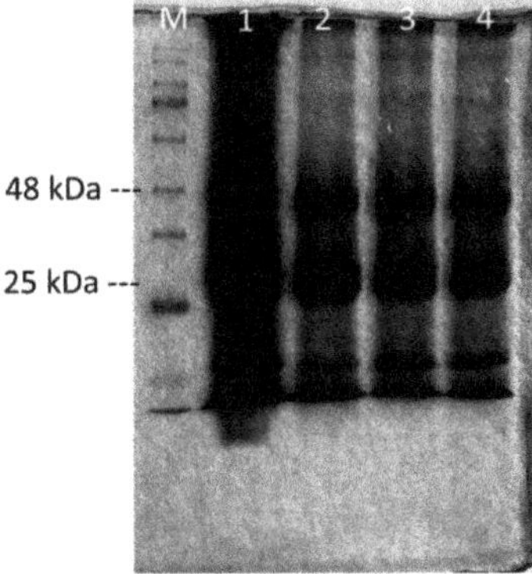

Figure 1. SDS–PAGE gel image of extracellular proteins produced by *B. halodurans*. Lane M is the marker (PageRuler prestained protein ladder), lane 1 is the extracellular protein induced by the combination of xylan and xylo-oligosaccharides, and lanes 2 and 4 are those induced by xylo-oligosaccharides.

Table 2. Concentration and activity of extracellular proteins induced by *B. halodurans* with xylan and xylo-oligosaccharides (DP2 and DP3).

Inducer (Xylan:DP2:DP3) *	1:0:0	1:0.15:0.04	1:0.28:0.04	1:1:0.04	1:1:0.14	1:1:0.26
Protein concentration (mg/mL)	1.03 ± 0.04	0.85 ± 0.02	0.88 ± 0.04	1.09 ± 0.25	1.27 ± 0.02	1.25 ± 0.08
Activity (U/mL)	50.8 ± 9.3	85.1 ± 4.5	89.1 ± 49.5	127.2 ± 54.4	135.6 ± 27.8	121.2 ± 15.8
Specific activity (U/mg)	49.6 ± 11.0	100.8 ± 7.5	102.1 ± 60.1	114.4 ± 41.1	106.1 ± 20.1	96.5 ± 11.6

* The mass ratio of xylan, DP2 and DP1 is based on 0.5% as a unit.

Recombinant endo-1,4-β-xylanase Xyn45 was produced using the bacterium harboring the plasmid-containing gene of Xyn45, *Escherichia coli* BL21(DE3)-pET29a(+)-xyn45. After induction with isopropyl β-D-1-thiogalactopyranoside (IPTG), the recombinant bacterium overexpressed the target protein Xyn45 and secreted it into the medium, suggesting that the original signal peptide from *B. halodurans* performed well in *E. coli*. When the recombinant *E. coli* was induced with 0.2 mM IPTG at 25 °C for 18 h, the harvested fermentation broth contained 0.21 mg/mL of protein and exhibited xylanase activity per unit volume of 52.1 U/mL, corresponding to specific activity of 240.2 U/mg.

2.2. XOS Production

XOS production was divided into two major steps. First, xylan was extracted from EFB through alkaline pretreatment. Second, the supernatant (the mixture of hemicellulose and lignin) obtained from the alkaline pretreatment was used to produce XOS through enzymatic hydrolysis. When the recombinant endo-1,4-β-xylanase Xyn45 produced by *E. coli* BL21 harboring the plasmid pET29a(+)-xyn45 was used as the catalyst, the concentration of xylobiose increased with the reaction time (Figure 2). When the xylotriose concentration increased at the beginning and then decreased in the fourth hour, it was degraded into xylobiose and xylose. After reaction for 48 h, the yield of XOS was 19.58% ± 0.52%, with a xylobiose-to-xylotriose ratio of 10:3.

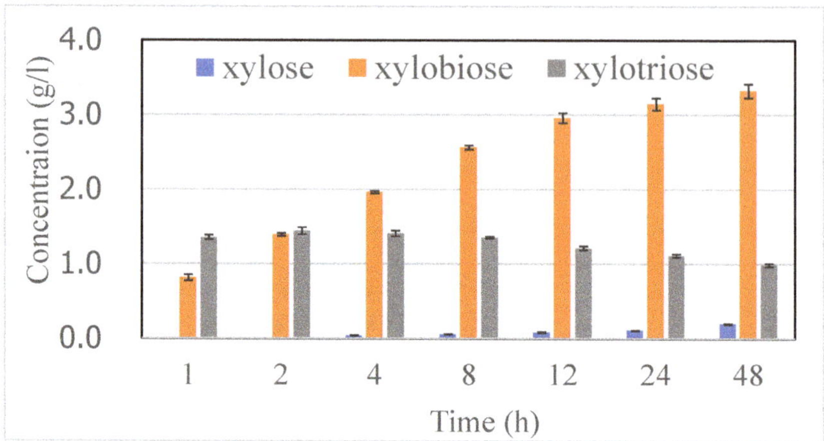

Figure 2. Time course of XOS production from EFB when using the recombinant endo-1,4-β-xylanase Xyn45. The enzyme dose was 0.5 U/mL. Each error bar represents the standard deviation of triplicate experiments.

When extracellular endo-1,4-β-xylanases (a mixture of Xyn45 and Xyn23) produced by wild-type *B. halodurans* were used on the basis of the same activity dose, the time course of XOS production was different. As illustrated in Figure 3, the concentrations of xylobiose and xylotriose increased with time from the beginning to the end of the reaction. A comparison of Figures 2 and 3 indicates that combined use of the endo-1,4-β-xylanases Xyn45 and Xyn23 was more efficient than use of the single endo-1,4-β-xylanase Xyn45. Because of its steric effect, Xyn45 could not cleave bonds in the internal xylan chain. However, Xyn23 could cleave bonds in the internal xylan chain. The highest yield of XOS (38.15% ± 1.30%) with a xylobiose-to-xylotriose ratio of 5:6 was obtained at 48 h through enzymatic action of the mixed endoxylanases Xyn45 and Xyn23.

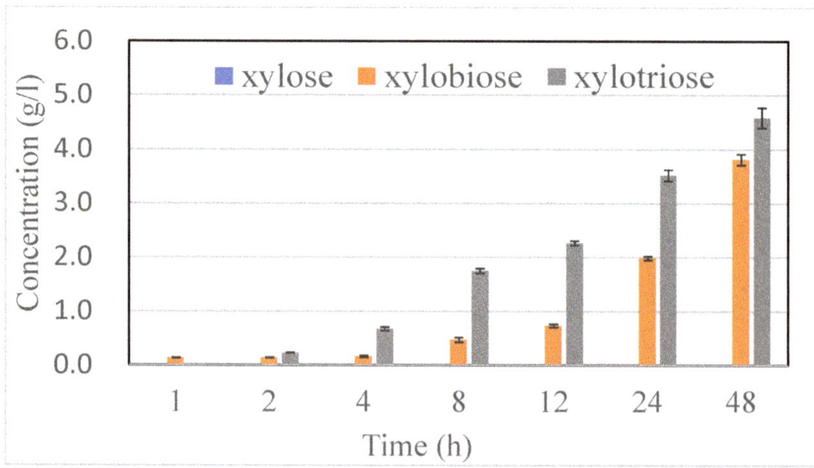

Figure 3. Time course of XOS production from EFB when using nonrecombinant endo-1,4-β-xylanases. The enzyme dose was 0.5 U/mL. Each error bar represents the standard deviation of triplicate experiments.

When a mixture of extracellular Xyn45 and Xyn23 produced by *B. halodurans* was used, the time course of XOS production varied depending on the dosage of the enzyme. When a low dose was used (0.5 U/mL), the concentrations of both xylobiose and xylotriose increased with time up to 48 h (Figure 3). However, use of a high dose (2 U/mL) resulted in higher total concentrations of xylobiose and xylotriose (Figure 4). As presented in Figure 4, the concentrations of xylobiose and xylose increased with time because soluble high-DP xylo-oligomers were gradually degraded into low-DP oligomers. The xylotriose concentration increased from the beginning of the reaction to the 24th hour and then decreased because it was degraded to produce xylobiose and xylose. A comparison of Figures 3 and 4 indicates that due to the higher dosage of the enzyme, the reaction speed of the latter (Figure 4) was faster than the former (Figure 3). The yield of XOS (xylobiose and xylotriose) was highest after 24 h of reaction, reaching 46.77% ± 1.64% (*w/w*). The ratio of xylobiose to xylotriose was 6:5.

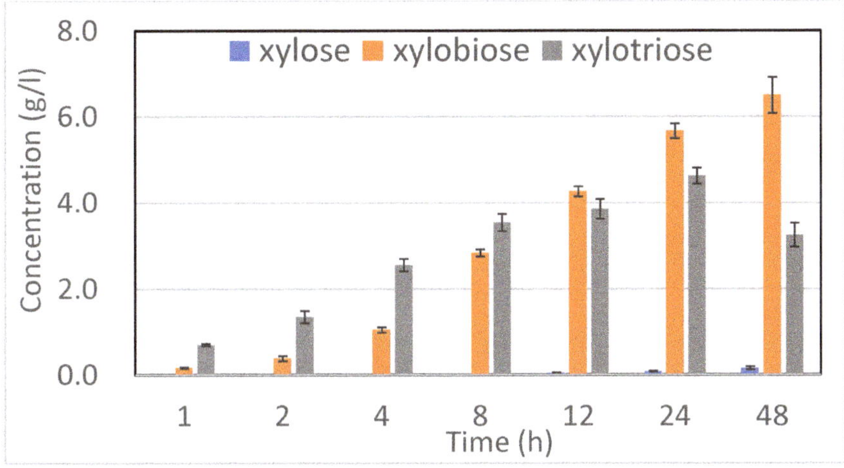

Figure 4. Time course of XOS production from EFB when using nonrecombinant endo-1,4-β-xylanases. The enzyme dose was 2 U/mL. Each error bar represents the standard deviation of triplicate experiments.

2.3. Promotion of the Growth and Metabolism of Bifidobacteria and Lactobacillus

The growth of *Bifidobacterium animalis*, *Bif. catenulatum* 14667, *Lactobacillus plantarum* 10069, and *L. acidophilus* NCFM on deMan, Rogosa, and Sharpe (MRS) medium was examined by measuring the optical density (OD) at 600 nm and the pH value. All *Bifidobacteria* and *Lactobacillus* species grew well on the MRS medium (Supplemental Table S1). *L. plantarum* 10069 had the highest growth rate on the MRS medium within 72 h, resulting in the maximum OD_{600} of 1.551 ± 0.008 and the lowest pH of 3.73 ± 0.02, followed by *Bif. catenulatum* 14667, *L. acidophilus* NCFM, and *Bif. animalis*.

Supplemental Table S2 presents the ability of *Bif. animalis*, *Bif. catenulatum* 14667, *L. plantarum* 10069, and *L. acidophilus* NCFM to ferment glucose as a carbon source in the MRS medium. *L. plantarum* 10069 had the highest growth rate on the MRS medium containing 2% (*w/v*) glucose in 72 h, resulting in a maximum OD_{600} of 1.543 ± 0.002 and the lowest pH of 3.41 ± 0.01, followed by *Bif. catenulatum* 14667, *L. acidophilus* NCFM, and *Bif. animalis*.

Supplemental Table S3 presents results obtained when XOS (originally 34.2 g/L with a xylobiose-to-xylotriose ratio of 6:5) from EFB instead of glucose in the MRS medium was fermented by *Bif. animalis*, *Bif. catenulatum* 14667, *L. plantarum* 10069, and *L. acidophilus* NCFM. *L. plantarum* 10069 had the highest growth rate on MRS medium containing 2% (*w/v*) XOS from EFB in 72 h, resulting in the maximum OD_{600} of 1.691 ± 0.007 and the lowest pH of 4.02 ± 0.01, followed by *Bif. catenulatum* 14667, *L. acidophilus* NCFM, and *Bif. animalis*.

Figure 5 presents the comparison of ΔOD_{600} and ΔpH for *Bifidobacteria* and *Lactobacillus* growth. The four bacterial strains grew well on the MRS medium containing XOS from EFB.

Compared with glucose as a supplemented carbon source, XOS resulted in a higher OD and lower pH. Among the four bacteria, *L. plantarum* 10069 exhibited the highest growth rate and maximum pH decrease on all the media, followed by *Bif. catenulatum* 14667, *L. acidophilus* NCFM, and *Bif. animalis*. The maximum ΔOD_{600} of 1.28 and ΔpH of -3.05 were obtained by cultivating *L. plantarum* 10069 on the MRS medium containing XOS from EFP for 72 h. The decline in pH in the culture of the probiotic strains was attributed to the production of short-chain fatty acids. The findings indicated that the prepared XOS enhanced the growth of all the examined probiotic stains.

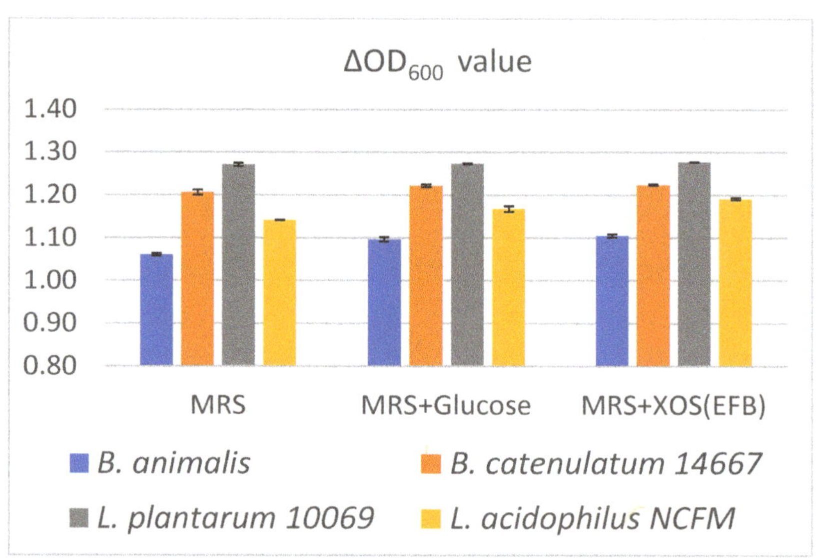

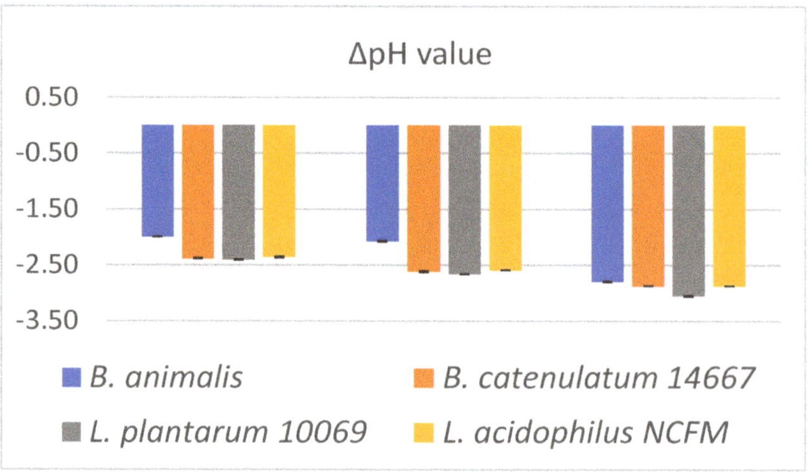

Figure 5. Biomass increment (ΔOD_{600}) and pH decline (ΔpH) of *Bifidobacteria* and *Lactobacillus* cultured in MRS medium without and with glucose or XOS for 72 h. Each error bar represents the standard deviation of triplicate experiments.

3. Discussion

Enzymes obtained from *Bacillus* strains are not only of academic but also industrial importance. The alkaliphilic *B. halodurans* C-125 was identified as a xylanase producer in 1985 [28]. The results of the present study revealed that wild-type *B. halodurans* BCRC 910501 produced two extracellular endo-1,4-β-xylanases, which were active under alkaline conditions and converted lignocellulosic xylan into xylo-oligomers. In *Bacillus* and other Gram-positive bacteria, the expression of extracellular xylanases is subject to carbon catabolite repression, a regulatory mechanism by which the expression of several catabolite genes and others are repressed in the present metabolizable carbon source, such as glucose, fructose, and mannose. For example, there is a report on glucose catabolite repression on the expression of xylanase gene (*xynA*) in *B. stearothermophilus* [29]. Expression of extracellular xylanase was repressed in a medium containing glucose, suggesting that carbon catabolite repression plays a role in regulating *xynA* [30]. Thus, in this study, we used a glucose-free medium for the production of endoxylanases by *B. halodurans*. Either xylan or XOS were present in the medium as the inducer for the expression of Xyn45 and Xyn23. The end-xylanases Xyn45 and Xyn23 with molecular weights of 45 and 23 kDa, respectively, were both secreted into the extracellular space by the wild-type bacterium *B. halodurans*. Short XOS, xylobiose (DP2) and xylotriose (DP3), effectively induced the expression of endoxylanases (Table 2). Moreover, the results revealed that the induction of xylobiose was slightly better than that of xylotriose. Although studies have extensively determined the utilization of xylan by cells and carbon catabolite repression in *B. subtilis* [31–34], the regulatory mechanisms of endoxylanase genes are not yet fully understood. For example, the presence of a cis-acting catabolite responsive element in endoxylanase genes in *B. halodurans* and the interaction of this element with the global transcription regulator should be evaluated.

Although both Xyn45 and Xyn23 catalyzed hydrolysis of the linear polysaccharide β-1,4-xylan, the product patterns were slightly different. The endoxylanase in the GH10 class is usually called XynA [35]. The GH10 endoxylanase from *B. halodurans* BCRC 910501, named Xyn45, was stable over a wide pH range, including under neutral and alkaline conditions, similar to the endoxylanases from *B. halodurans* TSEV1 [36] and *B. halodurans* S7 [37]. To obtain more benefits, the endoxylanase gene should not be overexpressed in its native host cell due to the carbon catabolite repression described previously. Thus, we constructed recombinant-strain *E. coli* BL21(DE3)-pET29a(+)-xyn45 to overexpress the endoxylanase Xyn45. Higher specific activity of xylanase was achieved using recombinant bacterial cells (240.2 U/mg) compared to wild-type *B. halodurans* (114.4 U/mg, the maximum value in Table 2).

EFB is a rich source of hemicellulose, which can be extracted through alkaline pretreatment. Endoxylanases were used to generate XOS by hydrolyzing xylan chains, that is, hemicellulosic polysaccharides, in EFB. When GH10 Xyn45 [recombinant protein produced by *E. coli* BL21(DE3)-pET29a(+)-xyn45] was used alone, due to its large molecular size, it mainly acted on the ends of xylan chains. Thus, the reaction began with the generation of xylobiose and xylotriose. This is similar to the finding that XynA (a GH10 endoxylanase) in *B. halodurans* S7 mainly resulted in the production of xylobiose and xylotriose unless the reaction was performed at high temperature, such as 65 °C [38]. With an increase in time, some xylotriose was further hydrolyzed into xylose and xylobiose, and more xylobiose was produced due to the hydrolysis of terminal xylan chains. Thus, the major reaction product was xylobiose. However, at the same activity dose (0.5 U/mL) of nonrecombinant endo-1,4-β-xylanases (a mixture of Xyn45 and Xyn23 produced by *B. halodurans*), the product distribution was considerably different. In addition to the cleavage of xylan chain ends by GH10 Xyn45 to yield xylobiose and xylotriose, the action of GH11 Xyn23 generated oligomers, including xylotriose, through cleavage in the middle of xylan chains. Therefore, xylotriose was the main reaction product. As revealed in the literature, a synergistic effect could be observed in a 48 h hydrolysis experiment, and the degradation efficiency (reducing sugar generation) of the double enzyme was higher than that of the single enzyme [39]. Because of the synergistic effect of Xyn45 and Xyn23, the overall yield of XOS (the sum of xylobiose and xylotriose) after reaction for 48 h was higher than that obtained when using a single enzyme (Xyn45). Using a mixture of enzymes nearly doubled the yield of XOS compared to using a single enzyme.

The results indicated that the reaction rate was proportional to the activity-based enzyme dose. When the dose of nonrecombinant endo-1,4-β-xylanases was increased from 0.5 to 2 U/mL, the formation of xylobiose and xylotriose was accelerated. At the 12th hour of the reaction, the concentration of xylobiose began to exceed the concentration of xylotriose, and this phenomenon was observed until the end of the reaction because xylotriose was hydrolyzed into xylose and xylotriose. At the end of the reaction, we discovered improved yield of XOS and more xylobiose than xylotriose. These experimental results indicate that by appropriately controlling the enzyme dosage and reaction time, the distribution of products can be manipulated and the yield of xylose as a byproduct can be kept very low. After 48 h of reaction, the concentrations of xylose were respectively 0.195 ± 0.004 g/L, 0 g/L, and 0.156 ± 0.034 g/L, corresponding to the data in Figures 2–4.

XOS are prebiotics that confer a health benefit on hosts by modulating microbiota. XOS can reduce intestinal pathogens, blood cholesterol, triglycerides, and glucose and can enhance antioxidant activity. Enhancement of the growth and metabolic activity of prebiotic bacteria can be quantified by an increase in OD and a decrease in pH. The decline in pH can be due to the production of short-chain fatty acids, the main metabolites produced by probiotics. The four probiotic bacteria investigated in this study—*Bif. animalis*, *Bif. catenulatum* 14667, *L. plantarum* 10069, and *L. acidophilus* NCFM—grew well on MRS medium containing XOS from EFB. Among them, *L. plantarum* 10069 exhibited the highest growth on the XOS-containing medium. Similarly, XOS were tested as a prebiotic for two probiotic bacteria, and XOS more efficiently enhanced the growth of *L. plantarum* WU-P19 than that of *Bif. bifidum* TISTR 2129 [23]. The ability of *Lactobacillus* strains to compete with other gut bacteria can be determined by examining their XOS utilization pattern [40]. In summary, XOS caused an increase in cell mass (OD) and a decrease in pH, suggesting that XOS enhanced the growth of all the studied probiotic strains. XOS from other types of lignocellulose waste have been demonstrated to exert similar prebiotic effects [41–43].

4. Materials and Methods

4.1. Preparation of Nonrecombinant Endoxylanases

Two types of enzymes were prepared in this study: (1) nonrecombinant endo-1,4-β-xylanases (a mixture of Xyn45 and Xyn23) produced by wild-type *B. halodurans* and (2) recombinant endo-1,4-β-xylanase Xyn45.

To produce the mixture of nonrecombinant endoxylanases, *B. halodurans* BCRC 910501 was grown on xylan-containing medium to induce the production of extracellular proteins. An inoculum was prepared by growing the bacterium in 5 mL of basal salt medium (composed of 1 g/L K_2HPO_4, 5 g/L NaCl, 0.2 g/L $MgSO_4 \cdot 7H_2O$, 5 g/L peptone, 2 g/L yeast extract, and 10 g/L glucose) at 37 °C for 16 h. Then, 5 mL of the precultured bacterial solution was poured into 100 mL of glucose-free Emerson medium (composed of 1 g/L K_2HPO_4, 0.2 g/L $MgSO_4 \cdot 7H_2O$, 5 g/L peptone, and 5.5 g/L yeast extract; pH 10) supplemented with xylan or a mixture of xylan and XOS from pineapple peel as the sole carbon source. The culture was grown at 37 °C with shaking at 175 rpm for 5 days. Then, the fermentation broth was centrifuged, and the obtained supernatant contained extracellular proteins secreted by bacterial cells into the culture medium.

Xylan was prepared from pineapple peel through alkaline pretreatment followed by ethanol precipitation. Briefly, pineapple peel waste was soaked in 4% sodium hydroxide solution at a solid-to-liquid ratio of 1:15 (*w/v*) in an incubator; the waste was incubated at 30 °C with shaking at 100 rpm for 24 h. Subsequently, after centrifugation at $2560 \times g$ to separate the precipitate out of the solution, the supernatant was adjusted to pH 7 by using concentrated hydrochloric acid. Three volumes of 95% ethanol were added to the supernatant, and the mixture was placed in a refrigerator at 4 °C. After 24 h, the sample was centrifuged to separate the precipitate (hemicellulose) from the solution. The precipitate (xylan) was dried in an oven and then used to induce the production of endoxylanases by *B. halodurans*.

Combinations of xylan and XOS containing certain amounts of xylobiose and xylotriose were also used for the induction of endoxylanases. To produce xylobiose and xylotriose, xylan obtained from pineapple peel was hydrolyzed using the recombinant endoxylanase Xyn45 (prepared in accordance with the protocol described in the next section). Hemicellulose (xylan) from pineapple peel was mixed with 100 mM Tris HCl buffer (pH 8.0) to obtain 2% (w/w) hemicellulose solution. Then, 2 U/mL Xyn45 was added, and the reaction was conducted at 50 °C in a water bath with shaking at 50 rpm for 24 h. Subsequently, the reaction mixture was placed in a water bath at 100 °C for 5 min to inactivate the enzyme and then centrifuged at 2560× g at 25 °C for 20 min. The supernatant was filtered using a 0.22-μm filter to yield XOS solution. After it was decolorized using the anion exchange resin HYDROLUX S5458, the XOS solution was ultrafiltered through a 1-kD hollow fiber membrane to a tenth of the original volume of the filtrate. High-DP oligomers and unhydrolyzed xylan were removed from the XOS solution. Finally, the filtrate was concentrated in a vacuum concentrator, and the concentration of each oligosaccharide in it was measured through high-performance liquid chromatography (HPLC) performed on a Rezex RSO-Oligosaccharide Ag+ 4% column (Phenomenex, Torrance, CA, USA).

4.2. Preparation of Recombinant Endoxylanase Xyn45

Recombinant *B. halodurans* endo-1,4-β-xylanase was prepared from *E. coli* BL21(DE3)-pET29a(+)-xyn45. To construct the recombinant plasmid, the gene encoding Xyn45 in *B. halodurans* was amplified through polymerase chain reaction (PCR) by using the forward primer 5′-GGA ATT CCA TAT GAT TAC ACT TTT TAG AAA GCC TTT TGT TGC TGG G-3′ and the reverse primer 5′-CCG CTC GAG CTA ATC AAT AAT TCT CCA GTA AGC AGG TTT CAC TCG-3′. The PCR-amplified product was purified and double-digested using the restriction endonucleases NdeI and XhoI. Furthermore, the vector plasmid pET-29a(+) was purified and double digested using the same restriction endonucleases. After ligation of the PCR-amplified product and vector plasmid, the resulting recombinant plasmid was transformed into *E. coli* DH5a. This recombinant plasmid was named pET-29a(+)-xyn45. Confirmed by DNA sequencing, it was transformed into *E. coli* BL21 (DE3) to yield a recombinant strain called *E. coli* BL21(DE3)-pET29a(+)-xyn45.

To produce the endoxylanase Xyn45, 5 mL of the recombinant bacterial preculture was inoculated into 50 mL of PM medium containing 0.05 mg/mL kanamycin. The bacterial culture was grown at 37 °C with shaking at 175 rpm. When absorbance at 600 nm reached 0.8–1.0, IPTG was added to induce recombinant protein expression for 18 h at 25 °C. After this induction, the bacterial solution was centrifuged at (4260× g) at 4 °C for 20 min, and the obtained supernatant was the extracellular protein solution.

4.3. Analysis of Enzyme Activity and Protein Assay

The activity of endo-1,4-β-xylanase was examined on the basis of the production of reducing sugars from xylan by using the 3,5-dinitrosalicylic acid (DNS) method. In two test tubes labeled A1 and A2, 50 μL of the enzyme was mixed with 450 μL of 1% birch xylan (in 100 mM Tris-HCl buffer, pH 8). The A2 test tube was placed in a water bath at 60 °C for 10 min, whereas simultaneously, the A1 test tube was placed in an ice bath for 10 min. Then, 1 mL of DNS reagent was added to A1 and A2. The A1 and A2 test tubes were placed in a water bath at 100 °C for 10 min to inactivate the enzyme. A spectrophotometer was used to measure the absorbance at 540 nm. The xylose calibration curve (y = 1.095x − 0.026) was used to convert the OD at 540 nm into the concentration of reducing sugars. Enzyme activity (U, in terms of μ mole/min) was calculated using the formula U = (A2 − A1)/Δt.

Extracellular proteins were analyzed through SDS–PAGE by using 12.5% acrylamide gel. The resolved protein bands were visualized through Coomassie brilliant blue staining. The amount of protein was quantified using the Bradford assay based on the use of Coomassie Brilliant Blue G-250, and bovine serum albumin was used as the standard protein.

4.4. XOS Production from EFB through Enzymatic Reaction

EFB was provided by Southern Palm (1978), Surat Thani, Thailand. The method developed by National Renewable Energy Laboratory (NREL) [44] was used to determine the composition of lignocellulose in raw materials. EFB was composed of 22.4% cellulose and 22.0% hemicellulose based on dry mass.

First, the raw EFB was washed with tap water to remove soil and other impurities from its surface. Then, it was dried in an oven at 60 °C for 24 h and pulverized to obtain powder with a particle size of 40–60 mesh. The EFB powder was soaked in 15% sodium hydroxide solution at a solid-to-liquid ratio of 1:10 (*w/v*) in an incubator and then incubated at 45 °C with shaking at 100 rpm overnight. Subsequently, the mixture was centrifuged ($2400\times g$ at 25 °C for 20 min) to separate the precipitate (cellulose) out of the solution. The supernatant (hemicellulose and lignin) was adjusted to pH 8 by using concentrated hydrochloric acid.

The solution was then treated with either recombinant *B. halodurans* endo-1,4-β-xylanase (from *E. coli* BL21-pET 29a-Xyn45) or extracellular endo-1,4-β-xylanases (from *B. halodurans* BCRC 910501) for the enzymatic hydrolysis of EFB-derived xylan in the solution. Enzyme preparations at different concentrations (0.5 U and 2 U for 1 mL of the sample) were used. The enzyme reaction was conducted in a water bath at 50 °C with shaking at 50 rpm for 48 h. Samples were removed at fixed intervals (1, 2, 4, 8, 12, 24, and 48 h), and the enzymatic reaction was inactivated by boiling the sample in a water bath at 100 °C for 10 min. After boiling, the sample was centrifuged ($2400\times g$ at 25 °C for 20 min) to separate out the precipitate (lignin) and keep the supernatant as the product (XOS). Finally, the supernatant was passed through a 0.22 μm filter by performing vacuum filtration. Then, the XOS concentration was determined through HPLC.

4.5. Fermentation of Probiotics on XOS

Ultrafiltration was performed using a hollow fiber membrane with a 1-kDa molecular weight cutoff to remove the enzyme and high-molecular-weight polysaccharides (DP5 and larger than DP5) from the XOS products after enzymatic hydrolysis. The permeate was concentrated by boiling to remove water.

Four bacterial strains were used in this study: *Bif. animalis*, *Bif. catenulatum* 14667, *L. plantarum* 10069, and *L. acidophilus* NCFM. The bacteria were cultured in MRS broth containing 34.15 g/L of Lactobacilli MRS broth, 0.5 g/L of cysteine monohydrochloride monohydrate, and 1 mL/L Tween #80. For each inoculum, 6 mL of MRS broth and 1 mL of mineral oil were added to a glass test tube. After being capped, the glass test tube was kept sterile at 121 °C for 20 min and then naturally cooled to room temperature. In a laboratory fume hood, 1 mL of lactic acid bacterial liquid (containing *Bifidobacterium or Lactobacillus*) was inoculated and stored at −80 °C in a glass test tube. Finally, the glass test tube was incubated anaerobically at 37 °C for 24 h.

Fermentation experiments were performed using different media (MRS broth, MRS broth containing 2% *w/v* glucose, and MRS broth containing 2% *w/v* of XOS from EFB). For each fermentation, 30 mL of MRS broth and 2.5 mL of mineral oil were added to a 50-mL sharp-bottomed centrifuge tube. After being capped, the sharp-bottom centrifuge tube was sterilized at 121 °C for 20 min and then naturally cooled to room temperature. The laboratory fume hood was used for inoculating 1 mL of the inoculum into the sharp-bottom centrifuge tube. Finally, the glass test tube was incubated anaerobically at 37 °C for 72 h. Bacterial growth was analyzed by measuring absorbance at 600 nm and pH at 0, 24, 48, and 72 h.

5. Conclusions

Both the recombinant and nonrecombinant endo-1,4-β-xylanases of alkaliphilic *B. halodurans* BCRC 910501 effectively converted the xylan within EFB into prebiotic XOS. The synergistic effect of the two endoxylanases Xyn45 and Xyn23—belonging to the GH10 and GH11 families of enzymes, respectively—enabled faster cleavage of xylan into small-molecule XOS. The yield of XOS was higher when a mixture of Xyn45 and Xyn23 was used compared with when a single enzyme, Xyn45, was used. The recombinant endoxylanase Xyn45 could be produced by recombinant *E. coli* BL21-pET 29a(+)-xyn45, while a mixture of nonrecombinant endoxylanases Xyn45 and Xyn23 could be obtained by culturing *B. halophilus* in a medium containing xylan and XOS. Xylobiose in XOS was more effective than xylotriose in inducing *B. halophilus* to secrete endoxylanases. The EFB-derived XOS promoted the growth and metabolism of probiotic strains, *Bifidobacteria* and *Lactobacilli*. The feasibility of the production of these two endoxylanases and their uses indicate their potential for industrial application.

Supplementary Materials: The following supporting information can be downloaded at: https://www.mdpi.com/article/10.3390/catal13010039/s1, Table S1: Cultivation of *Bifidobacteria* and *Lactobacillus* on MRS medium; Table S2: Cultivation of *Bifidobacteria* and *Lactobacillus* on MRS medium containing 2% (*w/v*) glucose; Table S3: Cultivation of *Bifidobacteria* and *Lactobacillus* on MRS medium containing 2% (*w/v*) XOS from EFB.

Author Contributions: Conceptualization, W.-C.L., N.L. and C.-Y.C.; data curation, C.T. and Y.-C.H.; methodology, C.T., Y.-C.H. and J.-M.S.; software, C.T. and Y.-C.H.; visualization, C.T.; writing—original draft, W.-C.L.; writing—review and editing, N.L. and C.-Y.C.; supervision, W.-C.L. All authors have read and agreed to the published version of the manuscript.

Funding: We thank the Taiwan Ministry of Science and Technology (Grant number: MOST 109-2221-E-194-014) for supporting this study.

Data Availability Statement: Data reported in this study were collected and stored by C.T and Y.-C.H.

Acknowledgments: We would like to express our gratitude to Kow-Jen Duan at Tatung University, Taiwan for providing the probiotic strains.

Conflicts of Interest: The authors declare no conflict of interest.

References

1. Takami, H.; Nakasone, K.; Takagi, Y.; Maeno, G.; Sasaki, R.; Masui, N.; Fuji, F.; Hirama, C.; Nakamura, Y.; Ogasawara, N.; et al. Complete genome sequence of the alkali-philic bacterium *Bacillus halodurans* and genomic sequence comparison with *Bacillus subtilis*. *Nucleic Acids Res.* **2000**, *28*, 4317–4331. [CrossRef] [PubMed]
2. Lee, Y.S.; Ratanakhanokchai, K.; Piyatheerawong, W.; Kyu, K.L.; Rho, M.S.; Kim, Y.S.; Om, A.; Lee, J.W.; Jhee, O.H.; Chon, G.H.; et al. Production and location of xylanolytic enzymes in alkaliphilic Bacillus sp. K-1. *J. Microbiol. Biotechnol.* **2006**, *16*, 921–926.
3. Honda, Y.; Kitaoka, M. A family 8 glycoside hydrolase from Bacillus halodurans C-125 (BH2105) is a reducing end xylose-releasing exo-oligoxylanase. *J. Biol. Chem.* **2004**, *279*, 55097–55103. [CrossRef] [PubMed]
4. Lin, Y.S.; Tseng, M.J.; Lee, W.C. Production of xylooligosaccharides using immobilized endo-xylanase of *Bacillus halodurans*. *Process Biochem.* **2011**, *46*, 2117–2121. [CrossRef]
5. Polizeli, M.L.; Rizzatti, A.C.; Monti, R.; Terenzi, H.F.; Jorge, J.A.; Amorim, D.S. Xylanases from fungi: Properties and industrial applications. *Appl. Microbiol. Biotechnol.* **2005**, *67*, 577–591. [CrossRef]
6. Tseng, M.J.; Yap, M.N.; Ratanakhanokchai, K.; Kyu, K.L.; Chen, S.T. Purification and characterization of two cellulase free xylanases from an alkaliphilic Bacillus firmus. *Enzyme Microb. Technol.* **2002**, *30*, 590–595. [CrossRef]
7. Chang, P.; Tsai, W.S.; Tsai, C.L.; Tseng, M.J. Cloning and characterization of two thermostable xylanases from an alkaliphilic Bacillus firmus. *Biochem. Biophys. Res. Commun.* **2004**, *319*, 1017–1025. [CrossRef] [PubMed]
8. Beg, Q.K.; Kapoor, M.; Mahajan, L.; Hoondal, G.S. Microbial xylanases and their industrial applications: A review. *Appl. Microbiol. Biotechnol.* **2001**, *56*, 326–338. [CrossRef]
9. Chen, H.; Wang, L. Enzymatic hydrolysis of pretreated biomass. In *Technologies for Biochemical Conversion of Biomass*; Elsevier: Amsterdam, The Netherlands, 2016; pp. 65–99.
10. Bhardwaj, N.; Kumar, B.; Verma, P. A detailed overview of xylanases: An emerging biomolecule for current and future prospective. *Bioresour. Bioprocess.* **2019**, *6*, 40. [CrossRef]
11. GeonomeNet Database: Bacillus halodurans C-125 DNA, Complete Genome. Available online: https://www.genome.jp/dbget-bin/www_bget?refseq+NC_002570 (accessed on 16 November 2022).
12. Wamalwa, B.M.; Zhao, G.; Sakka, M.; Shiundu, P.M.; Kimura, T.; Sakka, K. High-level heterologous expression of *Bacillus halodurans* putative xylanase xyn11a (BH0899) in Kluyveromyces lactis. *Biosci. Biotechnol. Biochem.* **2007**, *71*, 688–693. [CrossRef]
13. Prakash, P.; Jayalakshmi, S.K.; Prakash, B.; Rubul, M.; Sreeramulu, K. Production of alkaliphilic, halotolerent, thermostable cellulase free xylanase by Bacillus halodurans PPKS-2 using agro waste: Single step purification and characterization. *World J. Microbiol. Biotechnol.* **2012**, *28*, 183–192. [CrossRef] [PubMed]

14. Ho, A.L.; Carvalheiro, F.; Duarte, L.C.; Roseiro, L.B.; Charalampopoulos, D.; Rastall, R.A. Production and purification of xylooligosaccharides from oil palm empty fruit bunch fibre by a non-isothermal process. *Bioresour. Technol.* **2014**, *152*, 526–529. [CrossRef] [PubMed]
15. Aachary, A.A.; Siddalingaiya, P. Xylooligosaccharides (XOS) as an emerging prebiotic: Microbial synthesis, utilization, structural characterization, bioactive properties, and applications. *Compr. Rev. Food Sci. Food Saf.* **2010**, *10*, 2–16. [CrossRef]
16. Han, J.; Kim, J. Process simulation and optimization of 10-MW EFB power plant. *Comput. Aided Chem. Eng.* **2018**, *43*, 723–729.
17. Sudiyani, Y.; Styarini, D.; Triwahyuni, E.; Sudiyarmanto; Sembiring, K.C.; Aristiawan, Y.; Abimanyu, H.; Han, M.H. Utilization of biomass waste empty fruit bunch fiber of palm oil for bioethanol production using pilot—Scale unit. *Energy Procedia* **2013**, *32*, 31–38. [CrossRef]
18. Millati, R.; Wikandari, R.; Trihandayani, E.; Cahyanto, M.; Taherzadeh, M.; Niklasson, C. Ethanol from oil palm empty fruit bunch via dilute-acid hydrolysis and fermentation by Mucor indicus and Saccharomyces cerevisiae. *Agric. J.* **2011**, *6*, 54–59. [CrossRef]
19. Abdullah, N.; Sulaiman, F.; Gerhauser, H. Characterisation of oil palm empty fruit bunches for fuel application. *J. Phys. Sci.* **2011**, *22*, 1–24.
20. Baharuddin, A.S.; Yunos, N.S.H.; Mahmud, N.A.N.; Zakaria, R.; Yunos, K.F. Effect of high-pressure steam treatment on enzymatic saccharification of oil palm empty fruit bunches. *BioResources* **2012**, *7*, 3525–3538.
21. Palamae, S.; Dechatiwongse, P.; Choorit, W.; Chisti, Y.; Prasertsan, P. Cellulose and hemicellulose recovery from oil palm empty fruit bunch (EFB) fibers and production of sugars from the fibers. *Carbohydr. Polym.* **2017**, *155*, 491–497. [CrossRef]
22. Noorshamsiana, A.W.; Faizah, J.N.; Kamarudin, H.; Eliyanti, A.O.N.; Fatiha, I.; Astimar, A.A. Integrated production of prebiotic xylooligosaccharides and high value cellulose from oil palm biomass. *IOP Conf. Ser. Mater. Sci. Eng.* **2020**, *736*, 022044. [CrossRef]
23. Rathamat, Z.; Choorit, W.; Chisti, Y.; Prasertsan, P. Two-step isolation of hemicellulose from oil palm empty fruit bunch fibers and its use in production of xylooligosaccharide prebiotic. *Ind. Crops Prod.* **2021**, *160*, 113124. [CrossRef]
24. Morelli, L.; Callegari, M.L.; Patrone, V. *Prebiotics, Probiotics, and Synbiotics: A Bifidobacterial View*; Elsevier Inc.: Amsterdam, The Netherlands, 2018.
25. Li, Z.; Summanen, P.H.; Komoriya, T.; Finegold, S.M. In vitro study of the prebiotic xylooligosaccharide (XOS) on the growth of *Bifidobacterium* spp and *Lactobacillus* spp. *Int. J. Food Sci. Nutr.* **2015**, *66*, 919–922. [CrossRef] [PubMed]
26. Samanta, A.K.; Jayapal, N.; Jayaram, C.; Roy, S.; Kolte, A.P.; Senani, S.; Sridhar, M. Xylooligosaccharides as prebiotics from agricultural by-products: Production and applications. *Bioact. Carbohydr. Diet. Fibre* **2015**, *5*, 62–71. [CrossRef]
27. Huang, C.; Yu, Y.; Li, Z.; Yan, B.; Pei, W.; Wu, H. The preparation technology and application of xylo-oligosaccharide as prebiotics in different fields: A review. *Front Nutr.* **2022**, *9*, 996811. [CrossRef] [PubMed]
28. Honda, H.; Kudo, T.; Ikura, Y.; Horikoshi, K. Two types of xylanases of alkalophilic *Bacillus* sp. No. C-125. *Can. J. Microbiol.* **1985**, *31*, 538–542. [CrossRef]
29. Cho, S.G.; Choi, Y.J. Catabolite repression of the xylanase gene (xynA) expression in Bacillus stearothermophilus no. 236 and B. subtilis. *Biosci. Biotechnol. Biochem.* **1999**, *63*, 2053–2058. [CrossRef]
30. Shulami, S.; Shenker, O.; Langut, Y.; Lavid, N.; Gat, O.; Zaide, G.; Zehavi, A.; Sonenshein, A.L.; Shoham, Y. Multiple regulatory mechanisms control the expression of the Geobacillus stearothermophilus gene for extracellular xylanase. *J. Biol. Chem.* **2014**, *289*, 25957–25975. [CrossRef]
31. Hueck, C.J.; Hillen, W. Catabolite repression in Bacillus subtilis: A global regulatory mechanism for the gram-positive bacteria? *Mol. Microbiol.* **1995**, *15*, 395–401. [CrossRef]
32. Moreno, M.S.; Schneider, B.L.; Maile, R.; Weyler, W.; Saier, M.H., Jr. Catabolite repression mediated by the CcpA protein in Bacillus subtilis: Novel modes of regulation revealed by whole-genome analyses. *Mol. Microbiol.* **2001**, *39*, 1366–1381. [CrossRef]
33. Fujita, Y. Carbon Catabolite Control of the Metabolic Network in Bacillus subtilis. *Biosci. Biotechnol. Biochem.* **2009**, *73*, 245–259. [CrossRef]
34. Singh, K.D.; Schmalisch, M.H.; Stülke, J.; Görke, B. Carbon catabolite repression in Bacillus subtilis: Quantitative analysis of repression exerted by different carbon sources. *J. Bacteriol.* **2008**, *190*, 7275–7284. [CrossRef] [PubMed]
35. Wang, K.; Cao, R.; Wang, M.; Lin, Q.; Zhan, R.; Xu, H.; Wang, S. A novel thermostable GH10 xylanase with activities on a wide variety of cellulosic substrates from a xylanolytic *Bacillus* strain exhibiting significant synergy with commercial Celluclast 1.5 L in pretreated corn stover hydrolysis. *Biotechnol. Biofuels* **2019**, *12*, 48. [CrossRef] [PubMed]
36. Kumar, V.; Satyanarayana, T. Biochemical and thermodynamic characteristics of thermo-alkali-stable xylanase from a novel polyextremophilic *Bacillus halodurans* TSEV1. *Extremophiles* **2013**, *17*, 797–808. [CrossRef] [PubMed]
37. Mamo, G.; Hatti-Kaul, R.; Mattiasson, B. A thermostable alkaline active endo-beta-1-4-xylanase from *Bacillus halodurans* S7: Purification and characterization. *Enzyme Microb. Technol.* **2006**, *39*, 1492–1498. [CrossRef]
38. Faryar, R.; Linares-Pastén, J.A.; Immerzeel, P.; Mamo, G.; Andersson, M.; Stålbrand, H.; Mattiasson, B.; Karlsson, E.N. Production of prebiotic xylooligosaccharides from alkaline extracted wheat straw using the K80R-variant of a thermostable alkali-tolerant xylanase. *Food Bioprod. Process.* **2015**, *93*, 1–10. [CrossRef]
39. Yang, Y.; Yang, J.; Wang, R.; Liu, J.; Zhang, Y.; Liu, L.; Wang, F.; Yuan, H. Cooperation of hydrolysis modes among xylanases reveals the mechanism of hemicellulose hydrolysis by *Penicillium chrysogenum* P33. *Microb. Cell Fact.* **2019**, *18*, 159. [CrossRef] [PubMed]
40. Iliev, I.; Vasileva, T.; Bivolarski, V.; Momchilova, A.; Ivanova, I. Metabolic profiling of xylooligosaccharides by Lactobacilli. *Polymers* **2020**, *12*, 2387. [CrossRef] [PubMed]

41. Wang, J.; Sun, B.; Cao, Y.; Wang, C. In vitro fermentation of xylooligosaccharides from wheat bran insoluble dietary fiber by Bifidobacteria. *Carbohydr. Polym.* **2010**, *82*, 419–423. [CrossRef]

42. Madhukumar, M.S.; Muralikrishna, G. Fermentation of xylo-oligosaccharides obtained from wheat bran and Bengal gram husk by lactic acid bacteria and bifidobacteria. *J. Food Sci. Technol.* **2012**, *49*, 745–752. [CrossRef]

43. Zeybek, N.; Rastall, R.A.; Buyukkileci, A.O. Utilization of xylan-type polysaccharides in co-culture fermentations of Bifidobacterium and Bacteroides species. *Carbohydr. Polym.* **2020**, *236*, 116076. [CrossRef]

44. Sluiter, A.A.; Ruiz, R.; Scarlata, C.; Sluiter, J.; Templeton, D. *Determination of Extractives in Biomass: Laboratory Analytical Procedure (LAP)*; Technical Report NREL/TP-510-42619; National Renewable Energy Laboratory: Golden, CO, USA, 2008; pp. 1–9.

 *catalysts*

Article

Optimized Conditions for Preparing a Heterogeneous Biocatalyst via Cross-Linked Enzyme Aggregates (CLEAs) of β-Glucosidase from *Aspergillus niger*

Thiago M. da Cunha [1,2], Adriano A. Mendes [1,2,*], Daniela B. Hirata [1,2] and Joelise A. F. Angelotti [1,2,*]

[1] Graduate Program in Biotechnology, Federal University of Alfenas, Alfenas 37130-001, Brazil
[2] Institute of Chemistry, Federal University of Alfenas, Alfenas 37130-001, Brazil
* Correspondence: adriano.mendes@unifal-mg.edu.br (A.A.M.); joelise.angelotti@unifal-mg.edu.br (J.A.F.A.)

Abstract: This study mainly aims to find the optimal conditions for immobilizing a non-commercial β-glucosidase from *Aspergillus niger* via cross-linked enzyme aggregates (CLEAs) by investigating the effect of cross-linking agent (glutaraldehyde) concentration and soy protein isolate/enzyme ratio (or spacer/enzyme ratio) on the catalytic performance of β-glucosidase through the central composite rotatable design (CCRD). The influence of certain parameters such as pH and temperature on the hydrolytic activity of the resulting heterogeneous biocatalyst was assessed and compared with those of a soluble enzyme. The catalytic performance of both the soluble and immobilized enzyme was assessed by hydrolyzing ρ-nitrophenyl-β-D-glucopyranoside (ρ-NPG) at pH 4.5 and 50 °C. It was found that there was a maximum recovered activity of around 33% (corresponding to hydrolytic activity of 0.48 U/mL) in a spacer/enzyme ratio of 4.69 (mg/mg) using 25.5 mM glutaraldehyde. The optimal temperature and pH conditions for the soluble enzyme were 60 °C and 4.5, respectively, while those for CLEAs of β-glucosidase were between 50 and 65 °C and pH 3.5 and 4.0. These results reveal that the immobilized enzyme is more stable in a wider pH and temperature range than its soluble form. Furthermore, an improvement was observed in thermal stability after immobilization. After 150 days at 4 °C, the heterogeneous biocatalyst retained 80% of its original activity, while the soluble enzyme retained only 10%. The heterogeneous biocatalyst preparation was also characterized by TG/DTG and FT-IR analyses that confirmed the introduction of carbon chains via cross-linking. Therefore, the immobilized biocatalyst prepared in this study has improved enzyme stabilization, and it is an interesting approach to preparing heterogeneous biocatalysts for industrial applications.

Keywords: β-glucosidase; immobilization; CLEAs technique; stabilization

Citation: da Cunha, T.M.; Mendes, A.A.; Hirata, D.B.; Angelotti, J.A.F. Optimized Conditions for Preparing a Heterogeneous Biocatalyst via Cross-Linked Enzyme Aggregates (CLEAs) of β-Glucosidase from *Aspergillus niger*. *Catalysts* **2023**, *13*, 62. https://doi.org/10.3390/catal13010062

Academic Editors: Chia-Hung Kuo, Chwen-Jen Shieh and Hui-Min David Wang

Received: 29 November 2022
Revised: 23 December 2022
Accepted: 25 December 2022
Published: 28 December 2022

1. Introduction

Enzymes are biological catalysts having a wide range of industrial applications, as they can be employed more efficiently and sustainably than traditional chemical processes. However, their free forms (crude enzyme preparations, either soluble or powder enzyme extracts) have a few drawbacks if employed on an industrial scale, such as low physical and chemical stabilities and difficult separation from the reaction mixture for reusability purposes [1]. In order to overcome such drawbacks, different immobilization techniques have been proposed, since proper immobilization approaches can improve the thermal and operational stability of the resulting biocatalysts [2–6]. Moreover, a proper immobilization technique can improve the selectivity or specificity and reduce the inhibition effects and costs in downstream processes, in addition to increasing the flexibility of reactor configuration and design parameters [7,8]. In this context, several immobilization techniques have been used for preparing highly active and stable industrial biocatalysts, such as CLEAs (cross-linked enzyme aggregates) due to the fact that they are considered a cost-effective immobilization protocol requiring no supports, in addition to providing a platform for enzyme cascade reactions using multiple enzymes [8,9]. Moreover, this promising immobilization protocol offers a simultaneous purification of enzymes through a simple step that is well capable of reducing the total costs to prepare the heterogeneous biocatalyst [10–12].

CLEAs can be prepared through a couple of steps. Initially, enzymes undergo aggregation/precipitation induced by precipitating agents such as salts, water-miscible organic solvents or non-ionic polymers, among others. Afterward, the formed aggregates are cross-linked amino groups of lysine residues, and sulfhydryl groups of cysteine, phenolic OH groups of tyrosine, or imidazole groups of histidine present on the enzyme surface; they react with carbonyl groups at the extremity of a bifunctional reagent, i.e., glutaraldehyde in the majority of previous studies reported in the literature [9,11,13]. The CLEAs technique is an irreversible immobilization method to link enzymes together and form a complex three-dimensional structure composed of intermolecular cross-links between the enzyme and bi- or multifunctional reagents in order to make them insoluble in the reaction medium [14,15]. Given that the enzyme has a low surface density of amino groups, enzyme co-precipitation using an inert protein rich in lysine residues, such as bovine serum albumin (BSA), should contribute to form CLEAs having greater activity recovery and operational stability [11,16]. Although it is a widely explored technique involving several enzyme complexes, such as α-L-arabinosidase [17], invertase [18], naringinase [19], β-amylase [20], lipases [13,21], lipase and protease [22], glucose oxidase and catalase [23], there are few studies on the use of the CLEAs technique in β-glucosidase immobilization.

β-glucosidases (β-D-glucoside glucohydrolases, EC 3.2.1.21) are enzymes capable of hydrolyzing β-1,4 glycosidic bonds present in aryl-, amino-, alkyl-β-D-glucosides and cyanogenic glycosides, and oligo- and disaccharides [24]. It is also able to catalyze several reactions of industrial interest in a variety of segments, and it has attracted considerable attention in recent years due to its important roles in several biotechnological processes, such as the hydrolysis of exogenous glycolipids and glycosides and isoflavonoids glycosides, cell wall catabolism of cello-oligosaccharides, defense mechanisms, activation of conjugated phytohormones, the release of flavor compounds in plants, the release of aromatic compounds from flavor precursors used in beverage industries, the production of second-generation bioethanol from agricultural wastes, and so on [20,25–28]. In a study conducted by Zong et al. [26], β-glucosidase from plum seeds was immobilized via the CLEAs technique using a mixture of ethanol and isopropanol as precipitating agents and glutaraldehyde (20 mM) as a cross-linking agent. Ahumada et al. [17] co-immobilized β-glucosidase and arabinosidase from a commercial enzyme preparation (Rapidase®AR2000) via CLEAs technique to assess the effect of BSA (spacer) and glutaraldehyde concentrations on the enzyme immobilization process and the catalytic performance of the heterogeneous biocatalyst. These authors observed that the prepared CLEAs were more stable than the soluble enzyme. On the other hand, studies on the preparation of immobilized microbial β-glucosidase from *Aspergillus niger* via the CLEAs technique still are scarce in the literature.

Thus, this novel study mainly consists in preparing a heterogeneous biocatalyst through immobilizing β-glucosidase from *Aspergillus niger* using the CLEAs technique. For such a purpose, β-glucosidase was produced via solid-state fermentation using wheat bran as a substrate as it is an eco-friendly and economic alternative to produce industrial enzymes [29]. Enzyme precipitation of a crude extract obtained by solid-state fermentation was performed under optimal experimental conditions determined through a previous study using *n*-propanol as a precipitant agent [30]. The effect of relevant factors, such as spacer/enzyme ratio and glutaraldehyde (cross-linking agent) concentrations, was evaluated using the central composite rotatable design (CCRD), given that it is a well-known, robust and cost-effective statistical approach [31]. A low-cost spacer from soy protein isolate was used due to its cost-effectiveness and satisfactory performance in producing heterogeneous biocatalysts with enhanced catalytic performances [32]. The effect of certain factors (pH, and temperature) on the immobilized enzyme performance was evaluated and compared with its soluble form so as to enhance enzyme immobilization and consolidate the results of utilizing β-glucosidase in industrial bioprocesses. Thermogravimetric

(TG/DTG) and Fourier transform infrared (FT-IR) analyses were also conducted in order to characterize the resulting heterogeneous biocatalyst.

2. Results and Discussion

In this study, a heterogeneous biocatalyst of industrial interest was prepared via the CLEAs technique using β-glucosidase from *A. niger*. The effect of relevant factors such as spacer/enzyme ratio and cross-linking agent (glutaraldehyde) concentration on the catalytic performance of the enzyme was evaluated using a statistical tool (CCRD), as shown in Table 1. These parameters were selected due to their great relevance to the catalytic performance and stability of several industrial enzymes [8,9,12]. The levels of each parameter were selected based on preliminary studies conducted in our lab (data not shown). These tests were performed randomly so as to avoid system errors under fixed experimental conditions (50 °C, pH 4.5, 1 h of cross-linking time). Under such experimental conditions, recovered activity percentage values (response) ranged from 8.1% (Test #1) to 33.1% (Test #8)—see Table 1.

Table 1. CCRD experimental design of CLEAs of β-glucosidase from *A. niger* obtained after 60 min of reaction using glutaraldehyde as cross-linking agent. The response variable was the recovered activity percentage of the enzyme.

Tests	Coded Variables (Real Variables)		Response	
	Spacer/Enzyme Ratio (mg/mg)	Cross-Linking Agent (mM)	Recovered Activity (%)	Deviation (%)
1	−1 (1.37)	−1 (8.12)	8.1	−162.5
2	+1 (8.00)	−1 (8.12)	9.1	−34.1
3	−1 (1.37)	+1 (42.88)	21.5	−68.6
4	+1 (8.00)	+1 (42.88)	21.7	−25.3
5	−1.41 (0.00)	0 (25.50)	27.7	−38.3
6	+1.41 (9.37)	0 (25.50)	27.2	6.1
7	0 (4.69)	−1.41 (1.00)	9.1	33.6
8	0 (4.69)	+1.41 (50.00)	33.1	17.9
9	0 (4.69)	0 (25.50)	32.2	0.8
10	0 (4.69)	0 (25.50)	30.8	−3.7
11	0 (4.69)	0 (25.50)	32.8	2.7

Center points showed a slight variation, which indicates good process reproducibility. The CCRD analysis has revealed that the quadratic term of spacer/enzyme ratio (x_1^2) and linear (x_2) and quadratic terms of the cross-linking agent (x_2^2) were statistically significant at 10%, once the *p*-value < 0.1 (Table 2). Thus, non-significant terms were removed from the model and recalculated. Coded coefficients were then recalculated and used to write Equation (1).

$$\text{Recovered activity } (\%) = 31.93 - 4.45x_1^2 + 7.49x_2 - 7.71x_2^2 \tag{1}$$

Table 2. ANOVA of the CCRD using statistically significant regression coefficients at 10% significance.

Source of Variation	Coefficients	Degrees of Freedom	Mean Square	F-Calculated	*p*-Value
Regression	816.62	3	272.21	10.65	0.005319
Residual	179.00	7	25.57		
Lack of Fit	176.89				
Pure error	2.11				
Total	995.62	10			
$R^2 = 82.02\%$ $F_{tab.} = 3.07$					

The ANOVA (Table 2) was performed to identify statistically significant coefficients and higher $F_{calc.}$ (10.65) than $F_{tab.}$ (3.07) and a coefficient of determination (R^2) of 0.82

was found. Therefore, it is possible to use the obtained model equation to generate the response surface and analyze the studied variables (spacer/enzyme ratio and cross-linking agent concentration).

The model was used to generate the response surface (Figure 1A) and contour curves (Figure 1B). It was observed that the formation of an optimal region in which the highest values of β-glucosidase recovered activity was obtained using a spacer/enzyme ratio of 4.69 (mg/mg) at 25.5 mM of cross-linking agent concentration. Some factors may be accountable for affecting the performance of the CLEAs, among which cross-linking agent concentration stands out, as it is related to particle size and affects mass transfer (substrate or pH gradients) [9]. Moreover, cross-linking agent concentration is affected by the number of cross-links, i.e., the higher the concentration, the greater the number of links, therefore, leading to reduced enzyme release [5,12]. However, high concentration increases enzyme rigidity and causes catalytic activity loss. A combination of spacer/enzyme ratio (4.69) and glutaraldehyde (25.5 mM) provided optimal recovered activity percentage, as it increased the density of groups available to create cross-links using glutaraldehyde, thus preserving groups of β-glucosidase side chains which are important for its catalytic activity [33–35]. At the lowest glutaraldehyde concentrations (1.0 mM), there was no binding and the enzyme was leached, which explains such a low recovered activity value (only 9.1%—see Test #7 in Table 1). Furthermore, a high glutaraldehyde concentration (42.8 mM) increases enzyme rigidity and reduces its catalytic activity, which is in agreement with previous reports [2,12].

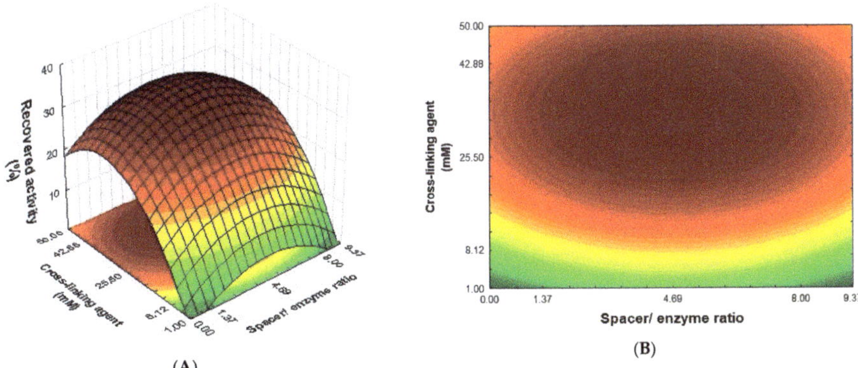

(A)　　　　　　　　　　　　　　(B)

Figure 1. β-glucosidase recovered activity as a function of cross-linking agent (glutaraldehyde) and spacer/enzyme ratio after 60 min of reaction time: (**A**) response surface and (**B**) contour curve.

The model was further validated by performing five experiments under optimal conditions, i.e., a spacer/enzyme ratio of 4.69 at 25.5 mM of cross-linking agent concentration, thus reaching β-glucosidase recovered activity of 31.12 ± 1.85% and demonstrating that the process has good reproducibility, as shown in Figure 2.

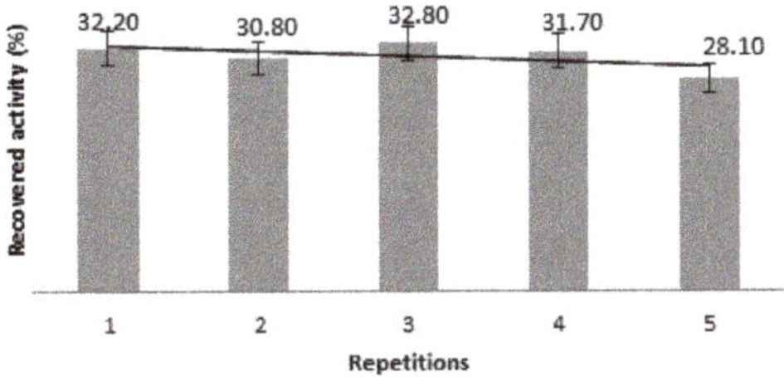

Figure 2. Optimized model validated for β-glucosidase immobilization via the CLEAs technique.

2.1. Study on the Effect of Cross-Linking Time

Once having optimized the parameters of cross-linking agent concentration and spacer/enzyme ratio, the effect of cross-linking time on the catalytic performance of the heterogeneous biocatalyst under optimal conditions determined by the response surface methodology was investigated. These results are illustrated in Figure 3.

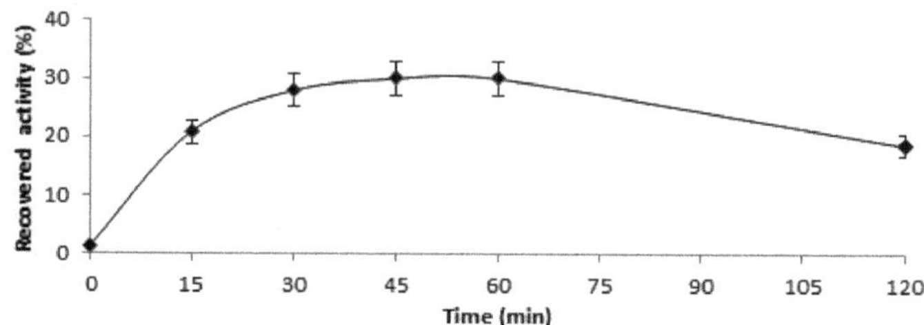

Figure 3. Effect of cross-linking time on β-glucosidase activity immobilized via the CLEAs technique (100% Relative activity corresponds to a hydrolytic activity of 1.48 U/mL).

According to the results, a gradual increase in the recovered activity percentage by increasing cross-linking time was observed. This phenomenon is possibly due to insufficient cross-links as reaction time was reduced, thus evidencing the leaching of free enzyme molecules while washing protein aggregates [34,35]. The maximum recovered activity value around 33% was found at a cross-linking time ranging between 45 min and 60 min, followed by a slight decrease after 120 min. For longer reaction times, an excessive number of cross-links can provide a more rigid heterogeneous biocatalyst having smaller pores in addition to hindering substrate access to its microenvironment [9]. In fact, further tests using immobilized β-glucosidase were performed using the biocatalyst prepared after 60 min of cross-linking time.

2.2. Biochemical Characterization

2.2.1. Optimum Temperature and pH Activity and Stability of Soluble and Immobilized β-Glucosidase

The highest catalytic activity was achieved at optimal temperature and pH conditions, i.e., 60 °C and pH 4.5 for soluble β-glucosidase (Figure 4A), and at temperatures ranging from 50 to 65 °C and pH from 3.5 to 4.0 for the CLEAs (Figure 4B). The heterogeneous biocatalyst used herein was the same as the one prepared in a spacer/enzyme ratio of 4.69, 25.5 mM of glutaraldehyde at 60 min of cross-linking time, i.e., conditions determined through the proposed CCRD.

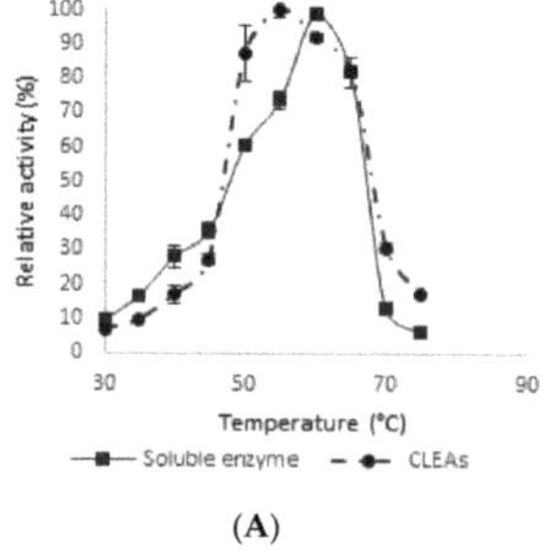

(**A**)

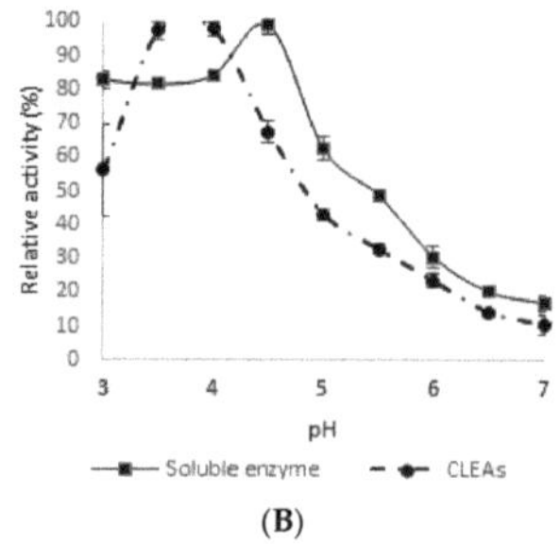

(**B**)

Figure 4. Hydrolytic activity profiles of soluble and immobilized β-glucosidase as a function of optimal temperature (**A**) and pH (**B**) conditions.

Optimal temperature shifts (Figure 4A), however, can be attributed to the lower flexibility of the enzyme's molecular structure within the structure of the CLEAs promoted by glutaraldehyde cross-linking, which makes the lateral bonds accountable for conferring conformational stability to the most stable and well-protected enzyme [36,37]. There were shifts in optimal pH conditions for the CLEAs activity (Figure 4B), probably due to a negatively charged microenvironment whose heterogeneity in size was comparable to that of soy protein isolate [38], thus leading to many different electrical charges in the internal microenvironment of the CLEAs. This might also explain the thermal stability of the CLEAs in comparison with that of the soluble enzyme (Figure 5A,B).

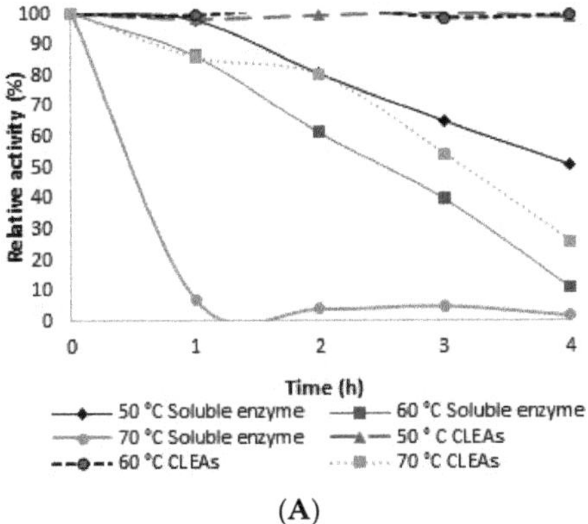

(A)

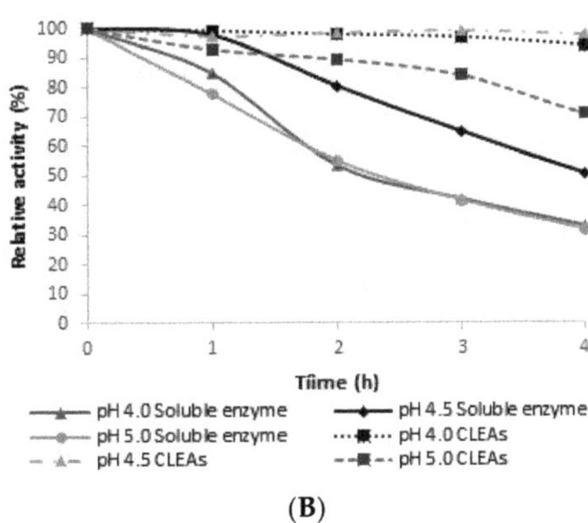

(B)

Figure 5. Hydrolytic activity profiles of soluble and immobilized β-glucosidase as a function of temperature (**A**) and pH (**B**) stability.

These results are in agreement with previous reports. Deng et al. [37] conducted assays using commercial β-glucosidase immobilized by CLEAs and found an increase by 0.5 of optimum pH for enzyme activity and a shift to a more alkaline range, i.e., from 5.0 to 5.5. In another study, Ahumada et al. [17] studied the co-immobilization of β-glucosidase and arabinosidase by the CLEAs technique and reported no changes in the biochemical profile of the enzyme, in addition to finding that the optimal temperature is 50 °C and pH 4.5 regarding enzyme activity using bovine serum albumin as a spacer. Different enzyme immobilization techniques possibly modify the biochemical profile of enzymes, and systems ought to be characterized thereof. Other authors also found no significant changes in the biochemical profile of β-glucosidase immobilized on various solid supports, such as chitosan and polyacrylamide [36], sodium alginate [39], silica gel and sol-gel [28].

The immobilized enzyme showed greater stability than the soluble enzyme at all studied temperatures. At 50 °C, the soluble enzyme retained around 50% of its original activity after 4 h of incubation. At 60 °C, the soluble enzyme maintained only 10% of its original activity after 4 h of incubation and it was completely inactivated after 1 h of the experiment at 70 °C. Nevertheless, the CLEAs were stable throughout the period tested at temperatures of 50 °C and 60 °C and maintained 100% of their activity. Moreover, it retained 54% of its initial activity after 3 h of incubation at 70 °C, as can be seen in Figure 5A. A thermal protective effect after enzyme immobilization is expected since there is an increase in rigidity and alteration in the enzyme's structural flexibility [36].

The CLEAs were also more stable at diverse pH conditions if compared to the soluble enzyme (Figure 5B). At pH 4.0 and after 4 h of incubation, the soluble enzyme maintained 50% of its initial activity, while the CLEAs presented only a 30% of loss of activity under such conditions. At pH values of 4.5 and 5.0, the soluble enzyme showed a retention of 30% of its original activity, while the CLEAs retained almost all of their original activity.

2.2.2. Storage Stability Tests

β-glucosidase immobilization via the CLEAs technique provided a derivative with greater storage stability if compared to the soluble enzyme, as can be seen in Figure 6.

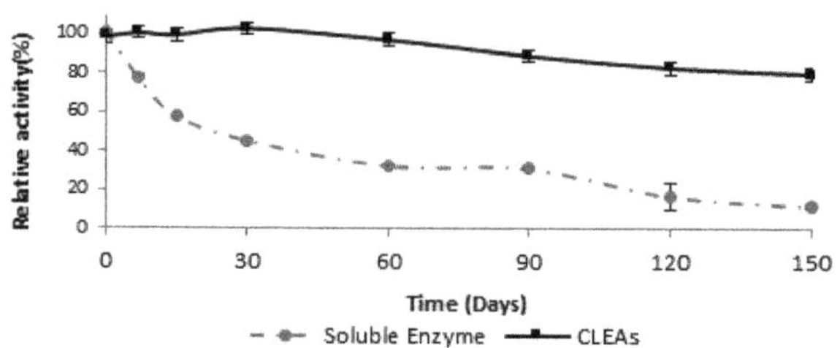

Figure 6. Storage stability of soluble and immobilized enzyme via the CLEAs technique after 150 days at 4 °C.

After 15 days of storage under refrigeration, the soluble enzyme lost 40% of its original activity, while the CLEAs still maintained 100% of their initial activity. After 150 days of storage, the soluble enzyme only retained 10% of its initial activity, while the CLEAs still maintained about 80% of their activity. During the immobilization process, enzyme crosslinking allows the enzyme to become more rigid and confined within a porous solid, in addition to offering protection against attacks from other proteolytic enzymes present in the fermented extract, thus preventing its degradation or decomposition [5,11,14] and improving storage stability. The profile of resulting CLEAs is more robust from an industrial process standpoint. Although there are mild pH and temperature fluctuations, there would be no yield alterations in enzymatic hydrolysis effectiveness.

2.3. Characterization of Biocatalysts by TGA and FT-IR Analyses

Fourier Transform Infrared spectroscopy (FT-IR) and thermogravimetric (TGA) analyses were performed for soluble and immobilized enzymes in order to confirm the structural changes occurring after the glutaraldehyde cross-linking step under optimal experimental conditions, as aforementioned. TGA was performed to assess whether modifications provided greater thermal stability to the heterogeneous biocatalyst prepared via the CLEAs technique.

The thermal stability of soluble and immobilized enzymes using TGA/DTG analysis was analyzed under a nitrogen atmosphere. Figure S1A,B shows that both enzyme forms (soluble—Figure S1A) and immobilized—Figure S1B) exhibited two main stage mass losses at temperatures ranging between 25 °C and 600 °C. It is observed that mass loss curves are very similar and there are two main thermal events occurring in the two samples by a loss of water through physical adsorption on the protein structure from the initial temperature to 100 °C, in addition to a thermal decomposition [40]. However, greater mass loss was found in CLEAs (39%—see Figure S1B) at around 300 °C if compared to the soluble enzyme (32%—see Figure S1A). These results may be due to the great compaction of the free enzyme chains, which in turn leads to greater thermal stability. The introduction of carbon chains after cross-linking using glutaraldehyde prevented such compaction and resulted in the formation of aggregates with enhanced porosity and, consequently, less stability in this temperature range [41]. The addition of functional groups can lead to a reduction in the effective compaction of these macromolecules, thus reducing their thermal stability in the temperature range evaluated in this study (from 25 °C and 600 °C), as they have a more porous structure [42], which explains the results obtained herein.

In this study, FT–IR analysis for soluble (Figure S2A) and immobilized β-glucosidase (Figure S2B) was performed to obtain the structural information after cross-linking using glutaraldehyde as a cross-linking agent. According to Figure S2A, an intense absorption band in the range of 3600–3000 cm^{-1} attributed to O–H and N–H stretching in the enzyme structure can be observed [43,44]. After cross-linking using glutaraldehyde, there was a reduction in the intensity of these bands (see Figure S2B), as this bifunctional agent interacts with different nucleophilic groups of the enzyme, such as O–H and N–H [45]. On the other hand, the band referring to the stretching of C–H (asymmetric) into –CH_2 (methyl groups), i.e., between 2923 and 2946 cm^{-1}, is intensified after biocatalyst preparation due to the cross-linking of nucleophilic groups described above, which confirms a chemical modification on the enzyme surface by glutaraldehyde molecules [2,41]. In cross-linked samples, the band referring to C–H (symmetric) ranges between 2870 and 2892 cm^{-1}, which is also intensified as carbon chains are introduced via glutaraldehyde cross-linking [46].

3. Materials and Methods

3.1. Materials

Aspergillus niger LBA 02 was obtained from the Culture Collection of Food and Biochemistry Laboratory, University of Campinas, Campinas-SP, Brazil and wheat bran (Nattuday, Formiga, MG, Brazil) was purchased from a local store. The glutaraldehyde solution (25% solution) and *n*-propanol were acquired from Dinâmica Química Contenporânea Ltd. (São Paulo, SP, Brazil). Bovine serum albumin (BSA) and ρ-nitrophenyl β-D-glucopyranoside (*rho*-NPG) were purchased from Sigma-Aldrich (St. Louis, MO, USA). Soy protein isolate (N4Natural, Santo André, SP, Brazil) was acquired from a local store. All other reagents and organic solvents were of analytical grade and acquired from Dinâmica and Synth® Ltd. (São Paulo, SP, Brazil).

3.2. β-Glucosidase Production via Solid-State Fermentation

The enzyme was produced by solid-state fermentation according to Angelotti et al. [29], but with a few modifications. *Aspergillus niger* strain belonging to the culture collection of the Biochemistry and Food Laboratory, Faculty of Food Engineering, State University of Campinas, Brazil, was kindly donated and used as a source of β-glucosidase. A culture medium of potato dextrose agar was placed in slant tubes for fungi growing, and tubes were covered with a protective layer of Vaseline during storage. The culture medium for enzyme production is composed of dry wheat bran (10 g) and 10 mL of distilled water, which was then poured into 500 mL Erlenmeyer flasks and sterilized through autoclaving (20 min, 121 °C); 10^5 spores/g of culture medium were inoculated in the flasks. Erlenmeyer flasks were incubated for 5 days at 30 °C.

3.3. Enzyme Extraction

Enzyme extraction was performed by adding 50 mL sodium acetate buffer 0.01 M pH 4.5 into the Erlenmeyer flasks containing the fermented medium. The flasks were shaken at 150 rpm for 20 min. The resulting suspension was then filtered with filter paper. The filtered solution was used as raw enzyme extract.

3.4. Determination of β-Glucosidase Activity

The β-glucosidase activity was determined in accordance with Matsuura et al. (1995), but with a few modifications [47]; 300 μL aliquots of a 5 mM solution of ρ-nitrophenyl β-D-glucopyranoside (ρ-NPG) in 0.05 M sodium acetate buffer at pH 4.5 were pre-incubated at 50 °C for 5 min, and 300 μL of enzyme extract was added afterward. Then, the mixture was incubated at 50 °C for 15 min and the reaction was stopped by adding 300 μL of a 0.5 M solution of sodium carbonate (pH 12). Synthetic substrate hydrolysis (ρ-NPG) was estimated by measuring absorbance at 410 nm using a UV/Vis spectrophotometer and a quartz cuvette. The amount of ρ-nitrophenol release was determined based on the standard curve of ρ-nitrophenol ranging from 5 to 300 μmol. A unit of enzyme activity was defined as the amount of enzyme required to release 1 μmol of ρ-nitrophenol per minute of reaction in the experimental conditions described previously.

3.5. Determination of Protein Concentration

Total protein concentration was determined through Bradford's protein assay [48] using BSA as the standard.

3.6. Immobilization of β-Glucosidase via CLEAs: Optimization by CCRD

The CLEAs were prepared according to Gupta and Raghava [49] with a few modifications. *n*-propanol was added to several protein solutions prepared in 100 mM sodium acetate buffer at pH 4.5 using a fixed volume ratio of 1:4 (protein-to-organic solvent). Under such experimental conditions, the complete precipitation of all initial proteins was determined by Bradford's method (see Section 3.5). After 30 min of precipitation in an ice bath, amounts of glutaraldehyde 25% (v/v) having final concentrations ranging from 8.0 to 50 mM (depending on the CCRD test), were added to the precipitated enzyme (cross-linking step). The aggregate suspension was incubated under gentle stirring (150 rpm) in a laboratory shaker at 4 °C for 1 h. After incubation, the suspension was centrifuged ($14,000 \times g$ for 5 min at 4 °C) and the precipitate was washed three times using 50 mM sodium acetate buffer pH 4.5 and re-suspended in the same buffer.

The effect of spacer/enzyme ratio (x_1) and cross-linking agent concentration (x_2) on the recovered activity percentage values of the immobilized enzyme was evaluated using a 2^2 central composite rotatable design (CCRD). Table 3 shows the CCRD with three center points and four axial points, totaling 11 tests and real and coded values for independent variables.

Table 3. Independent variables and their respective levels and real values used in the proposed CCRD approach.

Independent Variables		Levels				
		−1.41	−1	0	+1	+1.41
x_1	Spacer/enzyme ratio (mg/mg)	0.00	1.37	4.69	8.00	9.37
x_2	Cross-linking agent (mM)	1.00	8.12	25.50	42.88	50.00

The recovered activity percentage, determined as shown in Equation (2), of the CLEAs, obtained after 60 min of cross-linking time was used as a response in the experimental design. Results were analyzed at 10% statistical significance using independent variables coded to fit a second-order polynomial. The model was further validated by performing five experiments under optimal conditions.

$$\text{Recovered activity (\%)} = \left(\frac{\text{Hydrolytic activity of immobilized enzyme}}{\text{Hydrolytic activity of soluble enzyme}} \right) \times 100 \quad (2)$$

Effect of Cross-Linking Time

After optimizing the parameters evaluated in the experimental design, the effect of the cross-linking time (15–120 min) was assessed in order to prepare β-glucosidase CLEAs. The overall yield of immobilization was calculated by the ratio of enzyme activity of derivatives and the soluble enzyme under the same conditions described in Section 3.4.

3.7. Catalytic Properties of Soluble and Immobilized β-Glucosidase in Hydrolysis Reaction
3.7.1. Determination of Optimum Temperature

The effect of temperature on enzyme activity was determined for soluble and immobilized enzymes at temperatures ranging from 30 °C to 75 °C. Soluble and immobilized enzymes were incubated for 15 min in a 5 mM solution of ρ-NPG and 0.05 M sodium acetate buffer at pH 4.5. The reaction was stopped by adding 0.5 M sodium carbonate and enzyme activity was determined according to Section 3.4. Experimental results of the effect of temperature on enzyme activity were plotted in graphs in which the highest activity for each biocatalyst (soluble or immobilized β-glucosidase) was considered as 100%. All tests were performed in triplicate.

3.7.2. Determination of Optimum pH

The effect of pH on enzyme activity was determined at pH values ranging between 2.0 and 7.0 for both soluble and immobilized β-glucosidase (pH 3.0 at 5.0–100 mM sodium acetate buffer, pH 5.5 at 7.0–100 mM sodium phosphate buffer). Several ρ-NPG solutions at a fixed concentration (5 mM) were prepared and immersed in several buffer solutions (100 mM) at optimum activity temperature (55 °C for both soluble and immobilized β-glucosidase samples) after 15 min. Enzyme activity was determined as described in Section 3.4. Experimental results were also plotted in graphs and the maximum activity for each biocatalyst (soluble or immobilized β-glucosidase samples) was considered 100%.

3.7.3. pH and Thermal Stability Tests

Thermal stability tests were performed at the optimum pH activity of derivatives. In the absence of a substrate, both biocatalysts (soluble and immobilized enzyme) were immersed in a buffer solution (pH 4.5–100 mM sodium acetate buffer for a maximum period of 4 h) at temperatures ranging from 50 to 70 °C under static conditions. Samples were periodically collected (intervals of 1 h) to determine the residual catalytic activity of biocatalysts on ρ-NPG hydrolysis, as described in Section 3.4. The initial hydrolytic activity of each biocatalyst (3.33 U/mL for the soluble enzyme and 1.2 U/mL for the immobilized enzyme) was considered a control (100%).

pH stability tests were performed at optimum activity temperature (55 °C). In the absence of a substrate, soluble and immobilized enzymes were incubated for a maximum period of 4 h in solutions at pH values ranging from 4.0 to 5.0 under static conditions (sodium acetate buffer, 100 mM). In this set of experiments, samples were also periodically collected (intervals of 1 h) to determine the residual catalytic activity of biocatalysts, also on ρ-NPG hydrolysis (see Section 3.4). The initial hydrolytic activity of each biocatalyst (3.37 U/mL for the soluble enzyme and 1.02 U/mL for the immobilized enzyme) was also considered a control (100%).

3.7.4. Storage Stability Tests

Storage stability tests were also conducted for both biocatalyst forms (soluble and immobilized enzyme) for a maximum period of 5 months by immersion in a sodium acetate buffer pH 4.5 (100 mM) at 4 °C in a freezer under static conditions. In this study, samples were also collected every 15 days and the residual catalytic activity of biocatalysts was determined with ρ-NPG hydrolysis (see Section 3.4). The initial hydrolytic activity of each biocatalyst (1.15 U/mL for the soluble enzyme and 1.12 U/mL for the immobilized enzyme) was also considered a control (100%).

3.8. Characterization of Biocatalysts by TG/DTG and FT-IR Analyses

Thermogravimetric curves (TG/DTG) were obtained for the lyophilized soluble and immobilized enzyme using SII TG/DTA7300 Exstar in a temperature range of 25 °C to 600 °C at a heating rate of 10 °C/min under an inert atmosphere using nitrogen at a constant flow rate of 50 mL·min^{-1}. A Fourier Transform Infrared Spectroscopy (FT-IR) analysis was performed using Shimadzu equipment, model Affinity-1, coupled with a Pike Miracle ATR sampling accessory with ZnSe crystal plates at wavelengths ranging between 400 and 4000 cm^{-1} and 4 cm^{-1} of resolution.

4. Conclusions

The present results reveal that the model fits well with the data on the process of β-glucosidase immobilization via the CLEAs technique due to being statistically significant according to the experimental design. The spacer/enzyme ratio and cross-linking agent concentration were significant at the studied confidence level. Optimal conditions for

CLEAs production were the pinnacle of this study, as values of spacer/enzyme ratio of 4.69 and cross-linking agent of 25.5 mM were found. The generated model allowed plotting its response surface and explained 82% of the results. Furthermore, the heterogeneous biocatalyst is more robust than the soluble enzyme, whose optimal activity was achieved in a wider temperature range than the soluble enzyme (from 60 °C to 50–65 °C) and in a slightly more acidic pH range (from pH 4.5 to pH 3.5–4.0), thus offering greater stability in adverse temperature and pH conditions during a period of 4 h. TG/DTG and FT-IR analyses evidenced the effectiveness of cross-linking reaction and the insertion of carbon chains. The heterogeneous biocatalyst prepared in this study showed longer shelf life than that of the soluble enzyme, and residual activity of 80% was achieved after 150 days of storage at 4 °C if compared to 10% of residual activity reached by the soluble enzyme under the same conditions. These findings reveal that immobilized β-glucosidase via the CLEAs technique can be effectively used in industrial processes due to the enhanced pH activity and thermal stability, in addition to the improved storage time. Furthermore, this study offers new possibilities for further industrial applications of the heterogeneous biocatalyst prepared in this study in batch or continuous processes.

Supplementary Materials: The following supporting information can be downloaded at: https://www.mdpi.com/article/10.3390/catal13010062/s1, Figure S1: TG (blue curves) and DTG (red curves) analyses for soluble (A) and immobilized (B) β-glucosidase; Figure S2: FT-IR analysis for (A) and immobilized (B) β-glucosidase.

Author Contributions: Conceptualization, J.A.F.A. and T.M.d.C.; methodology, J.A.F.A., T.M.d.C. and D.B.H.; software, D.B.H.; validation, J.A.F.A., T.M.d.C. and D.B.H.; formal analysis, J.A.F.A., T.M.d.C., D.B.H. and A.A.M.; investigation, J.A.F.A., T.M.d.C. and D.B.H.; resources, J.A.F.A. and T.M.d.C.; data curation, J.A.F.A. and T.M.d.C.; writing—original draft preparation, J.A.F.A., T.M.d.C., D.B.H. and A.A.M.; writing—review and editing, J.A.F.A., D.B.H. and A.A.M.; visualization, J.A.F.A., T.M.d.C. and D.B.H.; supervision, J.A.F.A.; project administration, J.A.F.A.; funding acquisition, J.A.F.A. All authors have read and agreed to the published version of the manuscript.

Funding: The present study was financed in part by the Coordenação de Aperfeiçoamento de Pessoal de Nível Superior (CAPES)–Brazil–Finance Code 001.

Data Availability Statement: Not applicable.

Acknowledgments: Federal Center for Technological Education through a program to encourage personnel improvement.

Conflicts of Interest: The authors declare no conflict of interest.

References

1. Duan, F.; Sun, T.; Zhang, J.; Wang, K.; Wen, Y.; Lu, L. Recent Innovations in Immobilization of β-Galactosidases for Industrial and Therapeutic Applications. *Biotechnol. Adv.* **2022**, *61*, 108053. [CrossRef] [PubMed]
2. Xu, M.Q.; Wang, S.S.; Li, L.N.; Gao, J.; Zhang, Y.W. Combined Cross-Linked Enzyme Aggregates as Biocatalysts. *Catalysts* **2018**, *8*, 460. [CrossRef]
3. Xue, R.; Woodley, J.M. Process Technology for Multi-Enzymatic Reaction Systems. *Bioresour. Technol.* **2012**, *115*, 183–195. [CrossRef] [PubMed]
4. Rodrigues, R.C.; Berenguer-Murcia, Á.; Carballares, D.; Morellon-Sterling, R.; Fernandez-Lafuente, R. Stabilization of Enzymes via Immobilization: Multipoint Covalent Attachment and Other Stabilization Strategies. *Biotechnol. Adv.* **2021**, *52*, 107821. [CrossRef]
5. Sheldon, R.A.; van Pelt, S. Enzyme Immobilisation in Biocatalysis: Why, What and How. *Chem. Soc. Rev.* **2013**, *42*, 6223–6235. [CrossRef]
6. Figueira, J.A.; Sato, H.H.; Fernandes, P. Establishing the Feasibility of Using β-Glucosidase Entrapped in Lentikats and in Sol-Gel Supports for Cellobiose Hydrolysis. *J. Agric. Food Chem.* **2013**, *61*, 626–634. [CrossRef]
7. Bolivar, J.M.; Woodley, J.M.; Fernandez-Lafuente, R. Is Enzyme Immobilization a Mature Discipline? Some Critical Considerations to Capitalize on the Benefits of Immobilization. *Chem. Soc. Rev.* **2022**, *51*, 6251–6290. [CrossRef]
8. Sampaio, C.S.; Angelotti, J.A.F.; Fernandez-Lafuente, R.; Hirata, D.B. Lipase Immobilization via Cross-Linked Enzyme Aggregates: Problems and Prospects–A Review. *Int. J. Biol. Macromol.* **2022**, *215*, 434–449. [CrossRef]
9. Sheldon, R.A. Characteristic Features and Biotechnological Applications of Cross-Linked Enzyme Aggregates (CLEAs). *Appl. Microbiol. Biotechnol.* **2011**, *92*, 467–478. [CrossRef]

10. del Pilar Guauque Torres, M.; Foresti, M.L.; Ferreira, M.L. Cross-Linked Enzyme Aggregates (CLEAs) of Selected Lipases: A Procedure for the Proper Calculation of Their Recovered Activity. *AMB Express* **2013**, *3*, 25. [CrossRef]

11. Sheldon, R.A. Cross-Linked Enzyme Aggregates (CLEA®s): Stable and Recyclable Biocatalysts. *Biochem. Soc. Trans.* **2007**, *35*, 1583–1587. [CrossRef] [PubMed]

12. Amaral-Fonseca, M.; Kopp, W.; de Lima Camargo Giordano, R.; Fernández-Lafuente, R.; Tardioli, P.W. Preparation of Magnetic Cross-Linked Amyloglucosidase Aggregates: Solving Some Activity Problems. *Catalysts* **2018**, *8*, 496. [CrossRef]

13. Mohamad, N.R.; Marzuki, N.H.C.; Buang, N.A.; Huyop, F.; Wahab, R.A. An Overview of Technologies for Immobilization of Enzymes and Surface Analysis Techniques for Immobilized Enzymes. *Biotechnol. Biotechnol. Equip.* **2015**, *29*, 205–220. [CrossRef] [PubMed]

14. Cao, L.; van Langen, L.; Sheldon, R.A. Immobilised Enzymes: Carrier-Bound or Carrier-Free? *Curr. Opin. Biotechnol.* **2003**, *14*, 387–394. [CrossRef] [PubMed]

15. Talekar, S.; Joshi, A.; Joshi, G.; Kamat, P.; Haripurkar, R.; Kambale, S. Parameters in Preparation and Characterization of Cross Linked Enzyme Aggregates (CLEAs). *RSC Adv.* **2013**, *3*, 12485–12511. [CrossRef]

16. Shah, S.; Sharma, A.; Gupta, M.N. Preparation of Cross-Linked Enzyme Aggregates by Using Bovine Serum Albumin as a Proteic Feeder. *Anal. Biochem.* **2006**, *351*, 207–213. [CrossRef]

17. Ahumada, K.; Urrutia, P.; Illanes, A.; Wilson, L. Production of Combi-CLEAs of Glycosidases Utilized for Aroma Enhancement in Wine. *Food Bioprod. Process.* **2015**, *94*, 555–560. [CrossRef]

18. Mafra, A.C.O.; Beltrame, M.B.; Ulrich, L.G.; de Lima Camargo Giordano, R.; de Arruda Ribeiro, M.P.; Tardioli, P.W. Combined CLEAs of Invertase and Soy Protein for Economically Feasible Conversion of Sucrose in a Fed-Batch Reactor. *Food Bioprod. Process.* **2018**, *110*, 145–157. [CrossRef]

19. Ribeiro, M.H.L.; Rabaça, M. Cross-Linked Enzyme Aggregates of Naringinase: Novel Biocatalysts for Naringin Hydrolysis. *Enzym. Res.* **2011**, *2011*, 851272. [CrossRef]

20. Araujo-Silva, R.; Mafra, A.C.O.; Rojas, M.J.; Kopp, W.; De Campos Giordano, R.; Fernandez-Lafuente, R.; Tardioli, P.W. Maltose Production Using Starch from Cassava Bagasse Catalyzed by Cross-Linked β-Amylase Aggregates. *Catalysts* **2018**, *8*, 170. [CrossRef]

21. de Souza Sales Rocha, E.A.L.; de Carvalho, A.V.O.R.; de Andrade, S.R.A.; de Medeiros, A.C.D.; Trovão, D.M.B.M.; Costa, E.M.M. Potencial Antimicrobiano de Seis Plantas Do Semiárido Paraibano Contra Bactérias Relacionadas à Infecção Endodôntica. *Rev. Ciênc. Farm. Básica Apl.* **2013**, *34*, 351–355.

22. Furlani, I.L.; Amaral, B.S.; Oliveira, R.V.; Cassa, Q.B. Imobilização Enzimática: Conceito e Efeitos Na Proteólise. *Quim. Nova.* **2020**, *43*, 463–473. [CrossRef]

23. Mafra, A.C.O.; Ulrich, L.G.; Kornecki, J.F.; Fernandez-Lafuente, R.; Tardioli, P.W.; de Arruda Ribeiro, M.P. Combi-CLEAs of Glucose Oxidase and Catalase for Conversion of Glucose to Gluconic Acid Eliminating the Hydrogen Peroxide to Maintain Enzyme Activity in a Bubble Column Reactor. *Catalysts* **2019**, *9*, 657. [CrossRef]

24. Hildén, K.; Mäkelä, M.R. Role of Fungi in Wood Decay. In *Reference Module in Life Sciences*; Elsevier: Amsterdam, The Netherlands, 2018. [CrossRef]

25. Cairns, J.R.K.; Esen, A. β-Glucosidases. *Cell. Mol. Life Sci.* **2010**, *67*, 3389–3405. [CrossRef]

26. Chen, L.; Hu, Y.D.; Li, N.; Zong, M.H. Cross-Linked Enzyme Aggregates of β-Glucosidase from *Prunus domestica* Seeds. *Biotechnol. Lett.* **2012**, *34*, 1673–1678. [CrossRef]

27. Tu, M.; Zhang, X.; Kurabi, A.; Gilkes, N.; Mabee, W.; Saddler, J. Immobilization of β-Glucosidase on Eupergit C for Lignocellulose Hydrolysis. *Biotechnol. Lett.* **2006**, *28*, 151–156. [CrossRef]

28. Figueira, J.A.; Dias, F.F.G.; Sato, H.H.; Fernandes, P. Screening of Supports for the Immobilization of β-Glucosidase. *Enzym. Res.* **2011**, *2011*, 642460. [CrossRef]

29. Ohara, A.; dos Santos, J.G.; Angelotti, J.A.F.; Barbosa, P.D.P.M.; Dias, F.F.G.; Bagagli, M.P.; Sato, H.H.; de Castro, R.J.S. A Multicomponent System Based on a Blend of Agroindustrial Wastes for the Simultaneous Production of Industrially Applicable Enzymes by Solid-State Fermentation. *Food Sci. Technol.* **2018**, *38*, 131–137. [CrossRef]

30. Cunha, T.M.; Ferreira, C.F.M.L.; Angelotti, J.A.F. Avaliação de Agentes Precipitantes Para Produção de Agregados de Ligação Cruzada (Cleas) Da Enzima β-Glicosidase Produzida Por *Aspergillus niger*/Evaluation of Precipitating Agents for the Production of Cross-Linked Aggregates (Cleas) of the Enzyme β-Glucosidase Produced by *Aspergillus niger*. *Braz. J. Develop.* **2020**, *6*, 80538–80545. [CrossRef]

31. Prakasham, R.S.; Sathish, T.; Brahmaiah, P.; Subba Rao, C.; Sreenivas Rao, R.; Hobbs, P.J. Biohydrogen Production from Renewable Agri-Waste Blend: Optimization Using Mixer Design. *Int. J. Hydrog. Energy* **2009**, *34*, 6143–6148. [CrossRef]

32. Perkins, E.G. Composition of Soybeans and Soybean Products. *Pract. Handb. Soybean Process. Util.* **1995**, *1995*, 9–28. [CrossRef]

33. Guimarães, J.R.; de Lima Camargo Giordano, R.; Fernandez-Lafuente, R.; Tardioli, P.W. Evaluation of Strategies to Produce Highly Porous Cross-Linked Aggregates of Porcine Pancreas Lipase with Magnetic Properties. *Molecules* **2018**, *23*, 2993. [CrossRef] [PubMed]

34. Cui, J.D.; Liu, R.L.; Li, L.B. A Facile Technique to Prepare Cross-Linked Enzyme Aggregates of Bovine Pancreatic Lipase Using Bovine Serum Albumin as an Additive. *Korean J. Chem. Eng.* **2016**, *33*, 610–615. [CrossRef]

35. Guauque Torres, M.P.; Foresti, M.L.; Ferreira, M.L. Effect of Different Parameters on the Hydrolytic Activity of Cross-Linked Enzyme Aggregates (CLEAs) of Lipase from *Thermomyces lanuginosa*. *Biochem. Eng. J.* **2013**, *72*, 18–23. [CrossRef]

36. Abdel-Fattah, A.F.; Osman, M.Y.; Abdel-Naby, M.A. Production and Immobilization of Cellobiase from *Aspergillus Niger* A20. *Chem. Eng. J.* **1997**, *68*, 189–196. [CrossRef]
37. Deng, X.; He, T.; Li, J.; Duan, H.L.; Zhang, Z.Q. Enhanced Biochemical Characteristics of β-Glucosidase via Adsorption and Cross-Linked Enzyme Aggregate for Rapid Cellobiose Hydrolysis. *Bioprocess. Biosyst. Eng.* **2020**, *43*, 2209–2217. [CrossRef]
38. Blanch, H.W.; Clark, D.S. *Biochemical Engineering*; Dekker, M., Ed.; CRC Press: Boca Raton, FL, USA, 1996; ISBN 9780824700997.
39. Kumar, V.; Yadav, S.; Jahan, F.; Saxena, R.K. Organic Synthesis of Maize Starch-Based Polymer Using Rhizopus oryzae Lipase, Scale Up, and Its Characterization. *Prep. Biochem. Biotechnol.* **2013**, *44*, 321–331. [CrossRef]
40. Miguez, J.P.; Gama, R.S.; Bolina, I.C.A.; Melo, C.C.; Cordeiro, M.R.; Hirata, D.B.; Mendes, A.A. Enzymatic Synthesis Optimization of a Cosmetic Ester Catalyzed by a Homemade Biocatalyst Prepared via Physical Adsorption of Lipase on Amino-Functionalized Rice Husk Silica. *Chem. Eng. Res. Des.* **2018**, *139*, 296–308. [CrossRef]
41. Okura, N.S.; Sabi, G.J.; Crivellenti, M.C.; Gomes, R.A.B.; Fernandez-Lafuente, R.; Mendes, A.A. Improved Immobilization of Lipase from *Thermomyces lanuginosus* on a New Chitosan-Based Heterofunctional Support: Mixed Ion Exchange plus Hydrophobic Interactions. *Int. J. Biol. Macromol.* **2020**, *163*, 550–561. [CrossRef]
42. Sánchez, R.; Alonso, G.; Valencia, C.; Franco, J.M. Rheological and TGA Study of Acylated Chitosan Gel-like Dispersions in Castor Oil: Influence of Acyl Substituent and Acylation Protocol. *Chem. Eng. Res. Des.* **2015**, *100*, 170–178. [CrossRef]
43. Soares, A.M.B.F.; Gonçalves, L.M.O.; Ferreira, R.D.S.; de Souza, J.M.; Fangueiro, R.; Alves, M.M.M.; Carvalho, F.A.A.; Mendes, A.N.; Cantanhêde, W. Immobilization of Papain Enzyme on a Hybrid Support Containing Zinc Oxide Nanoparticles and Chitosan for Clinical Applications. *Carbohydr. Polym.* **2020**, *243*, 116498. [CrossRef] [PubMed]
44. Verma, R.; Kumar, A.; Kumar, S. Synthesis and Characterization of Cross-Linked Enzyme Aggregates (CLEAs) of Thermostable Xylanase from *Geobacillus thermodenitrificans* X1. *Process. Biochem.* **2019**, *80*, 72–79. [CrossRef]
45. Barbosa, O.; Ortiz, C.; Berenguer-Murcia, Á.; Torres, R.; Rodrigues, R.C.; Fernandez-Lafuente, R. Glutaraldehyde in Bio-Catalysts Design: A Useful Crosslinker and a Versatile Tool in Enzyme Immobilization. *RSC Adv.* **2014**, *4*, 1583–1600. [CrossRef]
46. Wang, X.; Chen, Z.; Li, K.; Wei, X.; Chen, Z.; Ruso, J.M.; Tang, Z.; Liu, Z. The Study of Titanium Dioxide Modification by Glutaraldehyde and Its Application of Immobilized Penicillin Acylase. *Colloids Surf. A Physicochem. Eng. Asp.* **2019**, *560*, 298–305. [CrossRef]
47. Matsuura, M.; Sasaki, J.; Murao, S. Studies on β-Glucosidases from Soybeans That Hydrolyze Daidzin and Genistin: Isolation and Characterization of an Isozyme. *Biosci. Biotechnol. Biochem.* **1995**, *59*, 1623–1627. [CrossRef]
48. Bradford, M.M. A Rapid and Sensitive Method for the Quantitation of Microgram Quantities of Protein Utilizing the Principle of Protein-Dye Binding. *Anal. Biochem.* **1976**, *72*, 248–254. [CrossRef]
49. Gupta, M.N.; Raghava, S. Enzyme Stabilization via Cross-Linked Enzyme Aggregates. *Methods Mol. Biol.* **2011**, *679*, 133–145. [CrossRef]

 *catalysts*

Communication

Bienzymatic Cascade Combining a Peroxygenase with an Oxidase for the Synthesis of Aromatic Aldehydes from Benzyl Alcohols

Yunjian Ma [1,2], Zongquan Li [2], Hao Zhang [2], Vincent Kam Wai Wong [1], Frank Hollmann [3,*] and Yonghua Wang [2,4,*]

[1] Neher's Biophysics Laboratory for Innovative Drug Discovery, State Key Laboratory of Quality Research in Chinese Medicine, Macau University of Science and Technology, Taipa, Macau 999078, China
[2] School of Food Science and Engineering, South China University of Technology, Guangzhou 510640, China
[3] Department of Biotechnology, Delft University of Technology, Van der Maasweg 9, 2629 HZ Delft, The Netherlands
[4] Guangdong Youmei Institute of Intelligent Bio-Manufacturing Co., Ltd., Foshan 528200, China
* Correspondence: f.hollmann@tudelft.nl (F.H.); yonghw@scut.edu.cn (Y.W.);
Tel.: +31-15-278-1957 (F.H.); +86-20-8711-3842 (Y.W.)

Abstract: Aromatic aldehydes are important aromatic compounds for the flavour and fragrance industry. In this study, a parallel cascade combining aryl alcohol oxidase from *Pleurotus eryngii* (*Pe*AAOx) and unspecific peroxygenase from the basidiomycete *Agrocybe aegerita* (*Aae*UPO) to convert aromatic primary alcohols into high-value aromatic aldehydes is proposed. Key influencing factors in the process of enzyme cascade catalysis, such as enzyme dosage, pH and temperature, were investigated. The universality of *Pe*AAOx coupled with *Aae*UPO cascade catalysis for the synthesis of aromatic aldehyde flavour compounds from aromatic primary alcohols was evaluated. In a partially optimised system (comprising 30 μM *Pe*AAOx, 2 μM *Aae*UPO at pH 7 and 40 °C) up to 84% conversion of 50 mM veratryl alcohol into veratryl aldehyde was achieved in a self-sufficient aerobic reaction. Promising turnover numbers of 2800 and 21,000 for *Pe*AAOx and *Aae*UPO, respectively, point towards practical applicability.

Keywords: aromatic aldehydes; flavour compounds; cascade catalysis; *Pe*AAOx; *Aae*UPO

Citation: Ma, Y.; Li, Z.; Zhang, H.; Wong, V.K.W.; Hollmann, F.; Wang, Y. Bienzymatic Cascade Combining a Peroxygenase with an Oxidase for the Synthesis of Aromatic Aldehydes from Benzyl Alcohols. *Catalysts* **2023**, *13*, 145. https://doi.org/10.3390/catal13010145

Academic Editors: Chia-Hung Kuo, Chwen-Jen Shieh and Hui-Min David Wang

Received: 19 December 2022
Revised: 4 January 2023
Accepted: 6 January 2023
Published: 8 January 2023

1. Introduction

Aromatic aldehydes are widely found in nature as secondary metabolites, e.g., in plants [1,2]. Commercially, aromatic aldehydes such as vanillin, anisaldehyde or cinnamaldehyde are popular flavour and fragrance ingredients [3–5].

Various chemical synthesis routes for aromatic aldehydes exist [6], but biocatalytic routes are highly desirable, as the products obtained from those are considered as 'natural' [7]. For the transformation of benzylic alcohols into the desired benzaldehyde derivates, a range of biocatalytic methods have been reported (Scheme 1). Alcohol dehydrogenases, for example, catalyse the NAD(P)$^+$-dependent oxidation of benzyl alcohols (Scheme 1A) [8]. Their nicotinamide cofactor-dependency, however, challenges the economic feasibility of these reactions and necessitates further efforts to ensure catalytic use of NAD(P)$^+$ and in situ regeneration. In the simplest scenario, this is achieved by simple administration of a sacrificial hydride acceptor such as acetone, which however complicates the reaction scheme and is less attractive from an environmental point-of-view due to the significant wastes generated by unreacted co-substrate and co-product accumulated. More elegantly, waste-free oxidation using O_2 or H_2O_2 as stoichiometric oxidants would result in environmentally more acceptable reaction schemes.

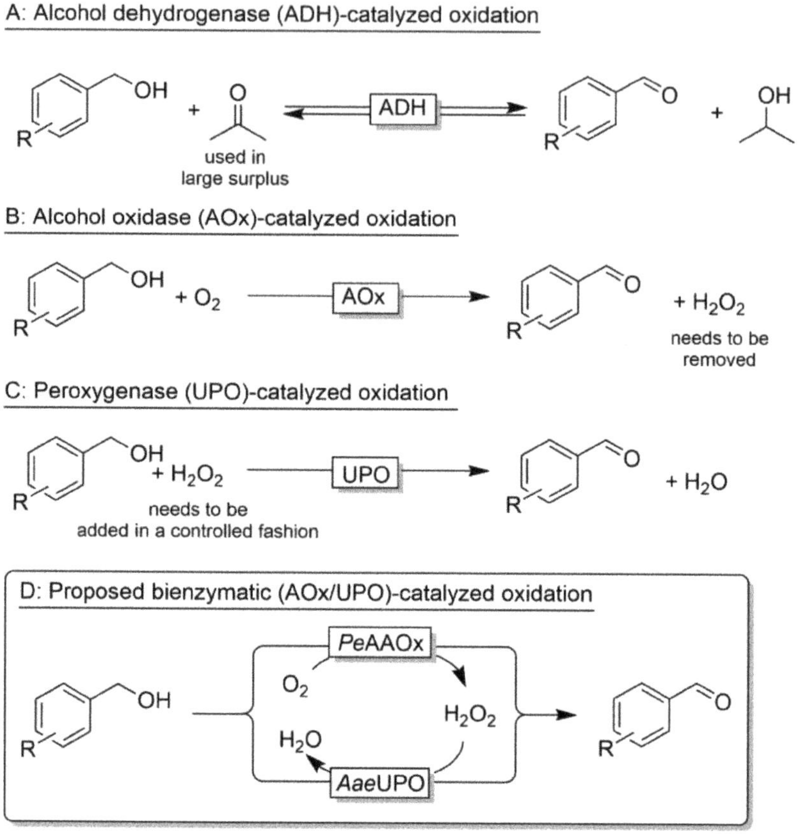

A: Alcohol dehydrogenase (ADH)-catalyzed oxidation

B: Alcohol oxidase (AOx)-catalyzed oxidation

C: Peroxygenase (UPO)-catalyzed oxidation

D: Proposed bienzymatic (AOx/UPO)-catalyzed oxidation

Scheme 1. Biocatalytic methods for the oxidation of benzylic alcohols. *Pe*AAOx: aryl alcohol oxidase from *Pleurotus eryngii*, *Aae*UPO: (unspecific) peroxygenase from *Agrocybe aegerita*.

Alcohol oxidases catalyse the aerobic oxidation of benzyl alcohols (Scheme 1B) [9]. Particularly, aryl alcohol oxidases (AAOx), have gained considerable interest in recent years [10–17]. An apparent drawback of using AAOx is the stoichiometric formation of H_2O_2 impairing the biocatalyts' robustness. Although H_2O_2 can easily be dimutated by catalase, this approach necessitates a second enzyme, thereby adding to the complexity of the reaction.

Even more recently, so-called unspecific peroxygenases (UPOs) have been reported to mediate the H_2O_2-dependent oxidation of alcohols (Scheme 1C) [18]. Again, H_2O_2 challenges the robustness of the overall reaction, therefore necessitating controlled provision with H_2O_2 [19,20].

The UPO- and AOx-catalysed reactions are co-substrate complementary, i.e., the by-product of the AAOx-catalysed oxidation serves as co-substrate for the UPO reaction. We therefore hypothesised that the combination of AAOx and UPOs may result in a synergistic parallel cascade for the oxidation of benzyl alcohols to the corresponding benzaldehydes (Scheme 1D). Another attractive feature of this system resides in the reduced waste formation of the proposed synergistic reaction scheme. Using an in situ H_2O_2 generation system requires the co-administration of a sacrificial co-substrate and results in the formation of a co-product. This not only negatively influences the environmental footprint of the overall reaction but may also complicate the reaction (e.g., by inhibitory effects of the co-reagents) and downstream processing.

Overall, an aerobic oxidation procedure yielding water as sole by-product was envisioned.

2. Materials and Methods

2.1. Chemical Reagents and Materials

All chemicals were purchased from Sigma-Aldrich (Louis, MO, USA), TCI (Tokyo, Japan), Acros (Morris Plains, NJ, USA) or Aladdin (Shanghai, China) with the highest purity available and used without further treatment.

2.2. Preparation of Enzyme

2.2.1. Preparation of *Aae*UPO

The unspecific peroxygenase from *Agrocybe aegerita* (*Aae*UPO) used in this study was obtained from a previous pilot-scale production of this enzyme [21].

2.2.2. Preparation of *Pe*AAOx

The plasmid pFLAG1-*Pe*AAOx reported previously [22] was kindly provided by Prof. Miguel Alcalde (CSIC, Madrid, Spain).

Cultivation Protocol

pFLAG1-*Pe*AAOx was transformed into *E. coli* BL21 star (DE3). After thermal activation, it was evenly coated with a coating rod on the LB solid medium containing ampicillin, and then it was incubated at 37 °C for 12–18 h at constant temperature until single colonies could be observed, of which one single colony was picked for further cultivation. Precultures of 25 mL LB-medium (10 g/L tryptone, 5 g/L yeast extract, 10 g/L NaCl, 50 mg/L kanamycin) were incubated overnight (12 h, 37 °C, 200 rpm) and used to inoculate the main cultures. The main cultures (500 mL TB-media) were mixed with the inoculum until an optical density of 0.01 was reached. They were then cultivated until an OD_{600} of 0.8 (4 h, 37 °C, 200 rpm) was obtained. Protein overexpression was induced by addition of isopropyl β-D-1-thiogalactopyranoside (IPTG; 1 mM final concentration). The induction time was 4 h and the induction temperature was 37 °C. After induction, cells were harvested by centrifugation (4000 rpm, 20 min, 4 °C). The resultant cell pellet was suspended in 20 mM sodium phosphate buffer, 500 mM NaCl, pH 7.5, and disrupted by sonication on ice. Soluble proteins were separated from cell fragments and insoluble proteins by centrifugation (10,000 rpm, 40 min, 4 °C). The supernatant was filtered through a 0.45 μm cellulose-acetate filter and further processed.

Refolding of PeAAOx from Inclusion Bodies

*Pe*AAOx was purified using an GE Chromatography system (Biorad). Initially, the crude enzyme was injected into a His Prep^{TM} FF16/10 column balanced by washing buffer A (20 mM sodium phosphate buffer, 500 mM NaCl, pH 7.5) at a flow rate of 5 mL min^{-1}. It was then equilibrated by washing buffer A and the binding protein was eluted using the elution buffer B (20 mM sodium phosphate buffer, 500 mM NaCl, 500 mM imidazole, pH 7.5) at a flow rate of 5 mL min^{-1}. Subsequently, the target protein was desalted on the column HiPrepTM 26/10 with the desalting buffer (20 mM sodium phosphate buffer, pH 7.5) at a flow rate of 5 mL min^{-1}. The purified protein was stored at 4 °C. An SDS-PAGE gel of *Pe*AAOx is shown in Figure S1.

2.3. Experimental Set-Up and Operating Conditions

The Agilent 7890B gas chromatography (GC) system (Agilent Technologies, Palo Alto, CA, USA) was used. The KB-FFAP gas chromatography column (30 m × 0.25 mm × 0.25 μm) was used for chromatographic separation. The analysis conditions were as follows: sample volume: 1 μL; solvent: ethyl acetate; injector temperature: 250 °C; split ratio: 30:1; detector: FID; detector temperature: 280 °C. The GC conditions were as follows: initial oven temperature was set at 60–120 °C for 6 min and ramped at 80 °C min^{-1}, and then increased up to 120–230 °C for 8 min and ramped at 20 °C min^{-1}.

Authentic standards of various substances were used to determine the retention time on the GC. Table S1 shows the retention times of various compounds. Standard solutions of different concentrations were prepared using the above standards; n-dodecane was used as the internal standard. The standard curve is prepared through gas detection for quantitative analysis. In the experiment, the product was not further separated and the conversion is specifically calculated by Formula (1).

$$\text{Conversion\%} = \frac{\text{Product concentration}}{\text{Initial substrate concentration}} \times 100\% \tag{1}$$

2.4. Experimental Procedures

Synthesis of Aromatic Aldehydes via Cascade Reaction of Aromatic Primary Alcohols Comparison of Catalytic Effects of PeAAOx Coupled with AaeUPO and PeAAOx Alone

(1)   PeAAOx coupled with AaeUPO catalytic cascade system experiment

Unless indicated otherwise, sodium phosphate buffer (50 mM, pH 7) was used. The buffer contained *Pe*AAOx (final concentration: 30 μM), *Aae*UPO (final concentration: 2 μM) and the substrate veratryl alcohol (final concentration: 50 mM, pre-dissolved in acetonitrile). The total volume of the reaction was 1 mL. The vessels (4 mL) were placed in self-contained round-hole reaction frames and thermostatted at 40 °C using an oil bath for 6, 12, 24 and 36 h under constant stirring (500 rpm). When the reaction is terminated, the reaction mixture was extracted with an ethyl acetate solution containing 25 mM n-dodecane internal standard, dried with anhydrous sodium sulfate and centrifuged at 12,000 rpm for 3 min. The upper organic phase was then transferred to the chromatographic bottle for GC detection.

(2)   PeAAOx catalysis alone experiment

The NaPi buffer (50 mM, pH 7), *Pe*AAOx enzyme solution (final concentration: 32 μM) and veratryl alcohol (final concentration: 50 mM, pre-dissolved in acetonitrile) were added to a 4 mL transparent glass reaction bottle. The total volume of the reaction was 1 mL. The reaction bottle was put on a self-contained round-hole reaction frame and then placed in a constant temperature oil bath for 6, 12, 24 and 36 h at 40 °C and a stirring speed of 500 rpm. The subsequent operation steps are identical to those described in *Pe*AAOx coupled with *Aae*UPO catalytic cascade system experiment.

Effect of PeAAOx Enzyme Dosage on the Oxidation of Veratryl Alcohol

The NaPi buffer (50 mM, pH 7), *Pe*AAOx enzyme solution (final concentrations: 10, 20, 30 and 40 μM, respectively), *Aae*UPO enzyme solution (final concentration: 2 μM) and veratryl alcohol (final concentration: 50 mM, pre-dissolved in acetonitrile) were added to a 4 mL transparent glass reaction flask. The total volume of the reaction was 1 mL. The reaction bottle was put on a self-contained round-hole reaction frame and then placed in a constant temperature oil bath for 3, 6, 9, 12, 24, 36 and 48 h at 30 °C and a stirring speed of 500 rpm. Subsequent operation steps are identical to those described in *Pe*AAOx coupled with *Aae*UPO catalytic cascade system experiment.

Effect of AaeUPO Enzyme Dosage on the Oxidation of Veratryl Alcohol

The NaPi buffer (50 mM, pH 7), *Pe*AAOx enzyme solution (final concentration: 30 μM), *Aae*UPO enzyme solution (final concentrations: 0.5, 1, 2 and 4 μM, respectively) and veratryl alcohol (final concentration: 50 mM, pre-dissolved in acetonitrile) were added to a 4 mL transparent glass reaction flask. The total volume of the reaction was 1 mL. The reaction bottle was put on a self-contained round-hole reaction frame and then placed in a constant temperature oil bath for 4, 8, 12, 24, 36 and 48 h at 30 °C and a stirring speed of 500 rpm. Subsequent operation steps are identical to those described in *Pe*AAOx coupled with *Aae*UPO catalytic cascade system experiment.

Factors Influencing Cascade Catalytic Oxidation of Veratryl Alcohol

(1)   Effect of temperature on cascade catalysis

The NaPi buffer (50 mM, pH 7), *Pe*AAOx enzyme solution (final concentration: 30 µM), *Aae*UPO enzyme solution (final concentration: 2 µM) and veratryl alcohol (final concentration: 50 mM, pre-dissolved in acetonitrile) were added to a 4 mL transparent glass reaction bottle. The total volume of the reaction was 1 mL. The reaction bottle was put on a self-contained round-hole reaction frame and then placed in a constant temperature oil bath for 24 h at 25, 30, 35, 40, 45 and 50 °C and a stirring speed of 500 rpm. Subsequent operation steps are identical to those described in *Pe*AAOx coupled with *Aae*UPO catalytic cascade system experiment.

(2)   Effect of pH on cascade catalysis

The buffer solution (50 mM, pH 5, 6, 7, 8, 9), *Pe*AAOx enzyme solution (final concentration: 30 µM), *Aae*UPO enzyme solution (final concentration: 2 µM) and then veratryl alcohol (final concentration: 50 mM, pre-dissolved in acetonitrile) were added to a 4 mL transparent glass reaction bottle. The total volume of the reaction was 1 mL. The reaction bottle was put on a self-contained round-hole reaction frame and then placed in a constant temperature oil bath for 24 h at 40 °C and a stirring speed of 500 rpm. Subsequent operation steps are identical to those described in *Pe*AAOx coupled with *Aae*UPO catalytic cascade system experiment.

2.5. Substrate Expansion

The NaPi buffer (50 mM, pH 7), *Pe*AAOx enzyme solution (final concentration: 30 µM), *Aae*UPO enzyme solution (final concentration: 2 µM) and veratryl alcohol, benzyl alcohol, 2-hydroxybenzyl alcohol, cinnamyl alcohol, p-methoxybenzyl alcohol or 4-hydroxy-3-methoxybenzyl alcohol (final concentration: 50 mM, pre-dissolved in acetonitrile) were added to a 4 mL transparent glass reaction flask. The total volume of the reaction was 1 mL. The reaction bottle was put on a self-contained round-hole reaction frame and then placed in a constant temperature oil bath for 24 h at 40 °C and a stirring speed of 500 rpm. Subsequent operation steps are identical to those described in *Pe*AAOx coupled with *Aae*UPO catalytic cascade system experiment.

3. Results and Discussion

In a first set of experiments we compared the catalytic performance of *Pe*AAOx alone with the envisioned bienzymatic cascade (Figure 1).

Pleasingly, the combination of *Pe*AAOx and *Aae*UPO proved to enable faster product formation compared to the single-enzyme catalyzed reaction system (Figure 1). Using *Pe*AAOx as oxidation catalyst alone, the product formation rate was somewhat slower than when using it in combination with *Aae*UPO, which we attribute to a positive effect of the double-catalyst usage postulated (Scheme 1).

Therefore, we further investigated the factors influencing the activity and robustness of the bienzymatic cascade. First, we systematically varied the concentration of either of the two enzymes (Figures 2, 3, S4 and S5). In general, the initial product formation rate of the overall reaction increased with increasing enzyme concentration. This trend was a bit more pronounced when varying the concentration of *Pe*AAOx, which may indicate that this represents the overall rate-limiting step of the cascade reaction. It is also interesting to note that the initial rate did not increase linearly with the *Pe*AAOx concentration. Possibly, at high *Pe*AAOx concentrations, diffusion of O_2 into the aqueous reaction mixture became overall rate-limiting. Finally, it should be noted that the reactions did not reach full conversion. We are currently lacking a plausible explanation for this observation but are convinced that further in-depth characterization of the reaction will reveal the current limitation.

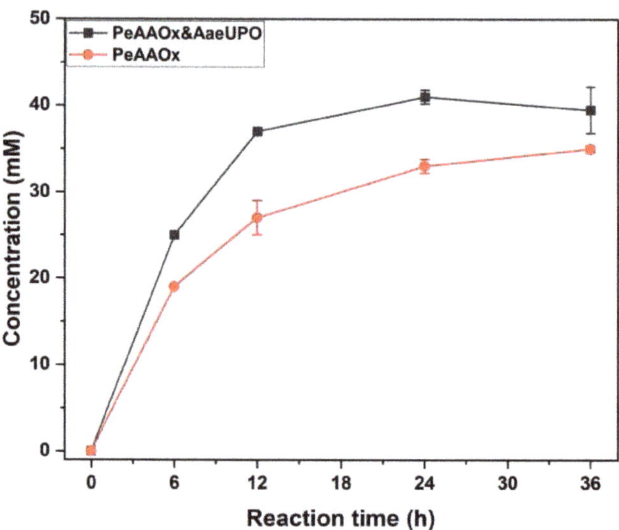

Figure 1. Comparison of catalytic effects of *Pe*AAOx catalysis alone (red) and *Pe*AAOx coupled with *Aae*UPO (black). (Reaction conditions: (1) [*Pe*AAOx] = 32 μM, [veratryl alcohol] = 50 mM (pre-dissolved in acetonitrile), 40 °C, pH 7 and 500 rpm); (2) [*Pe*AAOx] = 30 μM, [*Aae*UPO] = 2 μM, [veratryl alcohol] = 50 mM (pre-dissolved in acetonitrile), 40 °C, pH 7 and 500 rpm).

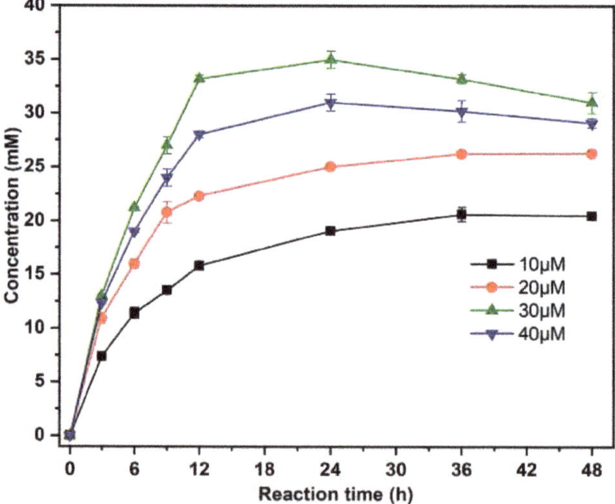

Figure 2. Effects of *Pe*AAOx dosage on the oxidation of veratryl alcohol. (Reaction conditions: [*Pe*AAOx] = 10, 20, 30 and 40 μM, [*Aae*UPO] = 2 μM, [veratryl alcohol] = 50 mM (pre-dissolved in acetonitrile), 30 °C, pH 7 and 500 rpm).

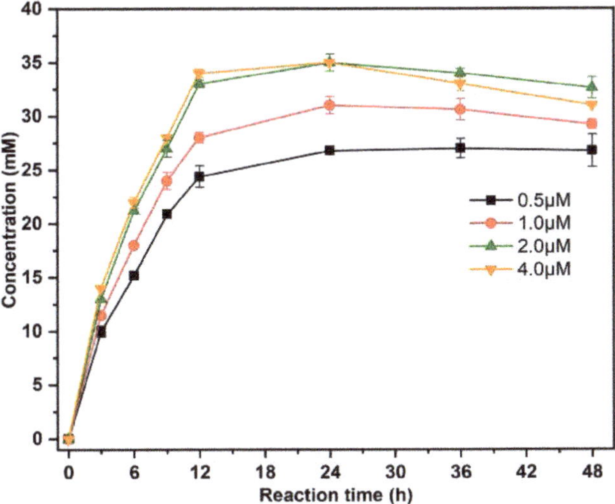

Figure 3. Effect of *Aae*UPO enzyme dosage on the oxidation of veratryl alcohol. (Reaction conditions: [*Pe*AAOx] = 30 μM, [*Aae*UPO] = 0.5, 1.0, 2.0 and 4.0 μM, [veratryl alcohol] = 50 mM (pre-dissolved in acetonitrile), 30 °C, pH 7 and 500 rpm).

Next, the influence of reaction temperature and pH on the overall oxidation rate was examined (Figure 4). Increasing the reaction temperature from 25 °C to 40 °C had only a minor influence on the overall product formation but increasing it to above 40 °C significantly decreased the conversion. We attribute this to the decreasing enzyme stability (Figure S2) and decreasing oxygen solubility at elevated temperatures. The pH profile was relatively broad with considerable activity between pH 5 and pH 8 and culminating at pH 7, which is consistent with the pH optimum of *Pe*AAOx (Figure S3) and *Aae*UPO [23]. Therefore, 40 °C and pH 7 were used for the following experiments.

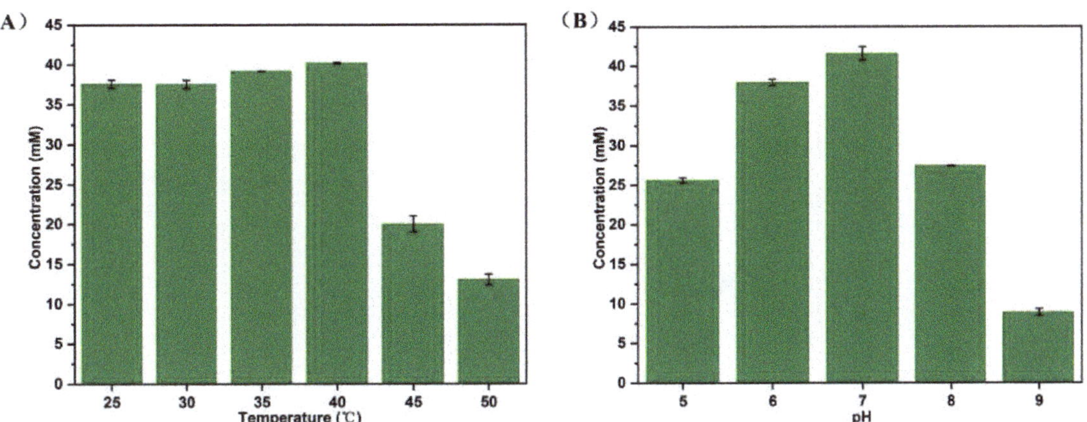

Figure 4. Factors influencing cascade catalytic oxidation of veratryl alcohol. (**A**) Temperature; (**B**) pH. (Reaction conditions: [*Pe*AAOx] = 30 μM, [*Aae*UPO] = 2 μM, [veratryl alcohol] = 50 mM (pre-dissolved in acetonitrile), 500 rpm and 24 h).

To evaluate the synthetic breadth of the proposed bienzymatic alcohol oxidation scheme, we explored the oxidation of further starting materials (Table 1 and Figure S6). A range of ring-substituted benzylic alcohols were converted in acceptable to good yields. Allylic alcohols such as cinnamyl alcohol were also converted, albeit at somewhat lower efficiency. The final product yields, however, correlated only poorly with the reported substrate spectrum of *Pe*AAOx [24]. This may partially be attributed to the initial rate measurements performed previously, which neglect possible inhibitory effects. Moreover, the *Aae*UPO substrate preference may interfere. In any case, a more extensive evaluation of the product scope of the proposed bienzymatic oxidation system will be desirable.

Table 1. Substrate expansion studies [1].

Entry	Substrate	Product	Product Concentration (mM)	Conversion (%)
1	H₃CO–, OCH₃ ring, –OH	H₃CO–, OCH₃ ring, –CHO	42.03	84.1
2	benzyl –OH	benzaldehyde –CHO	16.18	36.4
3	2-hydroxybenzyl –OH, –OH	2-hydroxy –CHO	5.06	10.1
4	cinnamyl alcohol –OH	cinnamaldehyde –O	12.27	24.5
5	H₃CO– ring, –OH	H₃CO– ring, –CHO	26.11	52.2
6	HO–, OCH₃ ring, –OH	HO–, OCH₃ ring, –CHO	16.20	32.4

[1] (Reaction conditions: [*Pe*AAOx] = 30 µM, [*Aae*UPO] = 2 µM, [substrate] = 50 mM (predissolved in acetonitrile), 40 °C, pH 7, 500 rpm, 24 h).

4. Conclusions

In the present study we have established a bienzymatic, parallel cascade combining aryl alcohol oxidases with peroxygenases to selectively oxidise benzylic alcohols into the corresponding aromatic aldehydes. Admittedly, many questions about the efficiency, robustness and scalability of the cascade remain to be answered. But the promising preliminary results obtained so far make us confident that the proposed approach may become a viable route to produce natural flavour compounds.

Supplementary Materials: The following supporting information can be downloaded at: https://www.mdpi.com/article/10.3390/catal13010145/s1, Figure S1: SDS-PAGE of *Pe*AAOx; Figure S2: Effect of temperature on *Pe*AAOx enzyme activity (A) and tolerance of *Pe*AAOx to temperature (B); Figure S3: Effect of pH on *Pe*AAOx enzyme activity (A) and tolerance of *Pe*AAOx to pH (B); Figure S4: Gas chromatograms of different *Pe*AAOx enzyme dosages used to catalyze the reaction of aromatic primary alcohols for 24 h; Figure S5: Gas chromatograms of different *Aae*UPO enzyme dosages used to catalyze the reaction of aromatic primary alcohols for 24 h; Figure S6: The gas chromatogram of the substrate expansion study (the specific peak time is shown in Table S1); Table S1: Retention time of different aromatic primary alcohols/aromatic aldehydes.

Author Contributions: Conceptualization, Y.M. and Z.L.; methodology, Z.L.; software, Z.L.; validation, Y.M., H.Z. and Z.L.; formal analysis, Y.M.; investigation, Z.L.; resources, Z.L.; data curation, Y.M. and Z.L.; writing—original draft preparation, F.H.; writing—review and editing, Y.M.; visualization, Y.M.; supervision, V.K.W.W. and Y.W.; project administration, Y.W.; funding acquisition, Y.M., V.K.W.W. and Y.W. All authors have read and agreed to the published version of the manuscript.

Funding: This research was funded by the Key Realm R&D Program of GuangDong Province (grant number 2022B0202050003), Key Program of Natural Science Foundation of China (grant number 31930084), National Natural Science Foundation of China (grant number 32001633), Guangzhou Science and technology planning project (grant number 202102020370), Macau Young Scholars Program (grant number AM2020024), Dr. Neher's Biophysics Laboratory for Innovative Drug Discovery from the Macau Science and Technology Development Fund (grant number 001/2020/ALC), and FDCT project (grant number 0033/2019/AFJ).

Data Availability Statement: All data are available from the corresponding author upon reasonable request.

Conflicts of Interest: The authors declare no conflict of interest. Guangdong Youmei Institute of Intelligent Bio-manufacturing Co., Ltd. provides some experimental instruments for testing during the whole experiment process. The company will not have any conflict of interest.

References

1. Chen, X.-M.; Kobayashi, H.; Sakai, M.; Hirata, H.; Asai, T.; Ohnishi, T.; Baldermann, S.; Watanabe, N. Functional characterization of rose phenylacetaldehyde reductase (PAR), an enzyme involved in the biosynthesis of the scent compound 2-phenylethanol. *J. Plant Physiol.* **2011**, *168*, 88–95. [CrossRef]
2. Torrens-Spence, M.P.; Chiang, Y.-C.; Smith, T.; Vicent, M.A.; Wang, Y.; Weng, J.-K. Structural basis for divergent and convergent evolution of catalytic machineries in plant aromatic amino acid decarboxylase proteins. *Proc. Natl. Acad. Sci. USA* **2020**, *117*, 10806–10817. [CrossRef]
3. Panten, J.; Surburg, H. Flavors and Fragrances, 2. Aliphatic Compounds. In *Ullmann's Encyclopedia of Industrial Chemistry*; Wiley-VCH: Weinheim, Germany, 2015. [CrossRef]
4. Panten, J.; Surburg, H. Flavors and Fragrances, 3. Aromatic and Heterocyclic Compounds. In *Ullmann's Encyclopedia of Industrial Chemistry*; Wiley-VCH: Weinheim, Germany, 2015. [CrossRef]
5. Panten, J.; Surburg, H. Flavors and Fragrances, 4. Natural Raw Materials. In *Ullmann's Encyclopedia of Industrial Chemistry*; Wiley-VCH: Weinheim, Germany, 2015. [CrossRef]
6. Punniyamurthy, T.; Velusamy, S.; Iqbal, J. Recent Advances in Transition Metal Catalyzed Oxidation of Organic Substrates with Molecular Oxygen. *Chem. Rev.* **2005**, *105*, 2329–2363. [CrossRef]
7. Fleige, C.; Meyer, F.; Steinbüchel, A. Metabolic Engineering of the Actinomycete Amycolatopsis sp. Strain ATCC 39116 towards Enhanced Production of Natural Vanillin. *Appl. Environ. Microbiol.* **2016**, *82*, 3410–3419. [CrossRef] [PubMed]
8. de Miranda, A.S.; Milagre, C.D.F.; Hollmann, F. Alcohol Dehydrogenases as Catalysts in Organic Synthesis. *Front. Catal.* **2022**, *2*, 900554. [CrossRef]
9. Urlacher, V.B.; Koschorreck, K. Pecularities and applications of aryl-alcohol oxidases from fungi. *Appl. Microbiol. Biotechnol.* **2021**, *105*, 4111–4126. [CrossRef] [PubMed]
10. Lappe, A.; Jankowski, N.; Albrecht, A.; Koschorreck, K. Characterization of a thermotolerant aryl-alcohol oxidase from Moesziomyces antarcticus oxidizing 5-hydroxymethyl-2-furancarboxylic acid. *Appl. Microbiol. Biotechnol.* **2021**, *105*, 8313–8327. [CrossRef] [PubMed]
11. Jankowski, N.; Koschorreck, K.; Urlacher, V.B. Aryl-Alcohol-Oxidase-Mediated Synthesis of Piperonal and Other Valuable Aldehydes. *Adv. Synth. Catal.* **2022**, *364*, 2364–2372. [CrossRef]
12. Jankowski, N.; Urlacher, V.B.; Koschorreck, K. Two adjacent C-terminal mutations enable expression of aryl-alcohol oxidase from Pleurotus eryngii in Pichia pastoris. *Appl. Microbiol. Biotechnol.* **2021**, *105*, 7743–7755. [CrossRef]
13. de Santos, P.G.; Lazaro, S.; Viña-Gonzalez, J.; Hoang, M.D.; Sánchez-Moreno, I.; Glieder, A.; Hollmann, F.; Alcalde, M. Evolved Peroxygenase–Aryl Alcohol Oxidase Fusions for Self-Sufficient Oxyfunctionalization Reactions. *ACS Catal.* **2020**, *10*, 13524–13534. [CrossRef]
14. Viña-Gonzalez, J.; Martinez, A.T.; Guallar, V.; Alcalde, M. Sequential oxidation of 5-hydroxymethylfurfural to furan-2,5-dicarboxylic acid by an evolved aryl-alcohol oxidase. *BBA-Proteins Proteom.* **2020**, *1868*, 140293. [CrossRef]
15. Serrano, A.; Sancho, F.; Viña-González, J.; Carro, J.; Alcalde, M.; Guallar, V.; Martínez, A.T. Switching the substrate preference of fungal aryl-alcohol oxidase: Towards stereoselective oxidation of secondary benzyl alcohols. *Catal. Sci. Technol.* **2019**, *9*, 833–841. [CrossRef]

16. de Almeida, T.P.; van Schie, M.M.C.H.; Ma, A.; Tieves, F.; Younes, S.H.H.; Fernández-Fueyo, E.; Arends, I.W.C.E.; Riul, A.; Hollmann, F. Efficient Aerobic Oxidation of trans-2-Hexen-1-ol using the Aryl Alcohol Oxidase from *Pleurotus eryngii*. *Adv. Synth. Catal.* **2019**, *361*, 2668–2672. [CrossRef]
17. van Schie, M.M.C.H.; de Almeida, T.P.; Laudadio, G.; Tieves, F.; Fernández-Fueyo, E.; Noël, T.; Arends, I.W.C.E.; Hollmann, F. Biocatalytic synthesis of the Green Note trans-2-hexenal in a continuous-flow microreactor. *Beilstein J. Org. Chem.* **2018**, *14*, 697–703. [CrossRef] [PubMed]
18. Hobisch, M.; Holtmann, D.; de Santos, P.G.; Alcalde, M.; Hollmann, F.; Kara, S. Recent developments in the use of peroxygenases—Exploring their high potential in selective oxyfunctionalisations. *Biotechnol. Adv.* **2021**, *51*, 107615. [CrossRef] [PubMed]
19. Burek, B.O.; Bormann, S.; Hollmann, F.; Bloh, J.Z.; Holtmann, D. Hydrogen peroxide driven biocatalysis. *Green Chem.* **2019**, *21*, 3232–3249. [CrossRef]
20. Grogan, G. Hemoprotein Catalyzed Oxygenations: P450s, UPOs, and Progress toward Scalable Reactions. *JACS Au* **2021**, *1*, 1312–1329. [CrossRef]
21. Tonin, F.; Tieves, F.; Willot, S.; van Troost, A.; van Oosten, R.; Breestraat, S.; van Pelt, S.; Alcalde, M.; Hollmann, F. Pilot-Scale Production of Peroxygenase from *Agrocybe aegerita*. *Org. Process. Res. Dev.* **2021**, *25*, 1414–1418. [CrossRef]
22. Ruiz-Dueñas, F.J.; Ferreira, P.; Martínez, M.J.; Martínez, A.T. In vitro activation, purification, and characterization of Escherichia coli expressed aryl-alcohol oxidase, a unique H2O2-producing enzyme. *Protein Expr. Purif.* **2006**, *45*, 191–199. [CrossRef]
23. Ullrich, R.; Nuske, J.; Scheibner, K.; Spantzel, J.; Hofrichter, M. Novel Haloperoxidase from the Agaric Basidiomycete *Agrocybe aegerita* Oxidizes Aryl Alcohols and Aldehydes. *Appl. Environ. Microbiol.* **2004**, *70*, 4575–4581. [CrossRef]
24. Francisco, G.; Angel, T.M.; Maria Jesús, M. Substrate specificity and properties of the aryl-alcohol oxidase from the ligninolytic fungus *Pleurotus eryngii*. *Eur. J. Biochem.* **1992**, *209*, 603–611.

catalysts

Article

Interaction of *Jania rubens* Polyphenolic Extract as an Antidiabetic Agent with α-Amylase, Lipase, and Trypsin: In Vitro Evaluations and In Silico Studies

Asmaa Nabil-Adam [1,*], Mohamed L. Ashour [2,3], Tamer M. Tamer [4], Mohamed A. Shreadah [1] and Mohamed A. Hassan [5,6,*]

[1]   Marine Biotechnology and Natural Products Laboratory, National Institute of Oceanography & Fisheries, Cairo 11516, Egypt
[2]   Department of Pharmacognosy, Faculty of Pharmacy, Ain-Shams University, Abbasia, Cairo 11566, Egypt
[3]   Department of Pharmaceutical Sciences, Pharmacy Program, Batterjee Medical College, P.O. Box 6231, Jeddah 21442, Saudi Arabia
[4]   Polymer Materials Research Department, Advanced Technologies and New Materials Research Institute (ATNMRI), City of Scientific Research and Technological Applications (SRTA-City), New Borg El-Arab City 21934, Alexandria, Egypt
[5]   Protein Research Department, Genetic Engineering and Biotechnology Research Institute (GEBRI), City of Scientific Research and Technological Applications (SRTA-City), New Borg El-Arab City 21934, Alexandria, Egypt
[6]   University Medical Center Göttingen, Georg-August-University, 37073 Göttingen, Germany
*   Correspondence: sama.biomarine@gmail.com (A.N.-A.); madel@srtacity.sci.eg (M.A.H.)

Citation: Nabil-Adam, A.; Ashour, M.L.; Tamer, T.M.; Shreadah, M.A.; Hassan, M.A. Interaction of *Jania rubens* Polyphenolic Extract as an Antidiabetic Agent with α-Amylase, Lipase, and Trypsin: In Vitro Evaluations and In Silico Studies. *Catalysts* 2023, 13, 443. https://doi.org/10.3390/catal13020443

Academic Editors: Chia-Hung Kuo, Chwen-Jen Shieh and Hui-Min David Wang

Received: 31 December 2022
Revised: 10 February 2023
Accepted: 13 February 2023
Published: 18 February 2023

Abstract: *Jania rubens* red seaweed has various bioactive compounds that can be used for several medicinal and pharmaceutical applications. In this study, we investigate the antidiabetic, anti-inflammatory, and antioxidant competency of *Jania rubens* polyphenolic extract (JRPE) by assessing their interactions with α-amylase, lipase, and trypsin enzymes. HPLC analysis revealed the dominance of twelve polyphenolic compounds. We performed computational analysis using α-amylase, lipase, and trypsin as target proteins for the polyphenols to explore their activities based on their predicted modes of binding sites following molecular modeling analysis. The molecular docking analysis demonstrated a good affinity score with a noticeable affinity to polyphenolic compositions of *Jania rubens*. The compounds with the highest affinity score for α-amylase (PDB: 4W93) were kaempferol, quercetin, and chlorogenic acid, with −8.4, −8.8 and −8 kcal/mol, respectively. Similarly, lipase (PDB: 1LPB) demonstrated high docking scores of −7.1, −7.4, and −7.2 kcal/mol for kaempferol, quercetin, and chlorogenic acid, respectively. Furthermore, for trypsin (PDB: 4DOQ) results, kaempferol, quercetin, and chlorogenic acid docking scores were −7.2, −7.2, and −7.1 kcal/mol, respectively. The docking findings were verified using in vitro evaluations, manifesting comparable results. Overall, these findings enlighten that the JRPE has antidiabetic, anti-inflammatory, and antioxidant properties using different diabetics' enzymes that could be further studied using in vivo investigations for diabetes treatment.

Keywords: *Jania rubens*; α-amylase; lipase; trypsin; antidiabetic; in silico analysis

1. Introduction

Obesity and diabetes (diabetes mellitus, DM) instigate several chronic diseases, including cardiovascular disease, kidney disease, eye disease, hypertension, osteoarthritis, and some forms of cancer. The worldwide rise of these conditions has become a major health concern. Excess fat mass accumulation is the defining feature of obesity, a complicated metabolic condition often accompanied by insulin resistance, elevated oxidative stress, and low-grade inflammation. Diabetes is a metabolic disorder caused by either a lack of pancreatic beta cells or a deficiency in insulin secretion and performance in terms of insulin resistance and sensitivity with different cell types. Genetic predisposition, a Western-style fast food diet, insufficient exercise, and socioeconomic standing are all thought to play a role in the epidemic of obesity. More than 600 million adults are overweight or obese, and an alarmingly rising percentage of infants are born overweight or obese in developing nations, according to research by the International Obesity Task Force.

By 2045, it is projected that 629 million people will have diabetes, up from 425 million in 2017. More than 85% of the diagnosed cases of diabetes were type 2 diabetes. Vascular diseases such as nephropathy, retinopathy, peripheral neuropathy, and stroke were present in many diabetic patients [1]. According to the International Diabetes Federation's (IDF) ninth-edition report on global diabetes (2019), the worldwide prevalence of diabetes was 9.3%, and approximately 463 million people were affected by diabetes worldwide (International Diabetes Federation 2019) [2]. Another study looking at the global prevalence of overweight and obesity between 1980 and 2015 found that, while the rate of obesity was highest in disadvantaged groups in high-income countries, it was highest in wealthy and urban families in low-income countries. The rapid changes in socioeconomic status and the acceptance of high-calorie, fat-rich foods and less active lifestyles have led to a considerable increase in obesity rates worldwide [3].

Obesity is currently treated with a variety of conventional medications; for instance, sibutramine (Meridia) and lorcaserin (Belviq). Despite their effectiveness, these drugs are rarely used due to concerns about their accessibility and safety. Moreover, recent studies reported the synthesis of new compounds to inhibit glycosidase enzymes implicated in various biomedical applications [4,5]. However, it is important to create accessible entities that are also secure, efficient, and cheap to use. It is widely accepted that medicines derived from plants should be used as the first line of defense in preventing illness and its complications [6,7]. Traditional synthetics have consistently provided a rich vein of novel chemical compounds from which to extract useful pharmaceuticals. Phytogenic herbal products account for over half of all FDA-approved prescriptions. Proteins, minerals, and vitamins can all be abundant in seaweeds, as can dietary fiber (non-starch polysaccharides) due to their low lipid content and reduced caloric value, as well as the fact that they disrupt increasing the other nutrients in your diet's bioavailability [8]. In addition, the dietary fiber found in seaweeds comes mostly from polysaccharides, including alginates, cellulose, fucans, and laminarins, all of which are indigestible to humans due to a lack of certain enzymes.

Marine biotechnology (also called blue biotechnology) involves the application of biological resources from the sea for industrial, medical, or environmental purposes [9–11]. On the other hand, enzymes are the dominant molecular targets for the major medicinal molecules introduced to the market. Moreover, they are considered a favorable target for new drug discovery due to their protein structures, which facilitate the exploration of diverse drugs with potential target validation. Clinical applications of enzyme inhibitors have suggested new avenues for enzyme implementation in various medical fields, including oncology, cardiology, diabetes, and neurology [12–14]. Trypsin and α-amylase inhibitors play vital roles in diabetes management since they hinder the digestion of dietary carbohydrates, reducing the risk of postprandial hyperglycemia [15,16]. Additionally, lipase enzymes are widely exploited in various biotechnological applications [17]. Importantly, carbohydrate and pancreatic lipase inhibition effectively impede weight gain and treat obesity through calorie restrictions [18,19]. Therefore, α-amylase, lipase, and trypsin inhibition assays are broadly applied in the screening of different plant extracts and natural products as a typical approach to the development of novel antidiabetic and anti-obesity medications. Polyphenols are one of the key plant compounds that have been demonstrated to possess considerable biomedical activity, such as their application in preventing cancer and heart disease and their significant role as natural antioxidants in the food industry.

Green algae (*Chlorophyta*), red algae (*Rhodophyta*), and brown algae (*Phaeophyta*) are the three main types of marine macroalgae based on their pigmentation [20,21]. Chemically speaking, macroalgae are characterized by a high percentage of water, carbohydrates, and proteins in addition to a low amount of lipids [22]. Rhodophyta, the red algae phylum, has the highest concentration of bioactive compounds, with over 1600 unique compounds accounting for 53% of all bioactive compounds found in algae [23]. Therefore, the current study investigates the biomedical application of the polyphenolic extract from Egyptian red algae (*Jania rubens*), JRPE, for the first time using different in vitro approaches, including antioxidant, anti-inflammatory, and antibacterial evaluations. Furthermore, the antidiabetic property of the JRPE was appraised through investigation for its ability to inhibit the pancreatic enzymes activity, such as α-amylase, lipase, and trypsin. Moreover, in silico computational drug screening studies were performed against these enzymes responsible for obesity and diabetes, which are targets for anti-diabetes treatment.

2. Results

2.1. The HPLC Analysis of JRPE

The HPLC analysis results of *Jania rubens* polyphenolic extract (JRPE) exhibited high concentrations of polyphenolic compositions, revealing the presence of 12 compounds, as depicted in Figure S1. Figure 1 illustrates the chemical structures of the identified polyphenolic compounds. It can be observed from the data in Table 1 that kaempferol, resveratrol, quercetin, and syringic acid are dominant products at concentrations of 140.68, 96.88, 67.48, and 49.60 mg/kg, respectively. In terms of the ratio of the polyphenolic compounds in JRPE, kaempferol revealed the highest percentage of 32.5%, followed by resveratrol (22%), quercetin (15%), syringic acid (11%), ferulic acid (6%), *o*-coumaric, vanillic, and caffeic acid (2%), while *p*-coumaric acid had the lowest concentration.

Table 1. Concentrations of polyphenolic compounds in JRPE based on HPLC analysis.

Polyphenolic Compounds	R. T/min	Con. (mg/kg)
p-Hydroxybenzoic acid	7.618	14.61616
Caffeic acid	9.954	6.93052
Catechin	9.124	3.99015
Chlorogenic	9.408	17.91833
Ferulic acid	15.715	25.81511
Kaempferol	24.757	140.68073
o-Coumaric acid	17.874	8.01992
p-Coumaric acid	13.526	1.71484
Quercetin	21.666	67.48636
Resveratrol	19.470	96.88487
Syringic acid	10.705	49.60852
Vanillic acid	15.40824	7.13708

2.2. The Antioxidant Properties of the JRPE

Figure 2A depicts the total antioxidant capacity (TAC) of JRPE in the presence of VitC as a reference, demonstrating that the TAC for JRPE is greater than VitC at all concentrations tested. However, the VitC demonstrated a slight increase in TAC compared to the JRPE at a concentration of 1000 µg/mL. In addition, the IC50 of TAC for JRE is 253.43 µg/mL, which is equivalent to VitC.

The antioxidant activity of the JRPE using DPPH methods compared with VitC as the standard antioxidant natural material revealed that JRPE had higher activity at concentrations from 300 to 700 µg/mL compared to VitC. Nevertheless, comparable antioxidant capacities for the JRPE and VitC were perceived at concentrations of 900 µg/mL and 1000 µg/mL. These findings substantiate those obtained from TAC. Furthermore, the IC50 values for the JRPE and VitC were 3.8772 µg/mL and 5601 µg/mL, respectively, as demonstrated in Figure 2B.

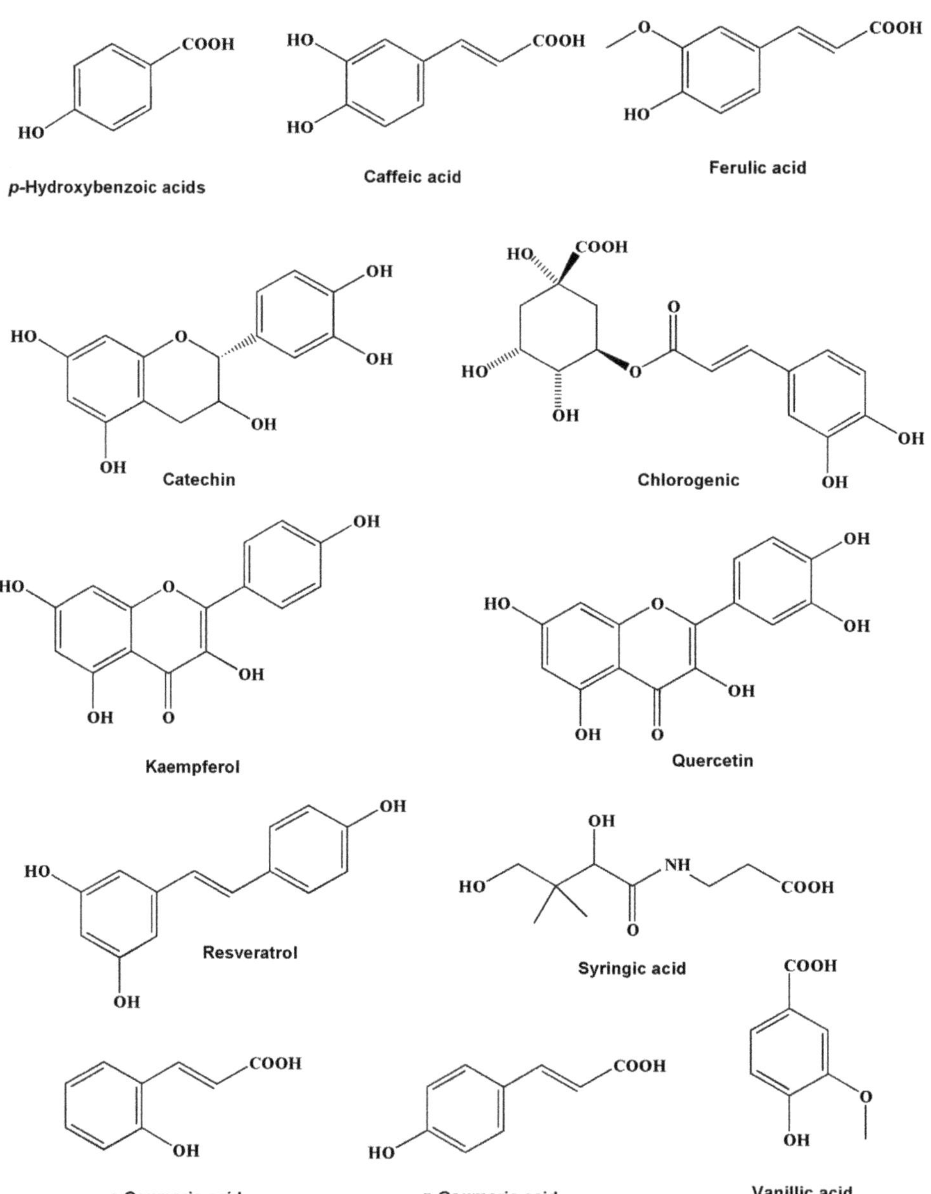

Figure 1. Chemical structures of the polyphenolic compounds in JRPE.

Figure 2C exhibited that the JRPE exerted the maximum ABTS$^{\bullet+}$ scavenging capacity of 98.91% at the highest concentration of 1000 μg/mL, whereas VitC manifested a scavenging capacity of 100% at a concentration of 700 μg/mL. In addition, the IC50 for JRPE was 1202.24 μg/mL, while the IC50 for VitC. was perceived at 345.40 μg/mL.

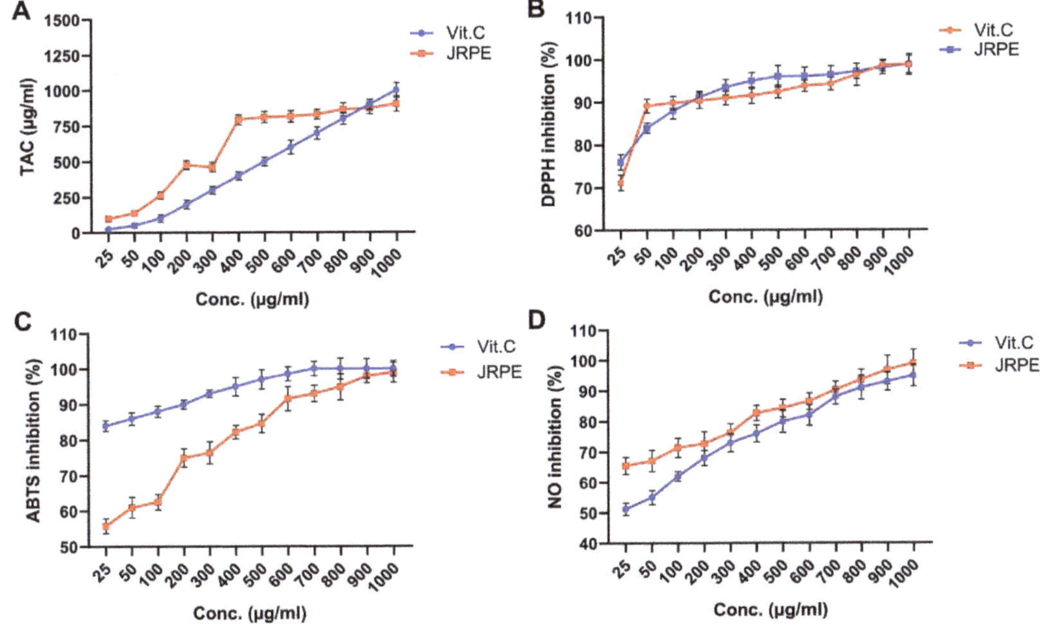

Figure 2. Antioxidant capacity of the JRPE using TAC (**A**), DPPH (**B**), ABTS$^{•+}$ (**C**), and anti-inflammatory using nitric oxide (**D**) compared with VitC as a standard drug. The results are expressed as mean ± SD.

2.3. The Anti-Inflammatory Activity Using Nitric Oxides Assay

The anti-inflammatory activity of the JRPE using the NO model reveals that the JRPE has higher anti-inflammatory activity at all levels compared to VitC with the highest activity of 98.92% at the highest concentration, as portrayed in Figure 2D. Furthermore, the IC50 values for JRPE and VitC in relation to the NO inhibition were reported to be 5326.53 and 5287.20, respectively.

2.4. The Antibacterial Activity of the JRPE

In this work, we also sought to determine whether the JRPE has antibacterial activity against gram-positive (*Streptococcus pyogenes*, *Staphylococcus aureus*, and *Enterococcus faecalis*) and gram-negative (*Escherichia coli* ATCC 8739, *Pseudomonas aeruginosa*, and *Klebsiella pneumonia*). From first sight, it could be discerned that the JRPE exerted significant growth hindrance in relation to the entire indicator bacteria, which was higher than the two reference antibiotics (ampicillin and amoxicillin), and even the later antibiotic is broad-spectrum. The first concentration of the extract (25 μg/mL) revealed low activity against all pathogenic bacteria, which significantly increased with the rise in the concentration of the extract, reaching full bacterial inhibition as presented in Figure 3. The antibacterial findings manifested that the JRPE has remarkable antibacterial properties, particularly against *S. pyogenes* (98.6%) and *S. aureus* (98.69%), which are higher than the antibacterial activities of the empirical antibiotics, ampicillin and amoxicillin. Notably, the JRPE demonstrated almost full bacterial growth inhibition of 99.8% in relation to *Enterococcus faecalis*, whereas ampicillin and amoxicillin exerted growth hindrance ratios of 87.7% and 95.98%, respectively, against the same bacteria. With regard to gram-negative bacteria, the JRPE exhibited a significant growth inhibition of 98.73% toward *E. coli*. By contrast, ampicillin and amoxicillin showed antibacterial rates of 91.68% and 97.05%, respectively, toward

E. coli. In the same manner, the antibacterial capacities of JRPE, ampicillin, and amoxicillin were perceived in relation to *P. aeruginosa* and *K. pneumonia.*

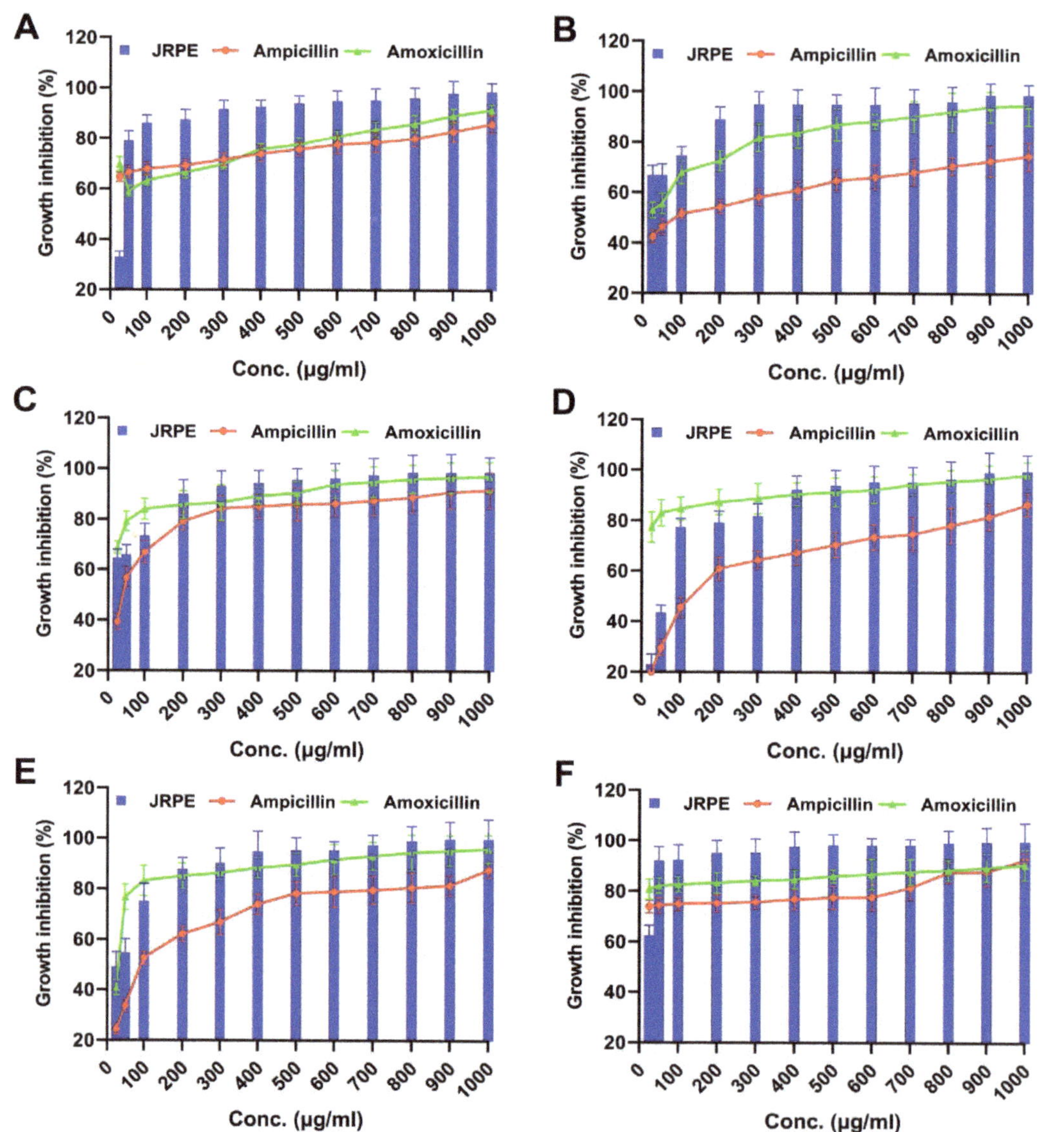

Figure 3. Antibacterial activity of the JRPE compared with two standard antibiotics (ampicillin and amoxicillin) against (**A**) *S. pyogenes*, (**B**) *S. aureus*, (**C**) *E. coli*, (**D**) *P. aeruginosa*, (**E**) *E. faecalis*, and (**F**) *K. pneumonia.* The results are shown as mean ± SD.

2.5. Inhibitory Effects of JRPE toward α-Amylase, Pepsin, Trypsin, and Lipase

The antidiabetic activity of JRPE was evaluated against three digestive enzymes (α-amylase, lipase, and trypsin) as illustrated in Figure 4. Alpha-amylase: considering the α-amylase inhibition after treatment with the JRPE, a minimum enzyme inhibition of 73.71% was detected at a JRPE concentration of 25 μg/mL. Additionally, the enzyme inhibition was augmented with the increase in JRPE concentration, reporting an inhibitory ratio of 97.43% at 1000 μg/mL. Furthermore, the IC50 value of JRPE against α-amylase was reported to be 2349.16 μg/mL. Lipase: in terms of the lipase enzyme, the minimum inhibition of lipolytic activity was perceived at 25 μg/mL with an inhibition ratio of 38.26%, while the maximum inhibition activity of lipase was 95.96% at 1000 μg/mL with an IC50 of 38.26 μg/mL. Trypsin: the minimum inhibitory activity of trypsin was 30.07% at a concentration of 25 μg/mL. On the other hand, the maximum inhibition of trypsin was 95.48% at 1000 μg/mL with an IC50 value of 517.9548 μg/mL.

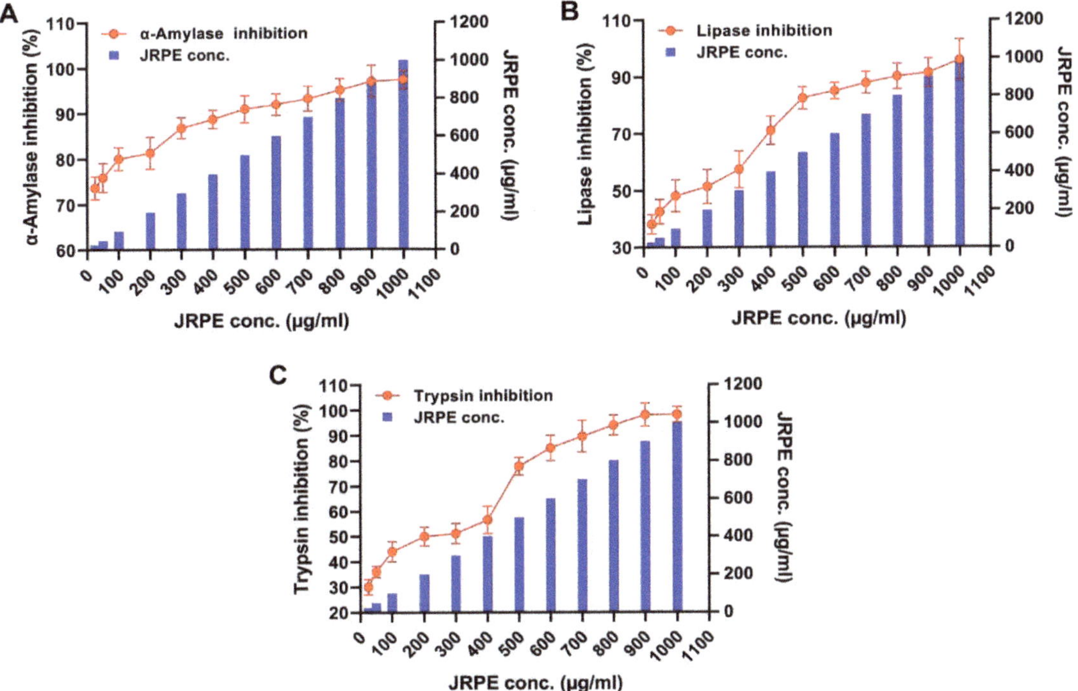

Figure 4. Inhibition properties of JRPE toward (**A**) α-amylase, (**B**) lipase, and (**C**) trypsin. The results are presented as mean ± SD.

2.6. Docking Studies of Polyphenolics in JRPE against Amylase, Lipase, and Trypsin

Docking studies were carried out to assess the interaction and potential binding model, affinity, and binding free energy (ΔG) of the polyphenolics in JRPE in relation to α-amylase, trypsin, and lipase as illustrated in Figures 5–7. The docking scores of the different polyphenolic compounds in the JRPE are enumerated in Table S1. Moreover, the rest of the docking results and the two-dimensional docking analyses for other polyphenolics are illustrated in Tables S2–S4.

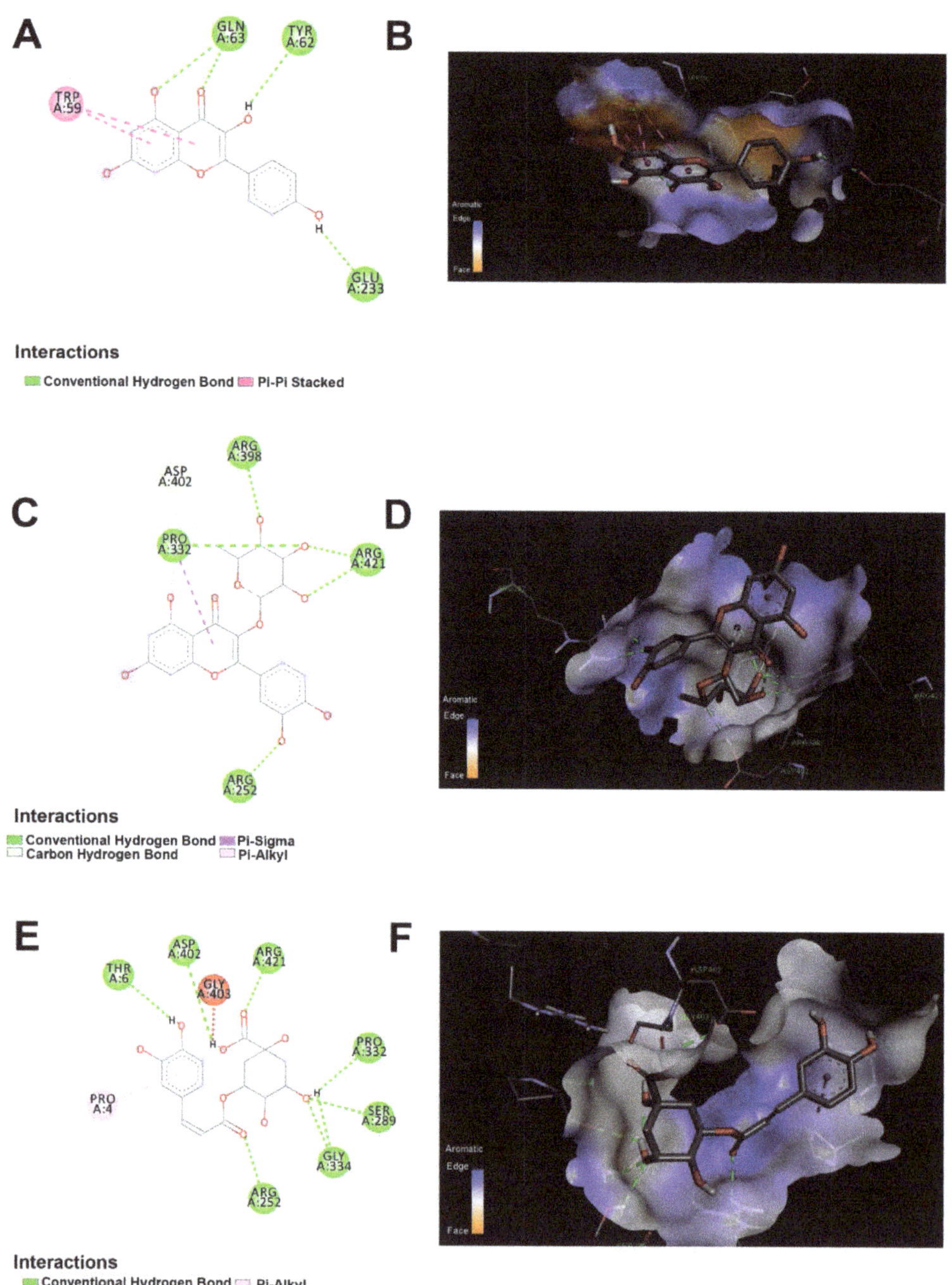

Figure 5. The 2D and 3D docking between α-amylase and polyphenolics compound in the JRPE. (**A**) 2D and (**B**) 3D of docking between kaempferol and α-amylase. (**C**) 2D and (**D**) 3D of docking between quercetin and α-amylase. (**E**) 2D and (**F**) 3D of docking analysis between chlorogenic acid and α-amylase.

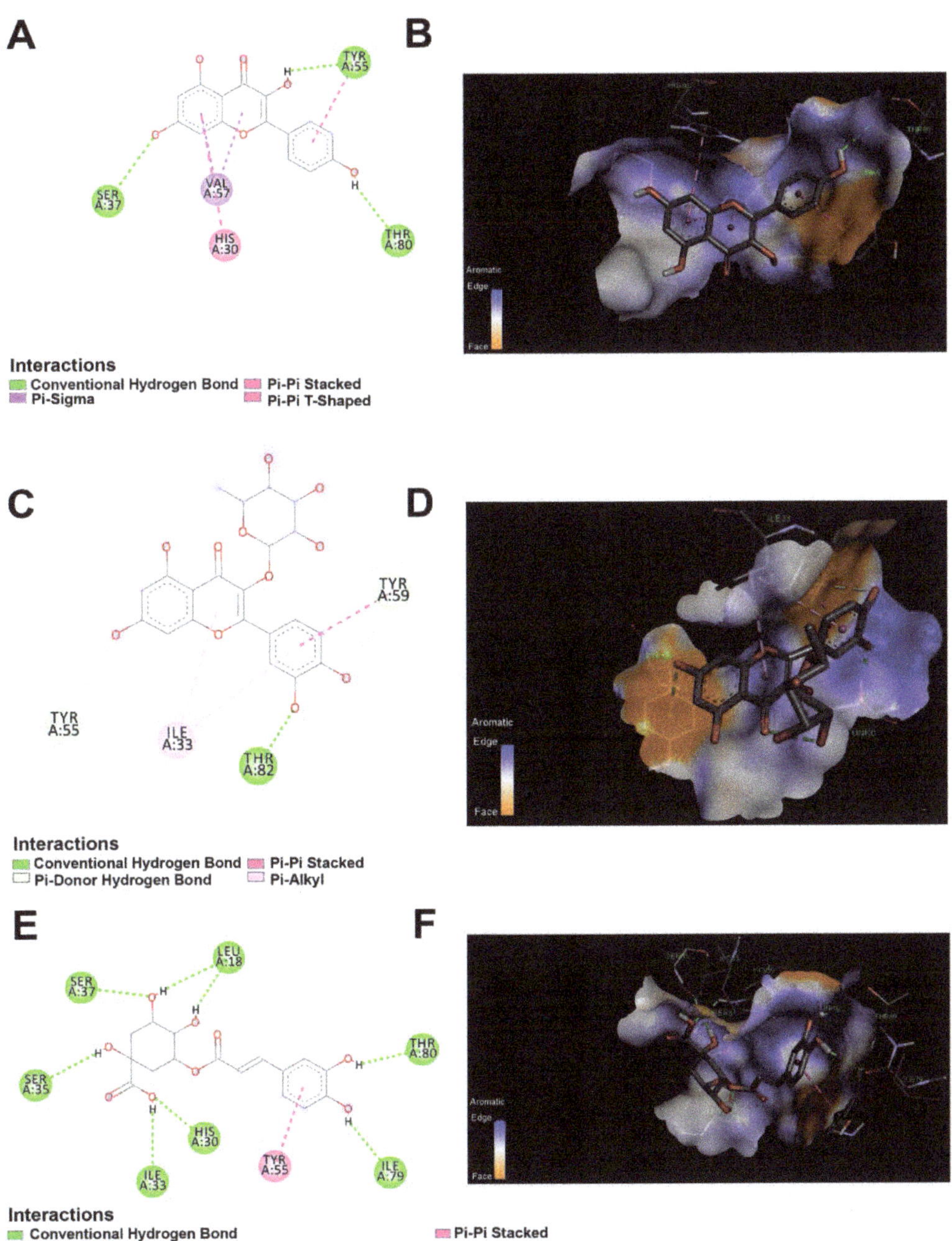

Figure 6. The 2D and 3D docking between lipase and polyphenolics compound in the JRPE. (**A**) 2D and (**B**) 3D of docking between kaempferol and lipase. (**C**) 2D and (**D**) 3D of docking between quercetin and lipase. (**E**) 2D and (**F**) 3D of docking between chlorogenic acid and lipase.

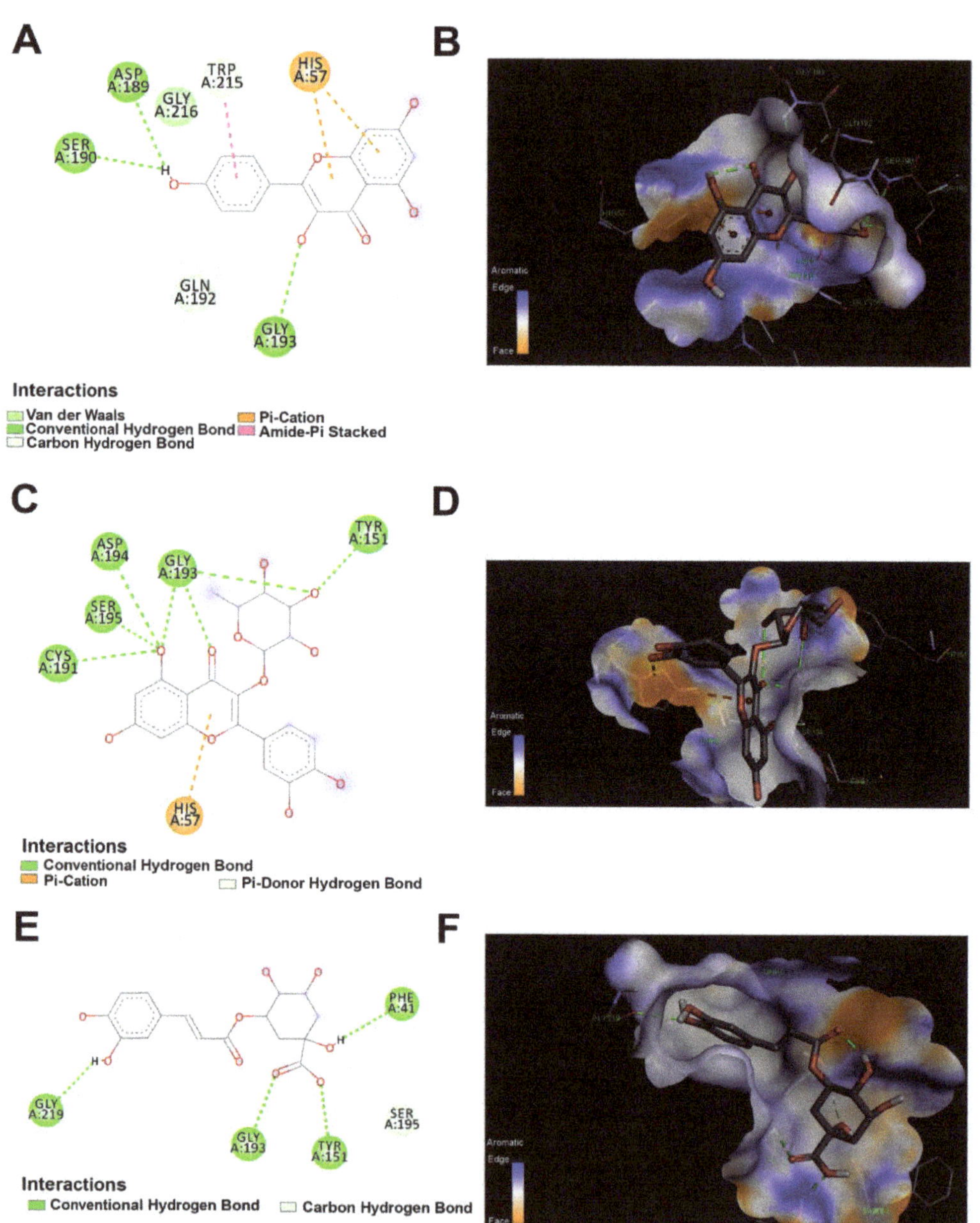

Figure 7. The 2D and 3D docking between trypsin and polyphenolics compound in the JRPE. (**A**) 2D and (**B**) 3D of docking between kaempferol and trypsin. (**C**) 2D and (**D**) 3D of docking between quercetin and trypsin. (**E**) 2D and (**F**) 3D of docking between chlorogenic acid and trypsin.

2.7. Interactions Assessment between the Twelve Polyphenolic Compounds and α-Amylase

According to the results obtained from in vitro studies, the computational docking analyses indicated that the twelve compounds could bind to the active site of α-amylase with the lowest binding energies of −4.5, −5.3, −6.43, and −5.7 kcal/mol, respectively, as presented in Table S1. Among all compounds, kaempferol, quercetin, and chlorogenic acid exhibited the highest docking scores of −8.4, −8.8, and −8 kcal/mol, respectively. As shown in Figure 5, all these residues are involved in the enzyme's binding to the docked compounds. Noticeably, the quercetin demonstrated the highest affinity score of −8.8 with α-amylase, revealing the interaction between the α-amylase and quercetin as shown in Figure 6. Furthermore, the amino acid residues (ARG A:398, PRO A:332, ARG A:421, and ARG A:252) exhibited a standard hydrogen bond with α-amylase. In addition, the ASP A:402 showed a carbon-hydrogen bond, whereas the PRO A:332 revealed p-alkyl and p-sigma bonds. The interaction of kaempferol with α-amylase exposed the second highest affinity score of −8.4, showing two types of interaction bond, including the conventional hydrogen bond (GLN A:63, TYR A:62, GLU A:233), and Pi–Pi stacked (TRP A:59). Chlorogenic acid demonstrated the third highest affinity score of −8, showing two types of interaction bonds, involving conventional hydrogen bonds (THR A:6, ASP A:402, ARG A:421, PRO A:332, SER A:289, GLY A334, ARG:252) and pi–Alkyl bonds (PRO A:4).

2.8. Interactions between the Twelve Polyphenolic Compounds and Lipase

Based on the findings of the in vitro investigations, the computational docking results exhibited that the twelve compounds attached to the active site of the lipase enzyme were catechin, *p*-coumaric acid, and vanillic acid with the lowest binding energies of −4.1, −4.3, and −4.3, kcal/mol, respectively. Out of the entire compounds, kaempferol, quercetin, and chlorogenic acid demonstrated the highest docking scores of −7.1, −7.4, and −7.2 kcal/mol, respectively. All these residues that participated in the enzyme binding to the docked compounds are represented in Table S1. The interaction between quercetin and lipase showed four types of interaction bonds, including the standard hydrogen bond (THR A:82), the pi–donor hydrogen bond (TYR A:59), (Pi–Pi stacked TYR A:59), and (Pi–Alkyl ILE A:33) as delineated in Figure 6. For kaempferol interaction bonds, it showed also four types of interaction bond, involving conventional (SER A:37, TYR A:55, THR A:80), pi–sigma (VALA: 57), Pi–Pi stacked (His A:30), and Pi–Pi shaped (TYR A:55). On the other hand, the chlorogenic acid had two types of interaction bonds: (SER A:37, SER A:35, LEU A:18, ILEA:33, HIS A:30, THR A:80, ILE A:79), and Pi–stacked (TYR A:55).

2.9. Interactions between the Twelve Polyphenolic Compounds and Trypsin

The computational docking results indicated that the twelve compounds bound to the active site of trypsin with the lowest binding energies of −4.1, −4.3, and −4.3, kcal/mol, for catechin, *p*-coumaric acid, and vanillic acid, respectively. As mentioned above in the docking analysis for lipase, the highest docking scores were observed for kaempferol, quercetin, and chlorogenic acid with affinity scores of −7.2, −7.2, and −7.1 kcal/mol, respectively. All these residues are contributed in the enzyme binding and the docked compounds are illustrated in Table S1. The interaction between lipase and quercetin is shown in Figure 7. Specifically, the interaction between quercetin and trypsin showed four types of interaction bonds, involving the conventional hydrogen bond (THR A:82), the pi–donor hydrogen bond (TYR A:59), Pi–Pi stacked (TYR A:59), and the pi–Alkyl bond (ILE A:33). Similarly, kaempferol docking with trypsin revealed conventional (SER A:37, TYR A:55, THR A:80), pi–sigma (VALA: 57), pi–pi stacked (His A:30), and pi–pi shaped (TYR A:55) interaction bonds. As observed above in the docking of lipase, chlorogenic acid displayed two types of interaction bonds: (SER A:37, SER A:35, LEU A:18, ILEA:33, HIS A:30, THR A:80, ILE A:79), and Pi–stacked (TYR A:55).

2.10. Pharmacodynamics and Pharmacokinetics of Polyphenolics Composition in JRPE

The pharmacokinetics, medicinal chemistry, drug-likeness, physicochemical properties, lipophilicity, and water solubility data are summarized in Table S5. According to the pharmacokinetic and ADMET properties, JRPE showed a high human intestinal absorption rate for almost all compounds, with the exception of chlorogenic acid and quercetin, which revealed a low intestinal absorption rate. Furthermore, polyphenolic compounds in the

JRPE exposed very low BBB permeability except for five compounds, involving *o*-coumaric, *p*-coumaric, ferulate, *p*-hydroxybenzoic acid and resveratrol. On the other hand, JRPE showed no effect on cytochrome P450 isomers for eight compounds, including caffeic, ferulate, quercetin, 4-hydroxybenzoic acid, chlorogenic acid, *p*-coumaric acid, *o*-coumaric acid, syringic acid, and vanillic acid. Most importantly, JRPE toxicity (non-mutagenic), hepatotoxicity, or skin sensitization were not perceived in the JRPE.

3. Discussion

In this study, the influence of *Jania rubens* polyphenolic extract (JRPE) on the activity of three common digestive enzymes, including α-amylase, lipase, and trypsin was investigated to comprehend their potential application as anti-obesity and anti-diabetic extract. Moreover, the antibacterial, anti-inflammatory, and antioxidant properties of JRPE polyphenols were examined. The polyphenolic contents bestow on JRPE the competency to deactivate various digestive enzymes. Thus, we investigated the inhibition of α-amylase, lipase, and trypsin activities by JRPE. After treatment with JRPE, the maximum inhibition ratios of -amylase, trypsin, and lipase were reported to be 97.43%, 98.20%, and 95.96%, respectively. This implies that the JRPE may have powerful anti-diabetic and anti-obesity properties. It is believed that polyphenol chemicals possess substantial antioxidant and antibacterial properties. Nevertheless, due to the binding of polyphenols and proteins, they are impounded into either soluble or insoluble complexes, which may frustrate the function of both polyphenols and proteins [11,24]. This makes seaweed extracts a promising candidate for the expansion of natural alternatives to synthetic compounds applied in food and cosmetic production [25–29].

The attachment of polyphenols to proteins can alter the structure, solubility, hydrophobicity, thermal stability, and isoelectric point of the protein by blocking specific amino acids, which certainly instigates conformational remodeling of the protein. Given the protein-phenolic complex, the digestibility and exploitation of dietary proteins, in addition to the activity of digestive enzymes, are altered [30,31]. Naturally occurring polyphenols have been shown to inhibit the activity of various digestive enzymes, including α-glycosidase, α-amylase, lipase, and trypsin [32]. This alters the nutrient availability and, in turn, the microbiota composition.

Considering the antibacterial activity of polyphenols, they have been evinced to have antibacterial effects by binding and inactivating essential bacterial proteins such as adhesins, enzymes, and cell envelope transport proteins. Previous studies reported that kaempferol and its derivatives could thwart the replication of *Streptococcus mutans* through disruption of a membrane enzyme identified as sortase A. Since this enzyme is vitally responsible for bacterial adherence and host cell invasion, it significantly contributes to the pathogenicity and even the virulence of the bacteria [33]. Moreover, recent studies evidenced the potent inhibitory impacts of quercetin, flavonoids, rutin, and phenolic acid against sortases A and B of *S. aureus* [34,35]. Inhibition of bacterial nucleic acid production is likely related to the B ring of the flavonoids intercalating or forming a hydrogen bond with the stacking of nucleic acid bases. Lou et al. [36] postulated two mechanisms to decipher the bactericidal activity of *p*-coumaric acid: (I) binding to bacterial genomic DNA, resulting in suppression of various metabolic pathways and, ultimately, cell death; and (II) disruption of bacterial cell membranes. It is worth mentioning that previous investigations highlighted the competency of chlorogenic acid to thwart biofilm production, swarming, and other virulence influences such as protease and elastase activity in *P. aeruginosa* along with the disordering of other mechanisms such as rhamnolipid and pyocyanin synthesis [37,38]. Likewise, quercetin effectively inhibited biofilm formation in *K. pneumoniae*, *P. aeruginosa*, and *Y. enterocolitica*, as well as other quorum-sensing-regulated attributes; for instance, inhibition of violacein and production of exopolysaccharide production as alginate [35]. Crucially, the swimming and swarming of *P. aeruginosa* and *Y. enterocolitica* were also drastically suppressed by quercetin [39,40]. Besides, resveratrol as another phenolic compound, it has the competency to disturb the physicochemical properties of the surface of *Lactobacillus paracasei*. Moreover, the adhesion, bacterial aggregation, and biofilm capabilities were also blocked as a consequence of *L. paracasei* treatment with resveratrol [41]. Overall, the substantial antibacterial activity of the JRPE compared to a broad-spectrum antibiotic, such as amoxicillin is most likely due to the synergistic influence of the polyphenolic compounds implicated in the extract.

In addition, we also studied the inhibitory impacts of the twelve polyphenolic compounds against digestive enzymes. The synergistic/antagonistic actions among bioactive features in JRPE may account for why it has a greater pancreatic digestive-inhibitory effect than other extracts with similar activity levels [42]. The systems were characterized by a predominance of the competitive part of mixed inhibition. These findings support the hypothesis that polyphenolic extract inhibits the activity of pancreatic α-amylase, lipase and trypsin through competitive mechanisms [43]. Additionally, the effect of quercetin on other pancreatic enzymes, including α-amylase, was previously investigated [44].

To determine the potential binding sites of polyphenolics with pancreatic enzymes (α-amylase, lipase, and trypsin), docking analyses were conducted employing the structures of all compounds. The competitive component of the polyphenolic compounds' mixed-inhibition may be explained by their interactions with residues close to the active site. Taking into account that the enzyme inhibitors interact with the substrate–enzyme, binding may shed light on the inhibitory process. To reiterate, the inhibitor would bind not to the substrate–enzyme combination itself but to the enzyme itself, and it would do so in close proximity to the substrate binding site.

The binding to the active site attains in a fashion that only influences on the catalytic cascades and not the substrate binding, which may explain why polyphenolic compound mixed-type inhibition is non-competitive. Zhu et al. [45] reported that polyphenolic substances could bind to a pancreatic lipase through hydrophobic interactions. For instance, pancreatic digesting enzymes and aromatic rings from polyphenolic substances generate π-stacking interactions. Our findings are in line with those of Swilam et al. [46], who observed that hydrophobic bonds were the primary form of interaction between polyphenolic chemicals and digestive enzymes. Furthermore, it was predicted by the docking study that the hydroxyl groups of polyphenolics and the polar groups of digestive enzymes could form a hydrophobic bond [47]. Compared to other polyphenolic compound–pancreatic lipase complexes, the quercetin–digestive enzyme complex had greater polar interactions. The larger size and rigid structure of quercetin may account for its predominant binding qualities (more polar contacts) and, by extension, its greater inhibitory competency in the digestive tract. Zhang et al. [48] also found that quercetin was more effective at inhibiting the target enzyme. Moreover, Ullah et al. [49] demonstrated that quercetin has a stronger affinity for proteins due to its structural features. The catechol structure in the B ring and the double bond between C2 and C3 are two of quercetin's distinguishing features. The capability of polyphenolics to bind proteins is predominantly correlated to structural features, including free hydroxyl groups and number of aromatic rings and [50].

4. Materials and Methods

4.1. Samples Collection

Jania rubens was collected during 2019 at a depth of 1–27 m in the Red Sea region, Hurghada (latitude: 27°11'37.5" and longitude: 33°50'48.4"), Egypt and immediately transported to the lab. The collected samples were identified at the National Institute of Oceanography and Fisheries (NIOF), Egypt.

4.2. Extraction and Preparation of Jania rubens Polyphenolic Extract (JRPE)

To extract the polyphenolic compounds from *Jania rubens*, the samples were dried in air before being ground to obtain the powder with a weight of 500 g. Afterward, the powder was immersed in ethyl acetate for 1 h, followed by sonication before being maintained overnight in the fridge at 4 °C in the dark bottle. The extraction process was conducted three times to maximize the yield of polyphenolic compounds. Following this, the solvent was eliminated by means of a rotary evaporator (R-300, Büchi Labortechnik GmbH, Essen, Germany) at 45 °C under a low pressure before being stored at 4 °C for further investigations [49,50].

4.3. High-Performance Liquid Chromatography (HPLC) Analysis of JRPE

To ascertain the phenolic and flavonoid compounds in the JRPE, HPLC analysis was performed by means of HPLC (Agilent 1260, Santa Clara, CA, USA) using a Kinetex® 5 μm EVO C18 column (100 mm × 4.6 mm) purchased from Phenomenex®, Torrance, CA, USA. The separation was accomplished utilizing a tertiary liner elution gradient with HPLC grade water, 0.2% H_3PO_4 (v/v), methanol, and acetonitrile. The injection volume was 20 μL, and the detection was performed using WWD at 284 nm.

4.4. Antioxidant Activity of JRPE

To assess the antioxidant properties of the JRPE, DPPH and ABST$^{\bullet+}$ assays were conducted using 2,2-diphenyl-1-picrylhydrazyl and 2,2′-azino-bis(3-ethylbenzothiazoline-6-sulfonic acid), respectively, following the previous protocols with minor adaptations [51,52]. All investigations were performed in triplicate, and the free radical scavenging was then calculated using Equation (1):

$$\text{Scavenging (\%)} = [(Ac - As)/Ac] \times 100 \tag{1}$$

where Ac and As point to the absorbance of the control and the sample after reaction, respectively.

To assess the total antioxidant capacity (TAC) of the JRPE, the phosphomolybdenum approach was performed following the procedures reported by Prieto et al. [53]. Vitamin C (VitC) was utilized as a standard, and the reactions were replicated three times.

4.5. Anti-Inflammatory Activity of JRPE

To estimate the anti-inflammatory properties of the JRPE, a nitric oxide scavenging approach was performed in accordance with the protocol delineated by Garrat [54]. Vit. C was applied as a standard drug in this assay, and the investigations were accomplished in triplicate. The inhibition ratio of nitric oxide was computed using Equation (2):

$$\text{Nitric oxide scavenging (\%)} = [(Ac - As)/Ac] \times 100. \tag{2}$$

where Ac is the absorbance of the control, while as indicates the absorbance of the sample after reaction.

4.6. Antibacterial Assessments of JRPE

The antibacterial properties of JRPE were evaluated toward three gram-positive bacteria (*Streptococcus pyogenes* ATCC 19615, *Staphylococcus aureus* ATCC 25923, *and Enterococcus faecalis* ATCC 29212) and three gram-negative bacteria (*Escherichia coli* ATCC 8739, *Pseudomonas aeruginosa* ATCC 15442, and *Klebsiella pneumoniae* ATCC 13883) in accordance with the resazurin assay using a microtiter plate (Sigma-Aldrich, Taufkirchen, Germany). Each bacterial strain was revitalized by growing overnight at 37 °C in LB medium, followed by adjustment of their turbidities at 600 nm by means of a spectrophotometer in compliance with the McFarland 0.5 standard [55,56]. A volume of 50 μL representing different concentrations of the JRPE from 25 to 1000 μg/mL) was loaded into a sterile microtiter plate (96-well), followed by the addition of 10 μL of resazurin indicator solution to each well. Following this, 30 μL of LB medium was added to the wells before being inoculated with 10 μL of bacterial suspension (5×10^6 CFU/mL). Ampicillin and amoxicillin (Sigma-Aldrich, Taufkirchen, Germany) were applied as reference antibiotics to the microtiter plate with concentrations comparable to the extract. To avoid the rehydration of bacteria, each plate was wrapped with cling film before being incubated overnight at 37 °C. The antibacterial evaluations were carried out in triplicate and the bacterial cultures in each microplate were measured at 520 nm by means of a microplate reader (SpectraMax i3x Multi-Mode, San Jose, CA, USA). The growth inhibition ratio of bacteria was quantified using Equation (3):

$$\text{Growth inhibition of bacteria (\%)} = [(Ac - As)/Ac] \times 100 \tag{3}$$

where Ac and As indicate the absorbance of untreated bacterial cultures and bacterial cultures treated with JRPE, respectively.

4.7. Anti-Diabetics and Anti-Obesity of JRPE Using Inhibition of Digestive Enzymes

The anti-diabetics and anti-obesity of JRPE were evaluated utilizing α-amylase, lipase, and trypsin (Loba chemie, Mumbai, India). To estimate the activity of pancreatic α-amylase activity, we utilized soluble starch as a substrate and the reducing sugars were then calorimetrically evaluated adopting the dinitrosalicylic acid method demonstrated by Miller [57]. The reaction was performed by adding 20 μL of α-amylase solution containing 20 μg of the enzyme to 280 μL of starch solution (1%, w/v) prepared in phosphate buffer (pH 7) supplemented with 20 mM $CaCl_2$. The reaction was commenced by incubation at 37 °C before being terminated by the addition of a DNSA reagent. To evaluate the ability of the JRPE to inhibit the α-amylase activity, different concentrations of the extract from 25 to 1000 μg/mL were added to 20 μL of α-amylase enzyme solutions before being incubated for 5 min. Following this, the enzyme activity was estimated using the DNS method and measured at 540 nm by means of a spectrophotometer.

To assess the lipase inhibition, a stock solution of pancreatic lipase (1 mg/mL) was prepared in 10 mM Tris-HCl buffer (PH 7.5). The lipase activity was assayed as previously described by Choi et al. [58] with slight modifications. DMPTB (2,3- mercapto-1-propanol tributyrate) was utilized as a substrate and dissolved in 50 mM Tris-HCl buffer (PH 7.2) complemented with 6% of Triton X-100, while the DTNB reagent (5,5-dithio-bis-(2-nitrobenzoic acid) at concentration of 40 mM was prepared in isobutanol. The lipase inhibition using the JRPE was conducted using different concentrations within a range from 25 to 1000 μg/mL. The reaction was performed by adding 200 μL from each concentration of JRPE to 100 μL of the enzyme solution, followed by addition of 700 μL of tris-HCl buffer (pH 7.4) to the mixture. After incubation of the tubes at 37 °C for 15 min, 100 μL of DMPTB substrate was added to the mixture for the enzyme assay. The activity of the lipase enzyme was evaluated at 405 nm employing a spectrophotometer.

For estimating the capacity of the JRPE to inhibit the activity of trypsin, 1.5 mL of trypsin solution prepared in Tris-HCl buffer (0.2 M, pH 8) was pre-incubated at 25 °C for 15 min with 1.5 mL of different concentrations of JRPE, ranging from 25 to 1000 μg/mL. Casein was used as a substrate and was prepared in Tris-HCl buffer (0.2 M, pH 8) supplemented with 20 mM $CaCl_2$ in a ratio of 1:1.8, respectively, before being incubated in a water bath at 37 °C for 20 min to fully dissolve. To estimate the enzyme activity, 2.8 mL of the substrate was thoroughly mixed with 200 μL of the enzyme and JRPE. After incubation for 20 min at 37 °C, the reaction was terminated by adding 6 mL of 2.5% trichloroacetic acid (TCA) and the tubes were then maintained on ice for 20 min. The tubes were centrifuged at 10,000 rpm for 10 min, and the supernatants were further measured at 280 nm using a spectrophotometer. The inhibition percentage was calculated using Equation (4):

$$\text{Enzyme inhibition (\%)} = [(Ac - At)/Ac] \times 100 \tag{4}$$

where Ac points to the absorbance of the control, which contains all reagents and 20% DMSO in the absence of the tested solution, while At indicates the absorbance of the examined sample.

The half-maximal inhibitory concentration (IC50) values were estimated following fitting inhibition parameters with standard log inhibitor vs. normalized response models using AAT Bioquest (https://www.aatbio.com/tools/ic50-calculator, accessed on 20 November 2022).

4.8. Molecular Docking Studies

The crystal structures of α-amylase (PDB ID: 4W93), lipase (PDB ID: 1LPB), and trypsin (PDB ID: 4DOQ) were retrieved from the Protein Data Bank website: http://www.rcsb.org (accessed on 20 November 2022) and the target proteins were prepared for docking analysis employing Pymol Opensource, Shirley, NY, USA. Furthermore, to prepare the ligands, the 3D structures of the investigated 12 polyphenolic compounds and the co-crystallized compound were obtained from the PubChem database (www.pubchem.ncbi.nlm.nih.gov (accessed on 20 November 2022)) and their chemical structures were then transformed into PDB using Pymol [59]. Afterward, the ligands were transformed into PDBQT format by means of AutoDock Tools, San Diego, CA, USA for molecular docking simulation. Before commencing the docking examinations, the docking procedure was verified by redocking the native inhibitors into the active site of the enzyme. The binding model with the minimum binding energy was superimposed on the retrieved co-crystallized inhibitor. Then, polyphenolics in JRPE were docked into the active sites of the enzymes employing the AutoDock Vina docking system. The prepared structures of enzymes were also imported, and the docking examination was commenced with all other parameters. The docked complexes were visualized by means of the Discovery Studio Visualizer, Shirley, NY, USA (V. 21) to explore and report the different molecular interactions.

4.9. In Silico Pharmacodynamics and Pharmacokinetics

To assess the drug-likeness of the 12 polyphenolic compounds in JRPE, in silico evaluation was attained adopting (http://www.swissadme.ch/index.php#, accessed on 20 November 2022) [60] based on specific properties of compounds, including absorption, distribution, metabolism, and excretion as assessments for pharmacokinetic features [61].

4.10. Statistical Analysis

All the experimental assays were carried out at least in triplicate, and results are shown as mean $\pm$ SD. The results were analyzed by means of GraphPad Prism (Version 8, GraphPad Software Inc., San Diego, CA, USA) and were considered to be significant at $p \leq 0.05$.

5. Conclusions

Twelve polyphenolic compounds were identified in the extract of *Jania rubens* (JRPE) collected from Egypt. The JRPE demonstrated remarkable antioxidant and antibacterial activities. Among the identified polyphenolic compounds, quercetin, kaempferol, and chlorogenic exposed the highest inhibition toward α-amylase, lipase, and trypsin with acceptable IC_{50}. This may be related to the antioxidant and anti-inflammatory properties of JRPE extract, which enable them to bind with digestive enzymes, forming a polyphenolic-enzyme complex. Furthermore, computational studies following molecular docking analysis for the twelve polyphenolic compounds in JRPE exhibited that the polyphenolics may develop a complex with digestive enzymes, demonstrating that all compounds could closely bind to the active site of digestive enzymes. Significantly, the in vitro studies substantiated the powerful inhibitory activity of JRPE, indicating the anti-obesity and anti-diabetes characteristics of the implicated polyphenolic compounds. Further in vivo investigations should be performed to evaluate the capacity of the JRPE to govern diabetes in animal models, and the polyphenolic compounds could be purified to assess their independent bioactivity.

Supplementary Materials: The following supporting information can be downloaded at: https://www.mdpi.com/article/10.3390/catal13020443/s1, Figure S1: HPLC chromatogram for the JRPE shows different polyphenolic compounds (phenolic and flavonoids); Table S1: The docking affinity scores of the different polyphenolics' compounds in the JRPE; Table S2: Affinity scores for docking analysis of α-amylase (PDB ID: 4W93) using twelve polyphenolic compounds identified from JRPE; Table S3: Affinity scores for docking analysis of lipase (PDB ID: 1LPB) using twelve polyphenolic compounds identified from JRPE; Table S4: Affinity scores for docking analysis of trypsin (PDB ID:

4DOQ) using twelve polyphenolic compounds identified from JRPE; Table S5: Pharmacodynamics and pharmacokinetics analysis of twelve polyphenolic compounds identified from JRPE.

Author Contributions: Conceptualization, A.N.-A., M.A.S. and M.A.H.; data curation, A.N.-A. and M.A.H.; investigation, A.N.-A. and M.A.H.; formal analysis, A.N.-A., M.L.A., T.M.T. and M.A.H.; writing—original draft, A.N.-A. and M.A.H.; writing—review and editing, A.N.-A., M.L.A., T.M.T., M.A.S. and M.A.H. All authors have read and agreed to the published version of the manuscript.

Funding: This research received no external funding.

Data Availability Statement: The datasets generated during the current study are available from the corresponding authors upon reasonable request.

Conflicts of Interest: The authors declare no conflict of interest.

References

1. Forouhi, N.G.; Wareham, N.J. Epidemiology of diabetes. *Medicine* **2014**, *42*, 698–702. [CrossRef]
2. Federation, I.D. *IDF Diabetes Atlas*, 9th ed.; International Diabetes Federation: Brussels, Belgium, 2019; Available online: https://www.diabetesatlas.org (accessed on 20 November 2022).
3. Galicia-Garcia, U.; Benito-Vicente, A.; Jebari, S.; Larrea-Sebal, A.; Siddiqi, H.; Uribe, K.B.; Ostolaza, H.; Martín, C. Pathophysiology of Type 2 Diabetes Mellitus. *Int. J. Mol. Sci.* **2020**, *21*, 6275. [CrossRef] [PubMed]
4. Rajasekaran, P.; Ande, C.; Vankar, Y.D. Synthesis of (5,6 & 6,6)-oxa-oxa annulated sugars as glycosidase inhibitors from 2-formyl galactal using iodocyclization as a key step. *ARKIVOC* **2022**, *2022*, 5–23.
5. Yang, L.-F.; Shimadate, Y.; Kato, A.; Li, Y.-X.; Jia, Y.-M.; Fleet, G.W.J.; Yu, C.-Y. Synthesis and glycosidase inhibition of N-substituted derivatives of 1,4-dideoxy-1,4-imino-d-mannitol (DIM). *Org. Biomol. Chem.* **2020**, *18*, 999–1011. [CrossRef] [PubMed]
6. Anand, U.; Jacobo-Herrera, N.; Altemimi, A.; Lakhssassi, N. A Comprehensive Review on Medicinal Plants as Antimicrobial Therapeutics: Potential Avenues of Biocompatible Drug Discovery. *Metabolites* **2019**, *9*, 258. [CrossRef] [PubMed]
7. Shreadah, M.A.; El Moneam, N.; El-Assar, S.A.; Nabil-Adam, A. Metabolomics and pharmacological screening of aspergillus versicolor isolated from *Hyrtios erectus* Red Sea sponge; Egypt. *Curr. Bioact. Compd.* **2020**, *16*, 1083–1102. [CrossRef]
8. Nabil-Adam, A.; Shreadah, A.M.; Abd El-Moneam, M.N.; El-Assar, A.S. Marine Algae of the Genus Gracilaria as Multi Products Source for Different Biotechnological and Medical Applications. *Recent Pat. Biotechnol.* **2020**, *14*, 203–228. [CrossRef] [PubMed]
9. AbdelMonein, N.M.; Yacout, G.A.; Aboul-Ela, H.M.; Shreadah, M.A. Hepatoprotective Activity of Chitosan Nanocarriers Loaded with the Ethyl Acetate Extract of a *Stenotrophomonas* sp. Bacteria Associated with the Red Sea Sponge Amphimedon ochracea in CCl4 Induced Hepatotoxicty in Rats. *Adv. Biosci. Biotechnol.* **2017**, *8*, 27–50. [CrossRef]
10. Abd El-Moneam, N.M.; Shreadah, M.A.; El-Assar, S.A.; Nabil-Adam, A. Protective role of antioxidants capacity of *Hyrtios aff. Erectus* sponge extract against mixture of persistent organic pollutants (POPs)-induced hepatic toxicity in mice liver: Biomarkers and ultrastructural study. *Environ. Sci. Pollut. Res.* **2017**, *24*, 22061–22072. [CrossRef]
11. Abd El-Moneam, N.M.; El-Assar, S.A.; Shreadah, M.A.; Nabil-Adam, A. Isolation, identification and molecular screening of Pseudomonas sp. metabolic pathways NRPs and PKS associated with the Red Sea sponge, *Hyrtios aff. Erectus*, Egypt. *J. Pure Appl. Microbiol.* **2017**, *11*, 1299–1311. [CrossRef]
12. Wang, C.C.L.; Hess, C.N.; Hiatt, W.R.; Goldfine, A.B. Clinical Update: Cardiovascular Disease in Diabetes Mellitus. *Circulation* **2016**, *133*, 2459–2502. [CrossRef]
13. Chaudhury, A.; Duvoor, C.; Reddy Dendi, V.S.; Kraleti, S.; Chada, A.; Ravilla, R.; Marco, A.; Shekhawat, N.S.; Montales, M.T.; Kuriakose, K.; et al. Clinical Review of Antidiabetic Drugs: Implications for Type 2 Diabetes Mellitus Management. *Front. Endocrinol.* **2017**, *8*, 6. [CrossRef] [PubMed]
14. Janssen, J.A.M.J.L. Hyperinsulinemia and Its Pivotal Role in Aging, Obesity, Type 2 Diabetes, Cardiovascular Disease and Cancer. *Int. J. Mol. Sci.* **2021**, *22*, 7797. [CrossRef]
15. Gong, L.; Feng, D.; Wang, T.; Ren, Y.; Liu, Y.; Wang, J. Inhibitors of α-amylase and α-glucosidase: Potential linkage for whole cereal foods on prevention of hyperglycemia. *Food Sci. Nutr.* **2020**, *8*, 6320–6337. [CrossRef] [PubMed]
16. Arbi, B.; Bouchentouf, S.; El-Shazly, M. Investigation Of The Potential Antidiabetic Effect Of *Zygophyllum* Sp. By Studying The Interaction Of Its Chemical Compounds With Alpha-Amylase And Dpp-4 Enzymes Using A Molecular Docking Approach. *Curr. Enzym. Inhib.* **2023**, *19*. [CrossRef]
17. Abol-Fotouh, D.; AlHagar, O.E.A.; Hassan, M.A. Optimization, purification, and biochemical characterization of thermoalkaliphilic lipase from a novel *Geobacillus stearothermophilus* FMR12 for detergent formulations. *Int. J. Biol. Macromol.* **2021**, *181*, 125–135. [CrossRef] [PubMed]
18. Wen, X.; Zhang, B.; Wu, B.; Xiao, H.; Li, Z.; Li, R.; Xu, X.; Li, T. Signaling pathways in obesity: Mechanisms and therapeutic interventions. *Signal Transduct. Target. Ther.* **2022**, *7*, 298. [CrossRef] [PubMed]
19. Kim, D.H.; Park, Y.H.; Lee, J.S.; Jeong, H.I.; Lee, K.W.; Kang, T.H. Anti-Obesity Effect of DKB-117 through the Inhibition of Pancreatic Lipase and α-Amylase Activity. *Nutrients* **2020**, *12*, 3053. [CrossRef] [PubMed]
20. Leandro, A.; Pereira, L.; Gonçalves, A.M.M. Diverse Applications of Marine Macroalgae. *Mar. Drugs* **2020**, *18*, 17. [CrossRef]

21. Daniotti, S.; Re, I. Marine Biotechnology: Challenges and Development Market Trends for the Enhancement of Biotic Resources in Industrial Pharmaceutical and Food Applications. A Statistical Analysis of Scientific Literature and Business Models. *Mar. Drugs* **2021**, *19*, 61. [CrossRef]

22. Sudhakar, K.; Mamat, R.; Samykano, M.; Azmi, W.H.; Ishak, W.F.W.; Yusaf, T. An overview of marine macroalgae as bioresource. *Renew. Sustain. Energy Rev.* **2018**, *91*, 165–179. [CrossRef]

23. Carpena, M.; Garcia-Perez, P.; Garcia-Oliveira, P.; Chamorro, F.; Otero, P.; Lourenço-Lopes, C.; Cao, H.; Simal-Gandara, J.; Prieto, M.A. Biological properties and potential of compounds extracted from red seaweeds. *Phytochem. Rev.* **2022**, 1–32. [CrossRef] [PubMed]

24. Lakey-Beitia, J.; Burillo, A.M.; La Penna, G.; Hegde, M.L.; Rao, K.S. Polyphenols as Potential Metal Chelation Compounds Against Alzheimer's Disease. *J. Alzheimer's Dis.* **2021**, *82*, S335–S357. [CrossRef]

25. Shreadah, M.; Abdel-El Moneam, N.; Al-Assar, S.; Nabil-Adam, A. The ameliorative role of a marine sponge extract against mixture of persistent organic pollutants induced changes in hematological parameters in mice. *Expert Opin. Env. Biol.* **2017**, *6*, 2. [CrossRef]

26. Abdel Moneam, N.; Al-Assar, S.; Shreadah, M.; Nabil-Adam, A. The hepatoprotective effect of *Hyrtios aff. Erectus* sponge isolated from the Red sea extract against the toxicity of Persistent organic pollutants (POPs) from Sediments of Lake Mariout. *J. Biotechnol. Biotechnol. Equip.* **2017**, *32*, 734–743. [CrossRef]

27. Shreadah, M.A.; El Moneam, N.M.A.; Al-Assar, S.A.; Nabil-Adam, A. Phytochemical and pharmacological screening of *Sargassium vulgare* from Suez Canal, Egypt. *Food Sci. Biotechnol.* **2018**, *27*, 963–979. [CrossRef]

28. Cotas, J.; Leandro, A.; Monteiro, P.; Pacheco, D.; Figueirinha, A.; Gonçalves, A.M.M.; da Silva, G.J.; Pereira, L. Seaweed Phenolics: From Extraction to Applications. *Mar. Drugs* **2020**, *18*, 384. [CrossRef]

29. Lomartire, S.; Cotas, J.; Pacheco, D.; Marques, J.C.; Pereira, L.; Gonçalves, A.M.M. Environmental Impact on Seaweed Phenolic Production and Activity: An Important Step for Compound Exploitation. *Mar. Drugs* **2021**, *19*, 245. [CrossRef]

30. Abd El Moneam, N.M.; Shreadah, M.A.; El-Assar, S.A.; De Voogd, N.J.; Nabil-Adam, A. Hepatoprotective effect of Red Sea sponge extract against the toxicity of a real-life mixture of persistent organic pollutants. *Biotechnol. Biotechnol. Equip.* **2018**, *32*, 734–743. [CrossRef]

31. Sęczyk, Ł.; Świeca, M.; Kapusta, I.; Gawlik-Dziki, U. Protein–Phenolic Interactions as a Factor Affecting the Physicochemical Properties of White Bean Proteins. *Molecules* **2019**, *24*, 408. [CrossRef]

32. Wojtunik-Kulesza, K.; Oniszczuk, A.; Oniszczuk, T.; Combrzyński, M.; Nowakowska, D.; Matwijczuk, A. Influence of In Vitro Digestion on Composition, Bioaccessibility and Antioxidant Activity of Food Polyphenols—A Non-Systematic Review. *Nutrients* **2020**, *12*, 1401. [CrossRef]

33. Susmitha, A.; Bajaj, H.; Madhavan Nampoothiri, K. The divergent roles of sortase in the biology of Gram-positive bacteria. *Cell Surf.* **2021**, *7*, 100055. [CrossRef]

34. Nguyen, T.L.A.; Bhattacharya, D. Antimicrobial Activity of Quercetin: An Approach to Its Mechanistic Principle. *Molecules* **2022**, *27*, 2494. [CrossRef] [PubMed]

35. Makarewicz, M.; Drożdż, I.; Tarko, T.; Duda-Chodak, A. The Interactions between Polyphenols and Microorganisms, Especially Gut Microbiota. *Antioxidants* **2021**, *10*, 188. [CrossRef]

36. Lou, Z.; Wang, H.; Rao, S.; Sun, J.; Ma, C.; Li, J. p-Coumaric acid kills bacteria through dual damage mechanisms. *Food Control* **2012**, *25*, 550–554. [CrossRef]

37. Wang, H.; Chu, W.; Ye, C.; Gaeta, B.; Tao, H.; Wang, M.; Qiu, Z. Chlorogenic acid attenuates virulence factors and pathogenicity of *Pseudomonas aeruginosa* by regulating quorum sensing. *Appl. Microbiol. Biotechnol.* **2019**, *103*, 903–915. [CrossRef] [PubMed]

38. Su, M.; Liu, F.; Luo, Z.; Wu, H.; Zhang, X.; Wang, D.; Zhu, Y.; Sun, Z.; Xu, W.; Miao, Y. The Antibacterial Activity and Mechanism of Chlorogenic Acid Against Foodborne Pathogen *Pseudomonas aeruginosa*. *Foodborne Pathog. Dis.* **2019**, *16*, 823–830. [CrossRef]

39. Mu, Y.; Zeng, H.; Chen, W. Quercetin Inhibits Biofilm Formation by Decreasing the Production of EPS and Altering the Composition of EPS in *Staphylococcus epidermidis*. *Front. Microbiol.* **2021**, *12*, 631058. [CrossRef]

40. Vipin, C.; Saptami, K.; Fida, F.; Mujeeburahiman, M.; Rao, S.S.; Athmika; Arun, A.B.; Rekha, P.D. Potential synergistic activity of quercetin with antibiotics against multidrug-resistant clinical strains of *Pseudomonas aeruginosa*. *PloS ONE* **2020**, *15*, e0241304. [CrossRef]

41. Al Azzaz, J.; Al Tarraf, A.; Heumann, A.; Da Silva Barreira, D.; Laurent, J.; Assifaoui, A.; Rieu, A.; Guzzo, J.; Lapaquette, P. Resveratrol Favors Adhesion and Biofilm Formation of *Lacticaseibacillus paracasei* subsp. *paracasei* Strain ATCC334. *Int. J. Mol. Sci.* **2020**, *21*, 5423. [CrossRef]

42. Kato, E.; Tsuruma, A.; Amishima, A.; Satoh, H. Proteinous pancreatic lipase inhibitor is responsible for the antiobesity effect of young barley (*Hordeum vulgare* L.) leaf extract. *Biosci. Biotechnol. Biochem.* **2021**, *85*, 1885–1889. [CrossRef] [PubMed]

43. Oluwagunwa, O.A.; Alashi, A.M.; Aluko, R.E. Inhibition of the in vitro Activities of α-Amylase and Pancreatic Lipase by Aqueous Extracts of *Amaranthus viridis*, *Solanum macrocarpon* and *Telfairia occidentalis* Leaves. *Front. Nutr.* **2021**, *8*, 772903. [CrossRef]

44. Oboh, G.; Ademosun, A.O.; Ayeni, P.O.; Omojokun, O.S.; Bello, F. Comparative effect of quercetin and rutin on α-amylase, α-glucosidase, and some pro-oxidant-induced lipid peroxidation in rat pancreas. *Comp. Clin. Pathol.* **2015**, *24*, 1103–1110. [CrossRef]

45. Zhu, W.; Jia, Y.; Peng, J.; Li, C.-M. Inhibitory Effect of Persimmon Tannin on Pancreatic Lipase and the Underlying Mechanism in Vitro. *J. Agric. Food Chem.* **2018**, *66*, 6013–6021. [CrossRef]

46. Swilam, N.; Nawwar, M.A.M.; Radwan, R.A.; Mostafa, E.S. Antidiabetic Activity and In Silico Molecular Docking of Polyphenols from *Ammannia baccifera* L. subsp. *Aegyptiaca* (Willd.) Koehne Waste: Structure Elucidation of Undescribed Acylated Flavonol Diglucoside. *Plants* **2022**, *11*, 452. [CrossRef]
47. Baruah, I.; Kashyap, C.; Guha, A.K.; Borgohain, G. Insights into the Interaction between Polyphenols and β-Lactoglobulin through Molecular Docking, MD Simulation, and QM/MM Approaches. *ACS Omega* **2022**, *7*, 23083–23095. [CrossRef]
48. Zhang, R.; Wei, Y.; Yang, T.; Huang, X.; Zhou, J.; Yang, C.; Zhou, J.; Liu, Y.; Shi, S. Inhibitory effects of quercetin and its major metabolite quercetin-3-O-β-D-glucoside on human UDP-glucuronosyltransferase 1A isoforms by liquid chromatography-tandem mass spectrometry. *Exp. Ther. Med.* **2021**, *22*, 842. [CrossRef]
49. Ullah, A.; Munir, S.; Badshah, S.L.; Khan, N.; Ghani, L.; Poulson, B.G.; Emwas, A.-H.; Jaremko, M. Important Flavonoids and Their Role as a Therapeutic Agent. *Molecules* **2020**, *25*, 5243. [CrossRef]
50. Cosme, P.; Rodríguez, A.B.; Espino, J.; Garrido, M. Plant Phenolics: Bioavailability as a Key Determinant of Their Potential Health-Promoting Applications. *Antioxidants* **2020**, *9*, 1263. [CrossRef] [PubMed]
51. Chakraborty, K.; Paulraj, R. Sesquiterpenoids with free-radical-scavenging properties from marine macroalga *Ulva fasciata* Delile. *Food Chem.* **2010**, *122*, 31–41. [CrossRef]
52. Tamer, T.M.; Sabet, M.M.; Alhalili, Z.A.H.; Ismail, A.M.; Mohy-Eldin, M.S.; Hassan, M.A. Influence of Cedar Essential Oil on Physical and Biological Properties of Hemostatic, Antibacterial, and Antioxidant Polyvinyl Alcohol/Cedar Oil/Kaolin Composite Hydrogels. *Pharmaceutics* **2022**, *14*, 2649. [CrossRef]
53. Prieto, P.; Pineda, M.; Aguilar, M. Spectrophotometric Quantitation of Antioxidant Capacity through the Formation of a Phosphomolybdenum Complex: Specific Application to the Determination of Vitamin E. *Anal. Biochem.* **1999**, *269*, 337–341. [CrossRef] [PubMed]
54. Garratt, D.C. The Quantitative Analysis of Drugs. In *The Quantitative Analysis of Drugs*; Garratt, D.C., Ed.; Springer US: Boston, MA, USA, 1964; pp. 1–669.
55. El-Samad, L.M.; Hassan, M.A.; Basha, A.A.; El-Ashram, S.; Radwan, E.H.; Abdul Aziz, K.K.; Tamer, T.M.; Augustyniak, M.; El Wakil, A. Carboxymethyl cellulose/sericin-based hydrogels with intrinsic antibacterial, antioxidant, and anti-inflammatory properties promote re-epithelization of diabetic wounds in rats. *Int. J. Pharm.* **2022**, *629*, 122328. [CrossRef]
56. Hassan, M.A.; Abd El-Aziz, S.; Elbadry, H.M.; El-Aassar, S.A.; Tamer, T.M. Prevalence, antimicrobial resistance profile, and characterization of multi-drug resistant bacteria from various infected wounds in North Egypt. *Saudi J. Biol. Sci.* **2022**, *29*, 2978–2988. [CrossRef] [PubMed]
57. Miller, G.L. Use of dinitrosalicylic acid reagent for determination of reducing sugar. *Anal. Chem.* **1959**, *31*, 426–428. [CrossRef]
58. Choi, S.-J.; Hwang, J.-M.; Kim, S.-I. A colorimetric microplate assay method for high throughput analysis of lipase activity. *BMB Rep.* **2003**, *36*, 417–420. [CrossRef] [PubMed]
59. O'Boyle, N.M.; Banck, M.; James, C.A.; Morley, C.; Vandermeersch, T.; Hutchison, G.R. Open Babel: An open chemical toolbox. *J. Cheminformatics* **2011**, *3*, 33. [CrossRef]
60. Daina, A.; Michielin, O.; Zoete, V. SwissADME: A free web tool to evaluate pharmacokinetics, drug-likeness and medicinal chemistry friendliness of small molecules. *Sci. Rep.* **2017**, *7*, 42717. [CrossRef]
61. Mbarik, M.; Poirier, S.J.; Doiron, J.; Selka, A.; Barnett, D.A.; Cormier, M.; Touaibia, M.; Surette, M.E. Phenolic acid phenethylesters and their corresponding ketones: Inhibition of 5-lipoxygenase and stability in human blood and HepaRG cells. *Pharmacol. Res. Perspect.* **2019**, *7*, e00524. [CrossRef]

 *catalysts*

Article

Improving Effects of Laccase-Mediated Pectin–Ferulic Acid Conjugate and Transglutaminase on Active Peptide Production in Bovine Lactoferrin Digests

Mingxia Xing [1,†], Ying Ji [1,†], Lianzhong Ai [1], Fan Xie [1], Yan Wu [2] and Phoency F. H. Lai [1,*]

[1] Shanghai Engineering Research Center of Food Microbiology, School of Health Science and Engineering, University of Shanghai for Science and Technology, Shanghai 200093, China
[2] School of Agriculture and Biology, Shanghai Jiao Tong University, Shanghai 200240, China
* Correspondence: plai856@hotmail.com; Tel.: +86-18516077898
† These authors contributed equally to this work.

Abstract: Bovine lactoferrin (bLf) is a multifunctional glycoprotein and a good candidate for producing diverse bioactive peptides, which are easily lost during over-digestion. Accordingly, the effects of laccase-mediated pectin–ferulic acid conjugate (PF) and transglutaminase (TG) on improving the production of bLf active peptides by in vitro gastrointestinal digestion were investigated. Using ultra-high-performance liquid chromatography tandem mass spectroscopy (UPLC-MS-MS), the digests of bLf alone, PF-encapsulated bLf complex (LfPF), and TG-treated LfPF complex (LfPFTG) produced by conditioned in vitro gastric digestion (2000 U/mL pepsin, pH 3.0, 37 °C, 2 h) were identified with seven groups of active peptide-related fragments, including three common peptides (VFEAGRDPYKLRPVAAE, FENLPEKADRDQYEL, and VLRPTEGYL) and four differential peptides (GILRPYLSWTE, ARSVDGKEDLIWKL, YLGSRYLT, and FKSETKNLL). The gastric digest of LfPF contained more diverse and abundant detectable peptides of longer lengths than those of bLf and LfPFTG. After further in vitro intestinal digestion, two active peptide-related fragments (FEAGRDPYK and FENLPEKADRDQYE) remained in the final digest of LfPFTG; one (EAGRDPYKLRPVA) remained in that of bLf alone, but none remained in that of LfPF. Conclusively, PF encapsulation enhanced the production of bLf active peptide fragments under the in vitro gastric digestion applied. TG treatment facilitated active peptide FENLPEKADRDQYE being kept in the final gastrointestinal digest.

Keywords: lactoferrin; pectin–ferulic acid conjugate; laccase; transglutaminase; active peptide; UPLC-MS-MS

Citation: Xing, M.; Ji, Y.; Ai, L.; Xie, F.; Wu, Y.; Lai, P.F.H. Improving Effects of Laccase-Mediated Pectin–Ferulic Acid Conjugate and Transglutaminase on Active Peptide Production in Bovine Lactoferrin Digests. *Catalysts* **2023**, *13*, 521. https://doi.org/10.3390/catal13030521

Academic Editors: Chia-Hung Kuo, Chwen-Jen Shieh and Hui-Min David Wang

Received: 31 December 2022
Revised: 20 February 2023
Accepted: 1 March 2023
Published: 3 March 2023

1. Introduction

Bovine lactoferrin (bLf) is a vital, multifunctional glycoprotein with diverse applications in foods [1] and pharmaceutics [2]. It is a "generally recognized as safe" (GRAS) ingredient at a maximal usage of 100–400 mg/100 g for milk products [3]. For foods, bLf is a promising nutraceutical and encapsulating agent for functional ingredients [1]. For pharmaceutics, bLf is a good candidate as an adjuvant or drug carrier in medicines such as antimicrobial, immunomodulatory, anticancer, anti-Parkinsonian, and anti-Alzheimer medicines [2,4]. For developing active peptides with improved activities, studies on bLf active peptides produced by in vitro gastric or gastrointestinal digestion are receiving more interest [5–9]. Several bLf active peptides have been discovered, e.g., antimicrobial peptides, bovine lactoferrampin (bLfampin) [5]; antihypertensive peptides, ENLPEKADRD [6]; and osteoblast-promoting peptides, ENLPEKADRDQYEL [7] and FKSETKNLL [8]. However, few pieces of evidence about whole active peptide distribution are reported for bLf digests, possibly due to over-digestion by the applied conditions or technical limitations for peptide identification [10].

Generally, enhancing active peptide production in bLf digests is essential for industrial concerns. For this purpose, several approaches for controlling bLf hydrolysis can be considered; examples include selecting suitable digestion conditions and encapsulation with food biopolymers. Natural pectins are multifunctional biopolymers widely applied in foods [11] and pharmaceuticals [12] and have potential for encapsulating bLf via electrostatic interactions [13,14]. Nonetheless, the encapsulation efficacy and stability of a bLf–natural pectin complex are limited [15] due to the high hydrophilicity of natural pectins. Recently, pectins have been receiving more interest for their ability to be conjugated with dietary phenolics (e.g., ferulic acid) [16–20], giving pectin–phenolic conjugates of improved hydrophobicity, solution viscosity, emulsion stability [16], and in vitro antioxidant activities [19,20]. In addition, a pectin–ferulic acid conjugate has been proven to encapsulate bLf better and give a lower particle size than its parent pectin [15]. The conjugate may be a good modulator for improving bLf active peptide production. However, the improvement remains to be proved.

Transglutaminase (TG), a popular modifier for food proteins [21], could be another choice for modulating bLf hydrolysis. The mechanism by which TG modifies protein functionality is well known to be the catalysis of an acyl transfer reaction between the γ-carboxyamide group in the glutamine residue and a primary amine, usually in lysine residue, leading to the formation of γ-glutamyl-lysine (Gln-Lys) bonds and cross-links between polypeptides or proteins [21]. This type of crosslink has been proposed for TG-treated bLf particles that increased the stability of Pickering emulsions [22]. The potential of TG in modulating bLf active peptide production during digestion is still unclear and under investigation.

Accordingly, this study aimed to investigate the improving effects on bLf active peptide production under conditioned in vitro gastrointestinal digestion accomplished using two approaches: encapsulation of laccase-mediated pectin–ferulic acid conjugate (PF) and TG treatment. The study was performed from the viewpoint of digestomics with UPLC-MS-MS [10,23]. Passion fruit pectin (PP) was purposedly applied for conjugation with ferulic acid, due to improved encapsulation ability toward bLf, giving the complex of significantly ($p < 0.05$) lessened effective particle sizes [15].

2. Results

2.1. Compositions and Characteristic Ratios of PP and PF

Table 1 indicates that PP possessed a total uronic acid content of 72.72% (w/w) on average, a total phenolic content (*TPC*) of 2.36 mg GAE/g, and a degree of esterification (*DE*) of 35.17%. In contrast to PP, PF showed a significantly ($p < 0.05$) lower total uronic acid content (63.19% (w/w)) but greater *TPC* (20.03 mg GAE/g) and *DE* (55.40%). For monosaccharide composition, PP was composed of GalA, Ara, Gal, Glc, and Rha at 53.42%, 14.73%, 18.50%, 9.11%, and 4.24%, respectively, in molar percentage. Comparatively, PF exhibited significantly ($p < 0.05$) lessened GalA (46.83%) and Rha (2.01%) compositions but greatened Glc% (16.02%). Five structural characteristic ratios for pectins were discovered. Homogalacturonan content (HG%) appeared to be 49.17% for PP, decreasing to 44.82% for PF. Rhamnogalacturonan type I composition (RG-I%) was estimated to be 41.72% for PP, slightly decreasing to 39.16% for PF. The ratio of GalA%/(Rha% + Ara% + Gal%), an index for structural linearity of the pectin chain, was 1.43 for PP, slightly lessening to 1.26 for PF. The ratio of Rha%/GalA%, an index for branching extent, was 0.08 for PP and was significantly reduced to 0.04 for PF. In addition, the ratio of (Ara% + Gal%)/Rha%, suggesting the side chain length of the RG-I fragment, was 7.87 for PP, significantly ($p < 0.05$) increasing to 17.53 for PF. FA seemed to bind selectively to the pectin molecules of the HG backbone with a low degree of branching and RG-I of long chain lengths.

Table 1. Approximate compositions, monosaccharide compositions, and ratios of PP and PF conjugates.

Parameter	Unit	PP	PF Conjugate
Total uronic acid content	(% w/w)	72.72 ± 2.53 [a,2]	63.19 ± 1.59 [b]
Total phenolic content (*TPC*)	(mg GAE/g) [1]	2.36 ± 0.01 [b]	20.03 ± 0.38 [a]
Degree of esterification (*DE*)	(%)	35.17 ± 1.04 [b]	55.40 ± 3.71 [a]
Monosaccharide composition			
Galacturonic acid (GalA)	(mol%)	53.42 ± 1.23 [a]	46.83 ± 0.64 [b]
Arabinose (Ara)	(mol%)	14.73 ± 0.18 [a]	15.16 ± 0.26 [a]
Galactose (Gal)	(mol%)	18.50 ± 1.28 [a]	19.98 ± 1.29 [a]
Glucose (Glc)	(mol%)	9.11 ± 0.27 [b]	16.02 ± 0.50 [a]
Rhamnose (Rha)	(mol%)	4.24 ± 0.32 [a]	2.01 ± 0.06 [b]
HG% = GalA% − Rha%	(mol%)	49.17 ± 1.46 [a]	44.82 ± 0.60 [b]
RG-I% = 2Rha% + Ara% + Gal%	(mol%)	41.72 ± 1.39 [a]	39.16 ± 0.92 [a]
GalA%/(Rha% + Ara% + Gal%)		1.43 ± 0.08 [a]	1.26 ± 0.05 [a]
Rha%/GalA%		0.08 ± 0.01 [a]	0.04 ± 0.00 [b]
(Ara% + Gal%)/Rha%		7.87 ± 0.59 [b]	17.53 ± 1.07 [a]

[1] GAE: gallic acid equivalent. [2] Means ± standard deviations ($n = 3$); means with different superscripts a, b in the same row differ significantly ($p < 0.05$).

2.2. NMR-Evidenced Structural Features of PP and PF

Figure 1 presents ^{13}C and ^{1}H-decoupled NMR spectra for structural identification of PP (A) and PF (B). Briefly, PP (Figure 1A) exhibited typical NMR signals of the HG backbone, accompanied by minor signals of RG-I branches, agreeing with its monosaccharide compositions (mainly ~49% HG% and partly ~42% RG-I of long-side-chain branches (Table 1)). According to our previous report [19] and those on passion fruit pectins [24] and pectic polysaccharides [25,26], the major peaks for PP could be assigned for the α-(1,4)-D-GalA residues of the HG backbone, where C1 δc = 100.40 ppm is associated with H1 δ_H = 5.03 ppm; C2 δc = 67.75 ppm with a shoulder peak C3 δc = 68.38 ppm, and both H2 and H3 δ_H = 3.75 ppm; C4 δc = 79.00 ppm, H4 δ_H = 3.94 ppm; C5 δc = 70.53 ppm, H5 δ_H ~ 5.05 ppm (associated with H1 δ_H); esterified carboxyl C6 δc = 170.67 ppm and free carboxyl C6 δc = 174.28 ppm (tiny peak), no δ_H; methoxyl C7 δc = 52.84 ppm, H7 δ_H ~ 3.75 ppm (coexisted with H2 and H3 δ_H); and Rha C6 δc = 16.51 ppm, δ_H = 1.16 ppm. In addition, tiny peaks at δc = 60.03, 71.27, 73–76, and 80–84 ppm could be attributed to branch Gal C6′, Gal C5′, Ara C3, and Ara C2 + C4, respectively. A peak at δc = 107.49 ppm was characteristic of terminal Ara C1 ($t − A1$) [25,26]. For the PF conjugate (Figure 1B), characteristic signals originating from FA were observed at δc = 43.71 ppm, δ_H = 2.69 and 2.82 ppm (-OCH$_3$ on phenyl ring), and δ_H = 7.45 and 7.51 ppm (parts of signals for methylene H on phenyl ring). The GalA C1 signal (originally δc = 100.37 ppm) likely shifted up to 98.00 ppm, accompanied by the other signals multiplied and of reduced intensities, possibly due to the shielding effect of bound FA. The above results confirm that FA is present in the PF conjugate, mostly via attaching to the nucleophilic carboxyl group of GalA residues. This kind of linkage is one of laccase-mediated FA redox reactions [18] and agrees with the evidence for citrus pectin–FA conjugates [17].

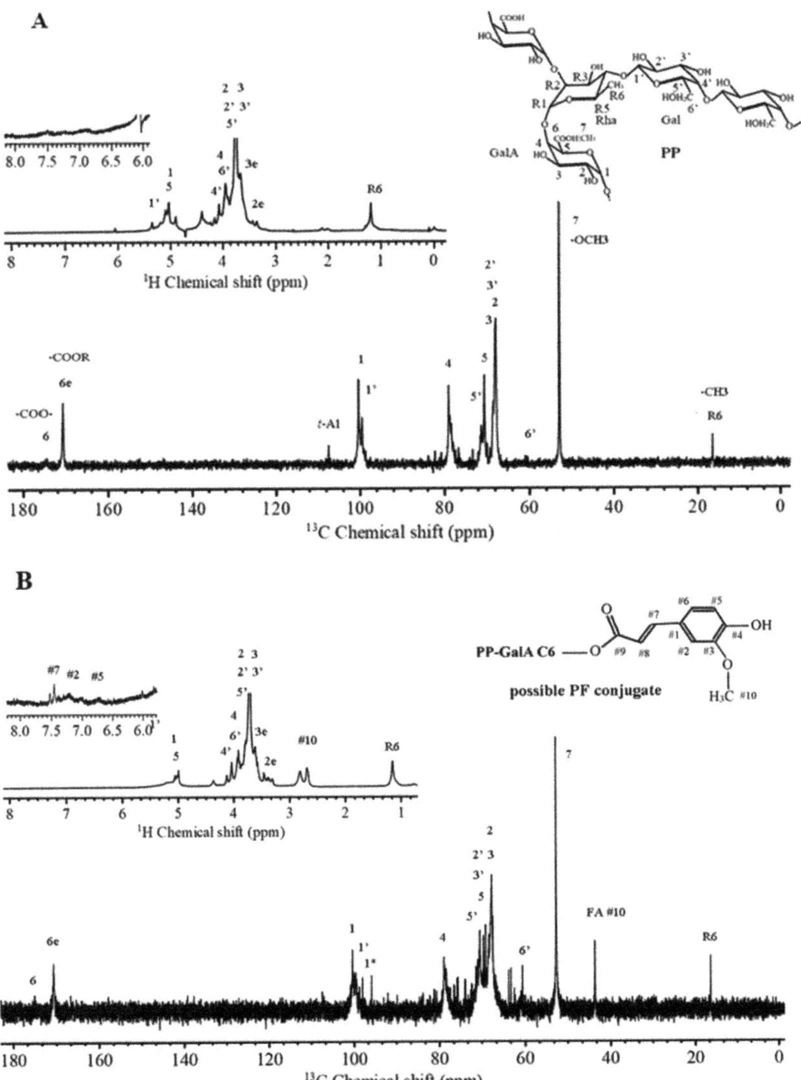

Figure 1. The 176 MHz ^{13}C and 700 MHz ^{1}H-decoupled NMR spectra of PP (**A**) and PF (**B**).

2.3. Particle Size Distributions of bLf, LfPF Complex, and LfPFTG Complex

Figure 2 shows that pure bLf aqueous solution (0.1% w/w) showed a bimodal distribution in particle size that peaked mainly at ~300 nm and minorly at ~2800 nm. After conjugation with PF, the LfPF complex (0.1% bLf + 0.025% PF) revealed a sharpened, bimodal distribution and shifted to small particle size regions, mainly peaking at 120 nm and associated with a small peak at ~25 nm. Additional treatment of 0.1% TG led to LfPFTG of a broad but monomodal distribution peaking at ~220 nm. The effective particle size and polydispersity index were, respectively, 269.1 ± 2.9 nm and 0.314 ± 0.076 for bLf alone, 157.0 ± 1.2 nm and 0.217 ± 0.006 for the LfPF complex, and 225.3 ± 7.3 nm and 0.231 ± 0.018 for the LfPFTG complex. In addition, the ξ-potential value was identified as +21.9 ± 0.42 mV for the pure bLf dispersed solution, changing to −20.5 ± 1.69 mV for the LfPF complex, implying that negative-charge PF tended to cover positive-charge, spherical bLf via electrostatic interactions. These changes are similar to those for a bLf complex with citrus low-methoxy pectin [14], which displayed a ξ-potential value (−43.6 mV) lower than that of LfPF [14] since that PF possessed a higher degree of esterification attributed to methoxy groups and esters formed between PP and FA (Figure 1). PF encapsulation on bLf was also confirmed by surface hydrophobicity and scanning electron micrograms [15].

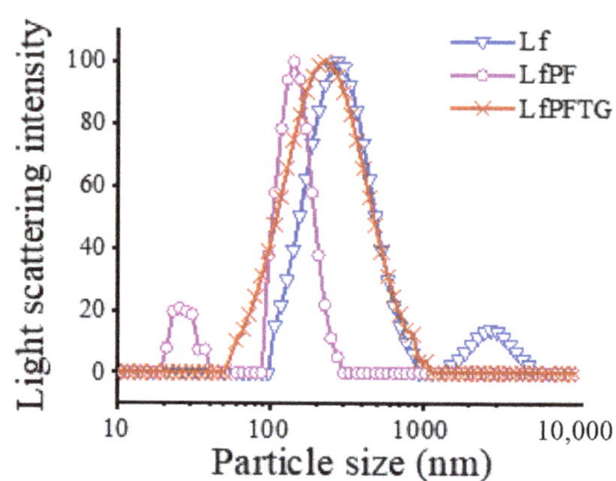

Figure 2. Particle size distributions of bLf, LfPF complex, and LfPFTG complex.

2.4. Differential Peptide Distributions after In Vitro Gastrointestinal Digestion

Figure 3 compares UPLC chromatograms for the peptide profiles in the digests of bLf, LfPF complex, and LfPFTG complex after in vitro gastrointestinal digestion. Obviously, in contrast to the chromatographic profile of bLf (A), those of in vitro gastric (pepsin) digests of LfPF (C) and LfPFTG (E) complexes were somewhat different. For the in vitro gastrointestinal (pepsin + trypsin) digests, the detected peak signals and peptide numbers lessened obviously in LfPFGI (D), but increased in LfPFTGGI (F), as compared with LfGI (B).

After MS-MS identification and homology matching with UniProt Knowledgebase (https://www.uniprot.org/uniprotkb/P24627/entry, accessed on 20 September 2022) and SWISS-MODEL (https://swissmodel.expasy.org/, accessed on 22 September 2022), the identified peptide fragments were tabulated (Supplementary Table S1). In total, there were 103 peptide fragments grouped into 19 peptide regions in the gastric digests studied (Table S2).

Briefly, the differential peptide numbers between samples are illustrated by Venn's plots (Figure 4). LfG, LfPFG, and LfPFTGG (Figure 4A) possessed total peptide numbers of 89, 91, and 77, respectively; individually, they possessed 4, 9, and 2 differential peptides, respectively, and they shared the same 66 peptides. There were 13, 6, and 3 same peptides between LfG and LfPFG, between LfG and LfPFTGG, and between LfPFG and LfPFTGG, respectively. For the gastrointestinal digests (Figure 4B), LfGI, LfPFGI, and LfPFTGGI contained total peptide numbers of only 7, 5, and 9, respectively; individually, they contained 4, 1, and 4 differential peptides, respectively, and they shared two same peptides. There were 0, 1, and 2 same peptides for LfGI vs. LfPFGI, LfGI vs. LfPFTGGI, and LfPFGI vs. LfPFTGGI, respectively.

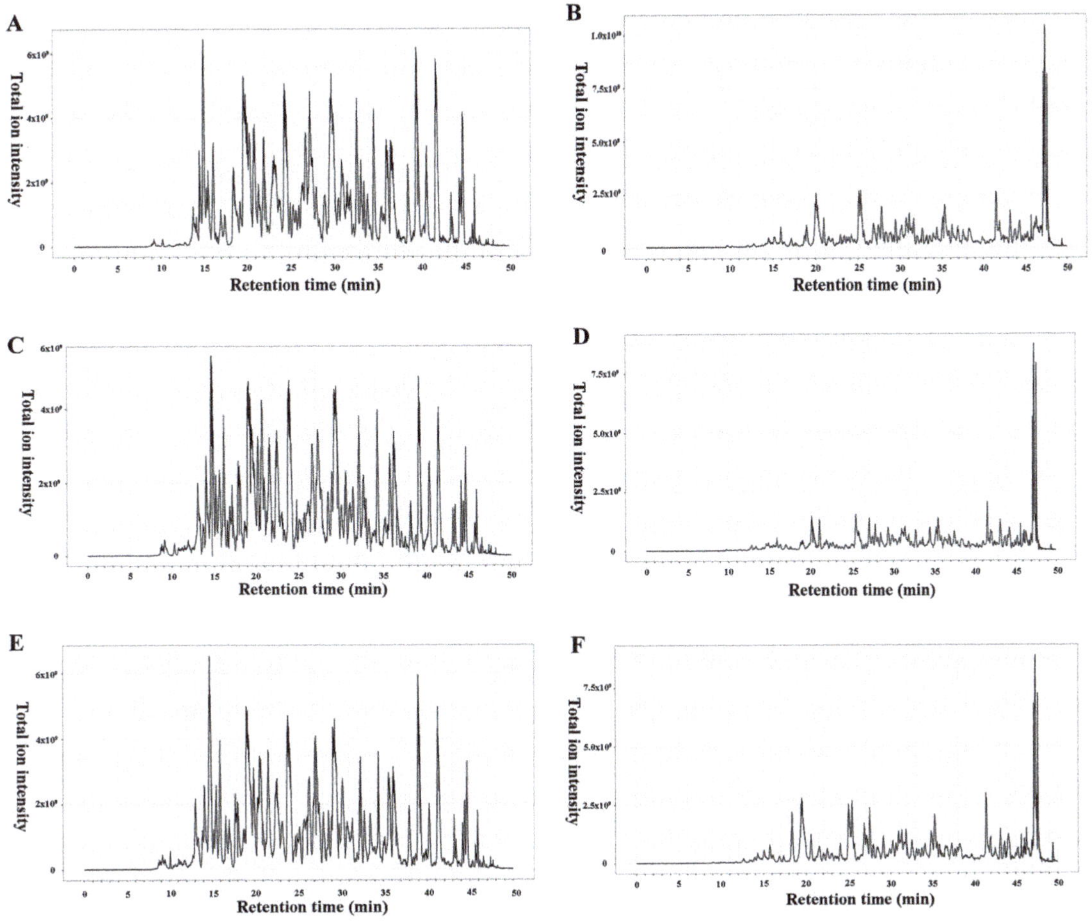

Figure 3. UPLC chromatograms for the peptide profiles in the digests of bLf (**A,B**), LfPF complex (**C,D**), and LfPFTG complex (**E,F**) after in vitro gastric (pepsin) digestion (**A,C,E**) and gastrointestinal (pepsin + trypsin) digestion (**B,D,F**).

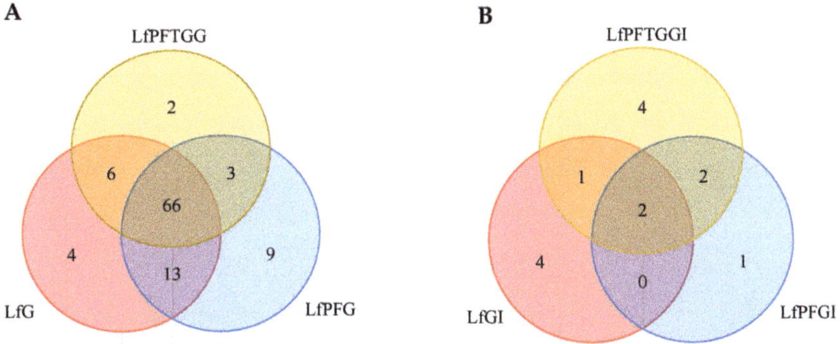

Figure 4. Venn's plots for differential peptide numbers detectable in the digests of bLf, LfPF complex, and LfPFTG complex after in vitro gastric digestion (**A**) and gastrointestinal digestion (**B**).

Figure 5 illustrates the histograms for peptide chain distributions detectable in the digests of bLf, LfPF, and LfPFTG by UPLC-MS-MS. The given LfG (A) showed detected peptide lengths in the range of 5–17 with abundances (number) topped at the lengths of 11–12, followed by 7–8. LfPFG (B) exhibited peptide lengths of 5–31 with abundance topped at 12, followed by 7–11. For LfPFTG (C), peptide lengths ranged from 6 to 16 with abundances maximized at 12, followed by 8. LfPFG exhibited high peptide abundance and long peptide chains (18–31) that were absent in both LfG and LfPFTGG. After further digestion by trypsin, the given LfGI (D), LfPFGI (E), and LfPFTGGI (F) displayed detectable peptide lengths in the ranges of 11–13, 7–14, and 6–16, respectively. Generally, total peptide abundance (peptide length × number) followed the order of LfPFG > LfG and LfPFTGG; LfPFGI > LfPFTGGI > LfGI, agreeing with the order in total purified peptide yield examined by BCA colorimetric assay, generally 0.98% (*w/w*) for both LfG and LfPFTGG, 1.12% (*w/w*) for LfPFG, 0.29% (*w/w*) for LfGI, 0.44% (*w/w*) for LfPFGI, and 0.34% (*w/w*) for LfPFTGGI. The conjugation with PF and TG modified the detectable peptide chain distributions in bLf digests. Further treatment by in vitro intestinal digestion notably reduced the lengths and abundances of detectable peptides in the final digests. Peptides of very short lengths (<5 amino acid residues), long polypeptides (length > 31), and hetero-oligopeptides cross-linked by disulfide bonds were undetectable by high-resolution MS [27] and not indicated in this study.

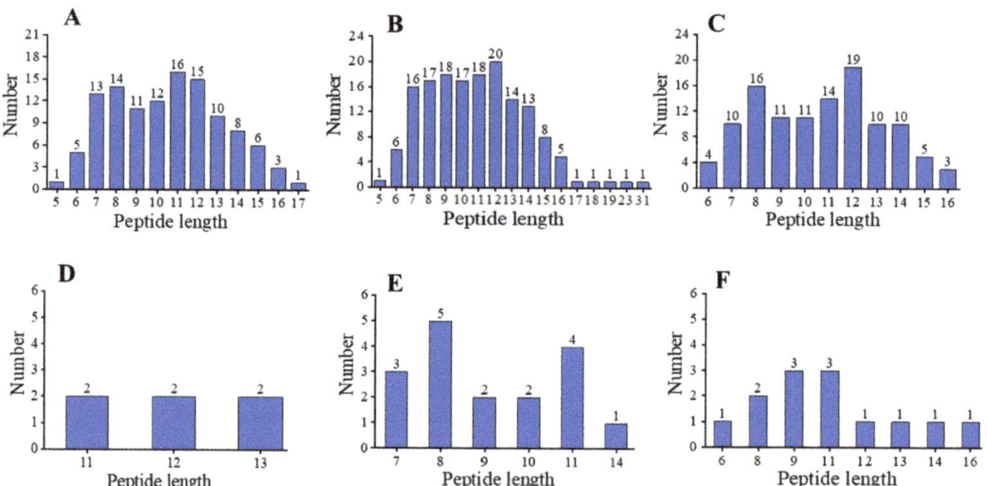

Figure 5. Detectable peptide profiles in the digests of bLf (**A,D**), LfPF complex (**B,E**), and LfPFTG complex (**C,F**) after in vitro gastric (pepsin) digestion (**A–C**) and gastrointestinal (pepsin + trypsin) digestion (**D–F**).

2.5. Differences in Active Peptide Fragments between Samples

Briefly, active peptide-related fragments are tabulated in Table 2; bioactive peptide sequences were determined from the milk bioactive protein database [28] and previous reports [5–9,29]. Seven groups of active peptide fragments with reported active peptide sequences were discovered in the gastric digests (LfG, LfPFG, and LfPFTGG): DGGMVFEAGRDPYKLRPVAAE, f(79–99); GILRPYLSWTE, f(149–159); FENLPEKADRDQYEL, f(234–248); ARSVDGKEDLIWKLLSK, f(276–292); YLGSRYLT, f(338–345); VLRPTEGYL, f(445–453); and LFKSETKNLL, f(650–659). For the final gastrointestinal digests, one active peptide type, EAGRDPYKLRPVA, f(85–97) and its related f(86–97), remained in LfGI. Two active peptides type, FEAGRDPYK, f(84–92), and FENLPEKADRDQYE, f(234–247) and related f(235–246), were discovered in LfPFTGGI. No active peptides remained in LfPFGI.

Table 2. Active peptide fragments in the digests of bLf, LfPF complex, and LfPFTG complex produced by in vitro gastrointestinal digestion at pH 3.0.

Targeted Peptide Fragment	Peptide Fragment	Sequence	LfG	LfPFG	Sample LfPFTGG	GI Digests
	VFEAGRDPYKLRPVAA	83–98	+	+	+	
	VFEAGRDPYKLRPVA	83–97	+	+	+	
	FEAGRDPYKLRPVAAE	84–99	+	+	+	
	FEAGRDPYKLRPVAA	84–98	+	+	+	
	FEAGRDPYKLRPVA	84–97	+	+	+	
	FEAGRDPYKLRPV	84–96	+	+	+	
	EAGRDPYKLRPVAAE	85–99	+	+	+	
	EAGRDPYKLRPVAA	85–98	+	+	+	
	EAGRDPYKLRPVA	85–97	+	+	+	LfGI (+)
	EAGRDPYKLRPV	85–96	+	+	+	
	AGRDPYKLRPVAAE	86–99	+	+	+	
	AGRDPYKLRPVAA	86–98	+	+	+	
	AGRDPYKLRPVA	86–97	+	+	+	LfGI (+)
DGGMVFEAGRDPYKLRPVAAE,	GRDPYKLRPVAA	87–98	+	+	+	
f(79–99)	GRDPYKLRPVA	87–97	+	+	+	
	RDPYKLRPVA	88–97	+	+	+	
	PYKLRPVA	90–97	+	+	+	
	YKLRPVA	91–97	+	+	+	
	AGRDPYKLRPV	86–96	+	+		
	PYKLRPVAAE	90–99	+	+		
	PYKLRPVAA	90–98	+	+		
	VFEAGRDPYKLRPV	83–96		+	+	
	EAGRDPYKLRP	85–95		+	+	
	DGGMVFEAGRDPYKLRPVA	79–97		+		
	VFEAGRDPYKLRPVAAE	83–99		+		
	GRDPYKLRPV	87–96		+		
	FEAGRDPYKLRP	84–95			+	
	DPYKLRPVA	89–97			+	
	GILRPYL	149–155	+	+	+	
	ILRPYLSW	150–157	+	+	+	
	ILRPYL	150–155	+	+	+	
GILRPYLSWTE, f(149–159)	RPYLSWT	152–158	+	+	+	
	RPYLSWTE	152–159	+	+		
	RPYLSW	152–157	+	+		
	LRPYLSWT	151–158	+			
	LRPYLSW	151–157		+		
	GILRPYLSW	149–157		+		
	FENLPEKADRDQYE	234–247	+	+	+	LfPFTGGI (+)
	FENLPEKADRDQ	234–245	+	+	+	
FENLPEKADRDQYEL,	ENLPEKADRDQYEL	235–248	+	+	+	
f(234–248)	ENLPEKADRDQYE	235–247	+	+	+	
	ENLPEKADRDQY	235–246	+	+	+	LfPFTGGI (+)
	FENLPEKADRDQY	234–246		+	+	
	ARSVDGKEDLIWKL	276–289	+	+	+	
ARSVDGKEDLIWKLLSK,	RSVDGKEDLIWKL	277–289	+	+	+	
f(276–292)	RSVDGKEDLIWKLLSK	277–292		+		
	SVDGKEDLIWKLLSK	278–292		+		
YLGSRYLT, f(338–345)	YLGSRYLT	338–345	+		+	
	YLGSRYL	338–344		+		
VLRPTEGYL, f(445–453)	VLRPTEGYL	445–453	+	+	+	
	VLRPTEGY	445–452	+	+		
LFKSETKNLL, f(650–659)	FKSETKNLL	651–659	+	+	+	
	LFKSETKNLL	650–659		+		

2.6. bLf Active Peptide Changes during In Vitro Gastrointestinal Digestion

The effects of PF encapsulation and TG treatment on the resultant bLf active peptide profiles in the digests are differentiated as depicted in Figure 6. According to the results of molecular docking, it is suggested that bLf whatever alone or in LfPF and LfPFTG was generally digested by pepsin on the N1, N2, C1, and C2 termini, involving 6, 7, and 8 peptide regions, respectively (indicated by dashed red frames). The obtained pepsin digests (LfG, LfPFG, and LfPFTGG) showed three common peptide fragments: f(83–99) (VFEAGRDPYKLRPVAAE), f(234–248) (FENLPEKADRDQYEL), and f(445–453) (VLRPT-EGYL). In addition, there were four differential peptides with lengths depending on the sample state: f(149–159) (GILRPYLSWTE), f(276–289) (ARSVDGKEDLIWKL), f(338–345) (YLGSRYLT), and f(650–659) (LFKSETKNLL). After further intestinal (trypsin) digestion, LfGI contained peptide fragment f(85–97) (EAGRDPYKLRPVA). LfPFTGGI contained two active peptides, f(84–92) (FEAGRDPYK) and f(234–247) (FENLPEKADRDQYE). These results imply that the f(85–97) (EAGRDPYKLRPVA) fragment on the N1 terminus was quite resistant to in vitro gastrointestinal digestion. The presence of TG protected the peptide f(234–247) (FENLPEKADRDQYE) in the N2 terminus against in vitro gastrointestinal digestion. All active peptide fragments in the C1 and C2 termini were feasibly hydrolyzed during in vitro intestinal digestion and absent in the final digests, regardless of the protective effect of PF during in vitro gastric digestion.

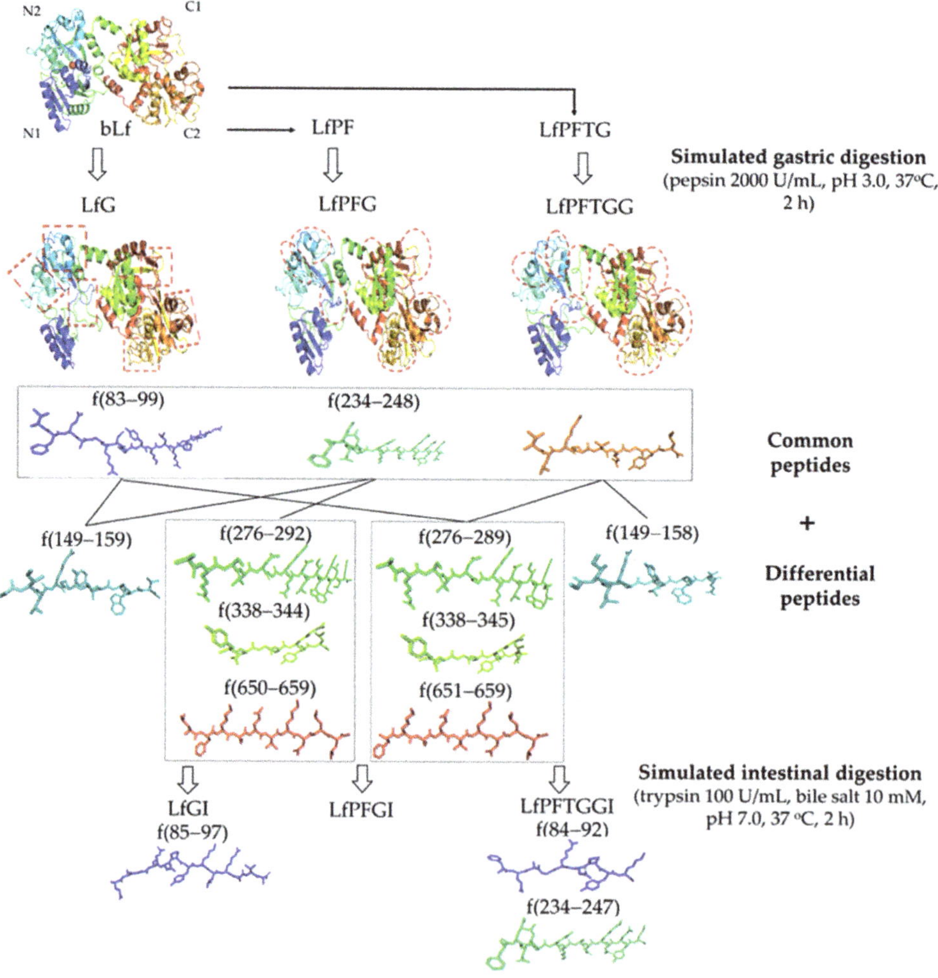

Figure 6. Active peptide production in the digests of bLf, LfPF complex, and LfPFTG complex during in vitro gastrointestinal digestion. Peptide fragments are in the same colors as in their parent bLf. Dashed frames indicate the action regions for pepsin.

3. Discussion

PP, as passion fruit low-methoxy pectin (*DE* = 35.17%), exhibits lower GalA%, HG%, and linearity ratio, but greater RG-I%, degree of branching, and RG-I side chain length compared to its high-methoxy counterpart (*DE* = 75.2%; GalA:Ara:Gal:Glc:Rha = 78.5:8.2:3.7:6.2:3.1; HG% = 74.3%; RG-I% = 18.1%; linearity ratio = 3.43; degree of branching = 0.04; RG-I side chain length = 3.9) [30]. In addition, PP displays similar GalA% and structural linearity, greater HG% and RG-I side chain length, but lower RG-I% and degree of branching when compared to citrus low-methoxy pectin (*DE* = 38.2%; GalA:Ara:Gal:Glc:Rha = 52.0:0.6:31.0:9.0:7.4; HG% = 44.6%; RG-I% = 46.4%; linearity ratio = 1.33; degree of branching = 0.14; RG-I side chain length = 4.2 [19]. The *TPC* of the PF conjugate formed between PP and FA in this study (20 mg GAE/g) is almost double those (9.6–12.3 GAE mg/g) of passion fruit high-methoxy pectin–FA or caffeic acid conjugates [20] and citrus low-methoxy pectin-caffeic acid conjugate [19]. Briefly, the enhanced *TPC* for the PF conjugate could be attributed to two pectin structural features that facilitate the binding of phenolic acids: low *DE* (alternatively, high proportion of free carboxyl groups) and high RG-I side chain length.

The laccase-mediated formation of the PF conjugate involves mostly the formation of an ester linkage between the GalA C6 free carboxyl group and the FA moiety, as suggested by NMR spectra (Figure 1). The proposed mechanism includes FA oxidation catalyzed by laccase in the presence of O_2 to generate quinones with free radicals [31], which in turn undergo cross-linking reactions with free carboxyl groups in pectins [16,19]. Similar results are reported in the cases of gum Arabic [32] and proteins [31]. In this study, the laccase-mediated effect endowed the resultant PF conjugate with a higher *DE* and hydrophobicity, resulting in the given PF-encapsulated bLf complex of lessened particle sizes, greatened total peptide yield, and longer peptides associated with higher peptide abundances in the digest LfPFG (Figure 5B), in contrast to bLf alone. The difference in peptide profile between LfPFG from LfG or LfPFTGG (Figure 5A–C) could be linked to PF interactions differentially with the N and C termini of bLf, since the N termini with positive charges [33] may interact with negatively charged PF more strongly than the C termini. However, more studies on the molecular mechanism for interactions between bLf and PF are necessary for understanding the controlled release of active bLf peptides by PF encapsulation approaches.

It is evident that seven types of active peptide fragments from bLf appeared in all gastric digests (LfG, LfPFG, and LfPFTGG) studied. The varieties of detected peptide fragments cover most of the bioactive peptides reported elsewhere. Accordingly, the sequences of bioactive peptide fragments reported are compared with those of the active peptide-related fragments in this study in Table 3. For antimicrobial peptides, bovine lactoferricin (bLfcin, f(17–41)) and lactoferrampin (bLfampin, f(268–288), f(271–288)) are well known, and bLfampin, rather than bLfcin, was identified in a digest using human gastric juice (pH 2.5) and duodenal juice (pH 7.0) [5]. The key sequences for antihypertensive bLf peptides discovered include the following: LIWKL, LFH, RPYL, and LNNSRAP fragments that are present in the gastric (pepsin) digest (pH 2.5, 4 h) of bLf [9]; ENLPEKADRD, present in the gastrointestinal (pepsin + trypsin) digest of bLf [6]; and LRP, DPYKLRP, PYKLRP, YKLRP, KLRP, and GILRP, present in the pepsin digest of Lf from yeast *Kluyveromyces marxianus* [29]. Osteoblast-promoting peptides of the sequences of ENLPEKADRDQYEL in the trypsin digest of bLf [7] and FKSETKNLL in the pepsin digest of bLf (pH 2.5, 4 h) [8] have been reported. In addition, the GSRY fragment, a sequence of potent SARS-CoV-2 inhibition, was found in the gastrointestinal (pepsin + trypsin) digest of bLf [34]. Accordingly, the peptides discovered comprising DGGMVFEAGRDPYKLRPVAAE, f(79–99); GILRPYLSWTE, f(149–159); YLGSRYLT, f(338–345); and VLRPTEGYL, f(445–453) (involving active fragments LRP, DPYKLRP, PYKLRP, YKLRP, YL, LRP, RPYL, or GILRP [29]), would be potent in antihypertensive activity as angiotensin-converting enzyme inhibitors (ACEIs). YLGSRYLT (containing GSRY) may be developed into a SARS-CoV-2 inhibitor. ARSVDGKEDLIWKLLSK, f(276–292), the fragment of bLfampin and composing antihypertensive fragment LIWKL [9], would be potent for antimicrobial and antihypertensive activities. FKSETKNLL and ENLPEKADRDQYEL are osteoblast-promoting peptides, as reported [7,8]. Additionally, ENLPEKADRDQYEL is also reported to have anticoagulant activity [7]. The obtained digests possessing seven types of potentially active bLf peptides could be applied for multifunction nutraceuticals [1] or purified in advance for nonantibiotic therapeutic agents [2] or targeting antibacterial vaccines [35].

Table 3. Active peptide fragments from various lactoferrins reported and in this study.

Source	Digestion Conditions	Active Peptides	Reference
Bovine Lf	Human gastric juice (HGJ, pH 2.5); human duodenal juice (HDJ, pH 7.0)	Antimicrobial peptide: bLfampin) f(268–288); no bLFcin, f(17–41), detected; 70% peptides from N terminus	[5]
Bovine Lf	Porcine pepsin (2540 U/mg, pH 2.5, 37 °C, 4 h)	Antihypertensive (ACE-inhibitory) peptides: LIWKL; LFH; RPYL; LNNSRAP	[9]
Kluyveromyces marxianusm Lf	Pepsin (0.02 mg/mL, pH 2.0, 37 °C, 90 min)	Antihypertensive (ACE-inhibitory) peptides: LRP, DPYKLRP, PYKLRP, YKLRP, KLRP, GILRP	[29]
Bovine Lf	Porcine pepsin (1:100 *w/w*), pH 2.0, 37 °C, 90 min; porcine pancreatin (E/S = 1%), pH 7.4, 3 h	ACE inhibitor, anticoagulant peptide: ENLPEKADRD	[6]
Bovine Lf	Porcine trypsin (3 U/mg, pH 8.0, 37 °C, 2 h)	Osteoblast-promoting peptide: ENLPEKADRDQYEL	[7]
Bovine Lf	Porcine pepsin (3%, 2500 U/mg, pH 2.5, 37 °C, 4 h)	Osteoblast-promoting peptide: FKSETKNLL	[8]
Bovine Lf	Porcine pepsin (2000 U/mL, pH 3.0, 37 °C, 2 h)	DGGMVFEAGRDPYKLRPVAAE, GILRPYLSWTE, VLRPTEGYL, YLGSRYLT, ARSVDGKEDLIWKLLSK, FKSETKNLL, ENLPEKADRDQYEL	This study

In this study, Lfampin, rather than bLfcin, fragments were discovered in three gastric digests, implying that the Lfampin fragment is more resistant to gastric digestion than Lfcin is. The incorporation of PF resulted in the LfPFG having a diverse peptide distribution (Table 2), with more long peptides (Figure 5B), and a higher total peptide abundance (protein coverage rate = 37.5%, in contrast to 32.5% for LfG). This points to the protection of PF encapsulation on bLf against in vitro gastric digestion, consistent with the case of bLf complexes with citrus low-methoxy pectins [13,14]. The fact that LfPFTGGI exhibited more peptide fragments than LfGI and LfPFGI could be attributed to the presence of TG (5 U/mL in this study), potentially via γ-glutamyl-lysine (Gln-Lys) cross-links between polypeptide chains [21,22].

4. Materials and Methods

4.1. Materials and Reagents

Passion fruit pectin (PP) was extracted from purple passion fruit peel, by using 0.48% (*w/v*) citric acid aqueous solution at 80 °C for 2.5 h. Anhydrous citric acid, anhydrous ethanol, ferulic acid (FA), bovine lactoferrin (bLf) with iron saturation = 20.8%, and bovine bile salt (bile acid > 60%) were obtained from Adamas Reagent Co. (Basel, Switzerland). Trypsin from bovine pancreas (50 U/mg), γ-glutamyl transferase (transglutaminase, TG) from bovine kidney (≥5 U/mg), artificial gastric electrolyte, and intestinal electrolyte were from Shanghai Yuanye Biotechnology Co. (Shanghai, China). Arabinose (Ara), galactose (Gal), galacturonic acid (GalA), glucose (Glc), rhamnose (Rha), laccase (0.5 U/mg, from *Aspergillus* sp.), and porcine pepsin (≥250 U/mg) were obtained from Sigma-Aldrich Co. LLC (St. Louis, MO, USA). Pierce BCA protein assay kit, trifluoroacetic acid, acetonitrile (LC/MS grade), formic acid (LC/MS grade), methanol (LC/MS grade), and water (LC/MS grade) were from ThermoFisher Scientific (Shanghai, China).

4.2. Preparation of PP-FA Conjugate (PF)

The preparation of PF conjugate was performed according to the methods of Karaki et al. [17] and our previous studies [15,19]. Five grams of PP was well dissolved in 425 mL phosphate buffer (50 mmol/L, pH 6.5) by heating (80 °C for 1 h) under stirring and cooling to room temperature. A portion of the given PP solution was then mixed with 0.2 mol/L ferulic acid (in anhydrous ethanol) at an equivalent volume, giving a final concentration of 1% (*w/w*) PP and 30 mmol/L FA (the molar ratio of free carboxy group on PP to FA = 1:1). The mixture was combined with 50 U laccase for reaction at 4 °C for 24 h in the dark. The sample was dialyzed (8–14 kDa) against deionized water for 48 h by refreshing water every other 6 h to remove free FA. The dialysate was concentrated in vacuo at ~55 °C and freeze-dried, yielding the PP-FA conjugate (namely PF).

4.3. Preparation of bLf-PF Complex (LfPF)

According to the results of our preliminary experiments and Niu et al. [14], the bLf preparation (0.2% (*w/w*), pH 5.5) was dissolved at ambient temperature under stirring (350 rpm) for 2 h. The PF preparation (0.05% (*w/w*) was dissolved in distilled water at 50 °C for 30 min and further stirred at room temperature for 1 h. bLf and PF preparations were then mixed at 1:1 *v/v* and kept at 50 °C for 10 min under stirring (350 rpm) for the spontaneous formation of the polyelectrolytic complex (LfPF). In addition, the LfPF dispersion was further treated with TG (5 U/mL, 0.1% (*w/w*)) for 10 min under mild stirring (350 rpm) at room temperature, yielding the LfPFTG complex. The complex product was collected by centrifugation (12,500× *g*, 10 min) to remove free bLf, followed by freeze-drying and storing in a desiccator before use.

4.4. Measurement of Uronic Acid Content

The uronic acid content assay was performed according to the *m*-hydroxybiphenyl method [36], using galacturonic acid (GalA) as a standard. The absorbance at 525 nm (A_{525}) of the reacted product was measured in a UV spectrophotometer (model 725, Shanghai Spectrum Instruments, Co., Ltd., Shanghai, China). The GalA content in the sample was derived by calibration with the standard curve of GalA: $y = 8.691x + 0.029$, $R^2 = 0.992$, where $y = A_{525}$ value and x = GalA concentration (0.01–0.09 mg/mL). All data were examined in triplicate.

4.5. Determination of Total Phenolic Content

Total phenolic content was determined by referring to the method of Sato et al. [37], using gallic acid as a standard. The absorbance at 720 nm (A_{720}) of the reacted product was measured. Total phenolic content in the sample was determined by calibration with the gallic acid standard curve: $y = 0.038x - 0.004$, $R^2 = 0.996$, where $y = A_{720}$ value and x = gallic acid concentration (0.05–0.25 mg/mL). The data were examined in triplicate and expressed as gallic acid equivalent (GAE) mg/g.

4.6. Measurement on Degree of Esterification

The degree of esterification (*DE*) was measured by titration with 0.05 mol/L NaOH, according to the method GB 25533-2010, National Food Safety Standard for food additive pectin, National Standard of People's Republic of China.

4.7. Assessment of Monosaccharide Composition

The sample (10 mg) was hydrolyzed with 0.2 mol/L trifluoroacetic acid (2 mL) at 110 °C for 2 h. The hydrolysate was successively cooled to room temperature, dried with nitrogen flux to remove trifluoroacetic acid, washed with methanol trice, dissolved in 2 mL deionized (18.25 MΩ·cm) water, microfiltered (0.22 μm), and subjected to high-performance anion exchange chromatography (HPAEC). A Thermo ICS-5000+ HPAEC (ThermoFisher Scientific Inc., Waltham, MA, USA) coupled with a PAD detector and guarded CarboPac PA20 analytic column (150 mm × 3 mm) (Dionex Co., Sunnyvale, CA, USA) at 30 °C was applied. The mobile phase was the mixture of four solutions: (A) deionized (18.25 MΩ·cm) water; (B) 25 mmol/L NaOH aqueous solution; (C) 1 mol/L sodium acetate aqueous solution; and (D) 200 mmol/L NaOH. The elution program for analysis was as follows: 0–20 min, A 75% + B 25%; 20.1 min, A 70% + B 25% + C 5%; 30 min, A 55% + B 25% + C 20%, in linear gradient. The elution rate was 0.5 mL/min. The injection volume of the sample was 25 μL. Data acquisition and processing and chromatographic analysis were managed with Chromeleon 7 software (ThermoFisher Scientific Inc., Waltham, MA, USA). The monosaccharide composition of the sample was quantified by calibration with monosaccharide standard curves: Ara: $y = 2098x + 1.481$, $R^2 = 0.992$; Rha: $y = 842.5x + 0.494$, $R^2 = 0.996$; Gal: $y = 1490x + 1.313$, $R^2 = 0.998$; Glc: $y = 1200x + 2.448$, $R^2 = 0.993$; GalA: $y = 1261x + 0.031$, $R^2 = 1.000$, where y = peak area (nC·min, nC = PAD detector signal) and x = monosaccharide concentration (1.25~12.5 μg/mL).

4.8. Nuclear Magnetic Resonance (NMR) Analysis

The pectin sample (40 mg) was subjected to H-D exchange by dissolving it in 1 mL of D_2O and freeze-drying it, and the procedure was repeated three times. The sample was then fully dissolved in 1 mL of D_2O and transferred to a 5 mm NMR dual tube. A 700 MHz AVANCE NEO NMR spectrometer (Bruker, Karlsruhe, Germany) was used to acquire 700.23 MHz ^{1}H-decoupled and 176.07 MHz ^{13}C spectra at 298.0 K. The spectral width, acquisition time, and scan number were 14.7 kHz, 1.114 s, and 128 for ^{1}H-decoupled spectra and 41.7 kHz, 0.393 s, and 3000 for ^{13}C spectra, respectively. The chemical shifts (δ) were expressed as parts per million (p.p.m.) in reference to external standard tetramethylsilane (TMS).

4.9. Examination of Particle Size Distribution

Particle size distribution of sample solution was assayed using a multiangle light scattering particle size analyzer (model 173 plus, Brookhaven Instrument Co., Holtsville, NY, USA) at an angle of 90° for a detection time of 120 s after equilibrium for 120 s at room temperature. Effective particle size was obtained with Particle Solution software. All data were measured in three replications.

4.10. ζ-Potential Analysis

The ζ-potential value of the sample solution was determined using a Zetasizer Nano ZS (Malvern Instruments Ltd., Malvern, UK). All data were measured in three replications.

4.11. Digestomics Analysis
4.11.1. In Vitro Simulated Gastrointestinal Digestion

Experimental conditions for in vitro simulated gastrointestinal digestion were mainly established according to the method of Minekus et al. [38] with slight modifications. A portion of bLf or its complex sample (10 mg/mL in deionized water, 10 mL) was mixed with concentrated artificial gastric electrolyte at 1:1 (v/v), followed by successively adjusting pH to 3.0 with diluted NaOH aqueous solution and adding $CaCl_2$ to a final concentration of 0.15 mmol/L $CaCl_2$ and pepsin to a final pepsin activity of 2000 U/mL. The mixture was kept at 37 °C for 2 h under shaking (95 rpm) for digestion, followed by centrifugation ($2599 \times g$, 5 min) at 4 °C to give the supernatant. The supernatant was lyophilized and stored at −18 °C for peptide identification.

For in vitro simulated intestinal digestion, a fresh supernatant of the gastric digest was mixed with concentrated simulated intestinal electrolyte at 1:1 (v/v), followed by consecutively adjusting pH to 7.0, adding bile salt solution to 10 mmol/L, mixing well, adding 0.04 mL $CaCl_2$ aqueous solution to 0.6 mmol/L $CaCl_2$, and adding bovine trypsin solution to 100 U/mL. The mixture was kept at 37 °C for 2 h under shaking (95 rpm) for digestion. The resultant digest was centrifuged ($6654 \times g$, 15 min) at 4 °C, yielding the supernatant for lyophilization and storage at −18 °C before peptide identification.

4.11.2. Peptide Purification

The lyophilized digested sample was dissolved in a portion of 8 mol/L urea, followed by microfiltration (molecular weight cut off = 10 kDa) to remove undigested residues and freeze-drying. The sample was successively re-dissolved in 0.1% (w/w) trifluoroacetic acid aqueous solution, desalted on an Oasis HLB 96-Well Plate column (30 μm) (Waters Co.,

Shanghai, China), freeze-dried, completely re-dissolved in the aqueous solution containing 50% (*v/v*) acetonitrile and 0.5% (*v/v*) trifluoroacetic acid, re-desalted on an Oasis MCX μElution Plate column (30 μm) (Waters Co., Shanghai, China), and freeze-dried. The resultant purified sample was detected for total peptide content using the Pierce BCA protein assay kit (ThermoFisher Scientific Co., Ltd.) and bovine serum albumin for the standard curve ($y = 2.025x + 0.720$, $R^2 = 0.999$, y = the absorbance at 480 nm, $x = 0.125$–1.00 μg/μL). The yield of total purified peptides was calculated on the basis of 100 mg bLf or its complex applied for gastrointestinal digestion.

4.11.3. UPLC-MS-MS Analysis for Peptide Distribution

According to total peptide content, sample solution (0.25 μg peptide/μL) was prepared in the aqueous solvent containing 2% (*v/v*) acetonitrile and 0.1% (*v/v*) formic acid for UPLC-MS-MS analysis. An EASY-nLC 1200 UPLC system coupled with a C18 column (75 μm × 25 cm, Thermo Scientific) and Q Exactive HF-X mass spectrometer (Thermo Scientific, Waltham, MA, USA) was employed. The elution was performed by mixing aqueous solutions A (2% acetonitrile + 0.1% formic acid) and B (80% acetonitrile + 0.1% formic acid) according to the following gradient program: 0 min, 5% B; 34 min, 23% B; 39 min, 29% B; 41 min, 38% B; 43–60 min, 100% B; linear gradient. The elution rate was 300 nL/min. Data were collected and managed with Thermo Xcalibur 4.0 software (Thermo Scientific).

For mass spectrometry, the MS scanning range (*m/z*) was 300–1500, and the acquisition mode was DDA. The primary fragments of the 20 strongest signals were subjected to secondary fragmentation. The conditions for primary fragmentation were as follows: resolution, 60,000; AGC target, 3×10^6; maximal injection time, 20 ms; fragmentation mode, HCD. For secondary fragmentation, the conditions were as follows: resolution, 15,000; AGC target, 1×10^5; maximum injection time, 50 ms; fixed first mass, 100 *m/z*; minimal AGC target, 8000; intensity threshold: 1.6×10^5; dynamic exclusion time, 18 s.

4.11.4. Homology Matching and Molecular Modeling

The obtained MS data were exported to the UniProt Knowledgebase (https://www.uniprot.org/uniprotkb/P24627/entry, accessed on 20 September 2022) and SWISS-MODEL (Swiss Institute of Bioinformatics, https://swissmodel.expasy.org/, accessed on 22 September 2022) for homology matching using PEAKS Studio 8.5 server and based on the amino acid sequence of bLf retrieved from UniProt Knowledgebase. The bLf 3D model was built by using SWISS-MODEL. Peptide 3D models were created using PyMol 2.3.0 (Schrödinger, New York, NY, USA).

4.12. Statistical Analysis

Significant differences between data at $p < 0.05$ were assessed by one-way analysis of variance (ANOVA) using SPSS software 25.0 (SPSS Inc., Chicago, IL, USA). OriginPro 2021b (Origin Lab, Northampton, MA, USA) was applied for plotting.

5. Conclusions

In this study, 103 bLf peptide fragments identified by UPLC-MS-MS, including seven groups of active peptide fragments, were obtained by using conditioned in vitro gastric digestion (pepsin activity 2000 U/mL, pH 3.0, 37 °C, 2 h). The discovered active peptide fragments related to three common active peptides (VFEAGRDPYKLRPVAAE, FENLPEKADRDQYEL, and VLRPTEGYL) and four differential active peptides (GILRPYLSWTE, ARSVDGKEDLIWKL, YLGSRYLT, and FKSETKNLL). Encapsulation by laccase-mediated pectin–ferulic acid conjugate (PF) improved the diversity, abundances, and long-chain species of bLf peptide fragments (including active peptides) in the gastric digest. Differently, moderate TG treatment on LfPF facilitated active peptides FENLPEKADRDQYE and FEAGRDPYK remaining in the final gastrointestinal digest. The LfPFG digest containing seven types of potentially active bLf peptides with enhanced abundances is of great potential for multipurpose use in nutraceuticals, clinical therapy, or targeting antibacterial vaccines. Based on the flow chart suggested for the production of active peptide fragments from bLf and its complexes with PF, more studies on cost-effective processes for isolating diverse bLf active peptides from the gastric digest are necessary for industrial applications. In addition, the optimal TG concentration and action sites for TG-induced crosslinking between bLf chains are interesting and under investigation.

Supplementary Materials: The following supporting information can be downloaded at: https://www.mdpi.com/article/10.3390/catal13030521/s1, Table S1: UPLC-MS-MS-identified molecular information for all peptide fragments in the digests of Lf, LfPF complex, and LfPFTG complex; Table S2: Peptide fragments found in the digests of Lf, LfPF complex, and LfPFTG complex.

Author Contributions: M.X.: Methodology, Investigation, Formal Analysis, Data Curation, Writing—Original Draft; Y.J.: Methodology, Investigation, Formal Analysis; L.A.: Supervision, Funding Acquisition, Review and Editing; F.X.: Methodology, Project Administration, Review and Editing; Y.W.: Funding Acquisition, Review and Editing; P.F.H.L.: Conceptualization, Supervision, Validation, Data Curation, Writing—Review and Editing. All authors have read and agreed to the published version of the manuscript.

Funding: This work was supported by Shanghai Science and Technology Development Fund, Domestic Science and Technology Cooperation Project (Grant No.: 21015800300).

Data Availability Statement: UPLC-MS-MS-identified information for peptides is available in the Supplementary Materials.

Acknowledgments: The authors thank Shanghai Majorbio Bio-pharm Technology Co., Ltd., Shanghai, China, for technical help in UPLC-MS-MS analysis and peptide identification.

Conflicts of Interest: The authors declare that they have no known competing financial interests.

References

1. Iglesias-Figueroa, B.F.; Espinoza-Sánchez, E.A.; Siqueiros-Cendón, T.S.; Rascón-Cruz, Q. Lactoferrin as a nutraceutical protein from milk, an overview. *Int. Dairy J.* **2019**, *89*, 37–41. [CrossRef]
2. Elzoghby, A.O.; Abdelmoneem, M.A.; Hassanin, I.A.; Abd Elwakil, M.M.; Elnaggar, M.A.; Mokhtar, S.; Fang, J.-Y.; Elkhodalry, K.A. Lactoferrin, a multi-functional glycoprotein: Active therapeutic, drug nanocarrier & targeting ligand. *Biomaterials* **2020**, *263*, 120355. [PubMed]
3. Ji, Y.; Ai, L.Z.; Xing, M.X.; Xie, F.; Lai, P. Advances in research on bovine lactoferrin-based technical innovation and application. *Food Ferment. Ind.* **2022**, *48*, 332–340. [CrossRef]
4. Agwa, M.M.; Sabra, S. Lactoferrin coated or conjugated nanomaterials as an active targeting approach in nanomedicine. *Int. J. Biol. Macromol.* **2021**, *167*, 1527–1543. [CrossRef]
5. Furlund, C.B.; Ulleberg, E.K.; Devold, T.G.; Flengsrud, R.; Jacobsen, M.; Sekse, C.; Holm, H.; Vegarud, G.E. Identification of lactoferrin peptides generated by digestion with human gastrointestinal enzymes. *J. Dairy Sci.* **2013**, *96*, 75–88. [CrossRef] [PubMed]
6. Tu, M.; Xu, S.; Xu, Z.; Cheng, S.; Wu, D.; Liu, H.; Du, M. Identification of dual-function bovine lactoferrin peptides released using simulated gastrointestinal digestion. *Food Biosci.* **2021**, *39*, 100806. [CrossRef]
7. Shi, P.; Liu, M.; Fan, F.; Chen, H.; Yu, C.; Lu, W.; Du, M. Identification and mechanism of peptides with activity promoting osteoblast proliferation from bovine lactoferrin. *Food Biosci.* **2018**, *22*, 19–25. [CrossRef]
8. Shi, P.; Fan, F.; Chen, H.; Xu, Z.; Cheng, S.; Lu, W.; Du, M. A bovine lactoferrin–derived peptide induced osteogenesis via regulation of osteoblast proliferation and differentiation. *J. Dairy Sci.* **2020**, *103*, 3950–3960. [CrossRef]
9. Ruiz-Giménez, P.; Salom, J.B.; Marcos, J.F.; Vallés, S.; Martínez-Maqueda, D.; Recio, I.; Torregrosa, G.; Alborch, E.; Manzanares, P. Antihypertensive effect of a bovine lactoferrin pepsin hydrolysate: Identification of novel active peptides. *Food Chem.* **2012**, *131*, 266–273. [CrossRef]
10. Picariello, G.; Mamone, G.; Nitride, C.; Addeo, F.; Peranti, P. Protein digestomics: Integrated platforms to study food-protein digestion and derived functional and active peptides. *TrAC-Trend Anal. Chem.* **2013**, *52*, 120–134. [CrossRef]
11. Basak, S.; Annapure, U.S. Trends in "green" and novel methods of pectin modification—A review. *Carbohydr. Polym.* **2022**, *278*, 118967. [CrossRef]
12. Li, D.-q.; Li, J.; Dong, H.-l.; Li, X.; Zhang, J.-q.; Ramaswamy, S.; Xu, F. Pectin in biomedical and drug delivery applications: A review. *Int. J. Biol. Macromol.* **2021**, *185*, 49–65. [CrossRef] [PubMed]
13. Bengoechea, C.; Jones, O.G.; Guerrero, A.; McClements, D.J. Formation and characterization of lactoferrin/pectin electrostatic complexes: Impact of composition, pH and thermal treatment. *Food Hydrocoll.* **2011**, *25*, 1227–1232. [CrossRef]

14. Niu, Z.; Loveday, S.M.; Barbe, V.; Thielen, I.; He, Y.; Singh, H. Protection of native lactoferrin under gastric conditions through complexation with pectin and chitosan. *Food Hydrocoll.* **2019**, *93*, 120–130. [CrossRef]
15. Ji, Y. Study on Preparation Technology, Formulation Optimization and Active Peptide of Lactoferrin-Pectin Complex. Master Thesis, University of Shanghai for Science & Technology, Shanghai, China, 2022.
16. Karaki, N.; Aljawish, A.; Muniglia, L.; Humeau, C.; Jasniewski, J. Physicochemical characterization of pectin grafted with exogenous phenols. *Food Hydrocoll.* **2016**, *60*, 486–493. [CrossRef]
17. Karaki, N.; Aljawish, A.; Muniglia, L.; Bouguet-Bonnet, S.; Leclerc, S.; Paris, C.; Jasniewski, J. Functionalization of pectin with laccase-mediated oxidation products of ferulic acid. *Enz. Microbial Technol.* **2017**, *104*, 1–8. [CrossRef] [PubMed]
18. Robert, B.; Chenthamara, D.; Subramaniam, S. Fabrication and biomedical applications of arabinoxylan, pectin, chitosan, soyprotein, and silk fibroin hydrogels via laccase—Ferulic acid redox chemistry. *Int. J. Biol. Macromol.* **2022**, *201*, 539–556. [CrossRef]
19. Gao, F.; Ai, L.Z.; Wu, Y.; Lai, F.X.; Zhang, H.; Xie, F.; Song, Z.B. Structural mechanism and structure-antioxidant activity relationship of low-methoxyl pectin-caffeic acid conjugate. *Food Sci.* **2022**, *43*, 19–29.
20. Gao, F.; Ding, N.; Ai, L.; Lai, P.; Zhang, H.; Song, Z. Molecular characteristics, in vitro antioxidant and immunological activities of high-methoxy pectin-phenolic acid derivatives from passion fruit. *Food Sci.* **2022**, *43*, 84–94. Available online: https://www.spkx.net.cn/article/2022/1002-6630/2022-43-17-010.html (accessed on 20 September 2022).
21. Miwa, N. Innovation in the food industry using microbial transglutaminase: Keys to success and future prospects. *Anal. Biochem.* **2020**, *597*, 113638. [CrossRef]
22. Xia, T.; Gao, Y.; Liu, Y.; Wei, Z.; Xue, C. Lactoferrin particle assembled via transglutaminase-induced crosslinking: Utilization in oleogel-based Pickering emulsions with improved curcumin bioaccessibility. *Food Chem.* **2022**, *374*, 131779. [CrossRef] [PubMed]
23. Martini, S.; Solieri, L.; Tagliazucchi, D. Peptidomics: New trends in food science. *Curr. Opin. Food Sci.* **2021**, *39*, 51–59. [CrossRef]
24. Lin, Y.; An, F.; He, H.; Geng, F.; Song, H.; Huang, Q. Structural and rheological characterization of pectin from passion fruit (*Passiflora edulis f. flavicarpa*) peel extracted by high-speed shearing. *Food Hydrocoll.* **2021**, *114*, 106555. [CrossRef]
25. Cui, L.; Wang, J.; Huang, R.; Tan, Y.; Zhang, F.; Zhou, Y.; Sun, L. Analysis of pectin from Panax ginseng flower buds and their binding activities to galactin-3. *Int. J. Biol. Macromol.* **2019**, *128*, 459–467. [CrossRef]
26. Liu, D.; Zhai, L.-Y.; Shi, Z.-H.; Hong, H.-L.; Liu, L.-Y.; Zhao, S.-R.; Hu, Y.-B. Purification and fine structural analysis of pectin polysaccharides from *Osmunda Japonica Thunb*. *J. Mol. Struct.* **2022**, *1269*, 133828. [CrossRef]
27. De Cicco, M.; Mamone, G.; Di Stasie, L.; Ferranti, P.; Addeo, F.; Picariello, G. Hidden "digestome": Current analytical approaches provide incomplete peptide inventories of food digests. *J. Agr. Food Chem.* **2019**, *67*, 7775–7782. [CrossRef] [PubMed]
28. Nielsen, S.D.; Beverly, R.L.; Qu, Y.; Dallas, D.C. Milk bioactive peptide database: A comprehensive database of milk protein-derived bioactive peptides and novel visualization. *Food Chem.* **2017**, *232*, 673–682. [CrossRef]
29. García-Tejedor, A.; Castelló-Ruiz, M.; Gimeno-Alcañíz, J.V.; Manzanares, P.; Salom, J.B. In vivo antihypertensive mechanism of lactoferrin-derived peptides: Reversion of angiotensin I- and angiotensin II-induced hypertension in Wistar rats. *J. Funct. Foods* **2015**, *15*, 294–300. [CrossRef]
30. Ding, N. Study on Extraction of Passion Fruit Peel Pectin, Derivatization and Functionality. Master Thesis, University of Shanghai for Science & Technology, Shanghai, China, 2020.
31. Li, X.; Li, S.; Liang, X.; McClements, D.J.; Liu, X.; Liu, F. Applications of oxidases in modification of food molecules and colloidal systems: Laccase, peroxidase and tyrosinase. *Trends Food Sci. Tech.* **2020**, *103*, 78–93. [CrossRef]
32. Vuillemin, M.E.; Muniglia, L.; Linder, M.; Bouguet-Bonnet, S.; Poinsignon, S.; Morais, R.D.S.; Simard, B.; Paris, C.; Michaux, F.; Jasniewski, J. Polymer functionalization through an enzymatic process: Intermediate products characterization and their grafting onto gum Arabic. *Int. J. Biol. Macromol.* **2021**, *169*, 480–491. [CrossRef]
33. Moreno-Expósito, L.; Illescas-Montes, R.; Melguizo-Rodríguez, L.; Ruiz, C.; Ramos-Torrecillas, J.; de Luna-Bertos, E. Multifunctional capacity and therapeutic potential of lactoferrin. *Life Sci.* **2018**, *195*, 61–64. [CrossRef] [PubMed]
34. Zhao, W.; Li, X.; Yu, Z.; Wu, S.; Ding, L.; Liu, J. Identification of lactoferrin-derived peptides as potential inhibitors against the main protease of SARS-CoV-2. *LWT* **2022**, *154*, 112684. [CrossRef] [PubMed]
35. Schryvers, A.B. Targeting bacterial transferrin and lactoferrin receptors for vaccines. *Trends Microbiol.* **2022**, *30*, 820–830. [CrossRef] [PubMed]
36. Blumenkrantz, N.; Hansen, G.A. New method for quantitative determination of uronic acids. *Anal. Biochem.* **1973**, *54*, 484–489. [CrossRef]
37. Sato, M.; Ramarathnam, N.; Suzuki, Y.; Ohkubo, T.; Takeuchi, M.; Ochi, H. Varietal differences in the phenolic content and superoxide radical scavenging potential of wines from different sources. *J. Agr. Food Chem.* **1996**, *44*, 37–41. [CrossRef]
38. Minekus, M.; Alminger, M.; Alvito, P.; Balance, S.; Bohn, T.; Bourlieu, C.; Carrière, F.; Boutrou, R.; Corredig, M.; Dupont, D.; et al. A standardised static in vitro digestion method suitable for food—an international consensus. *Food Funct.* **2014**, *5*, 1113–1124. [CrossRef] [PubMed]

 *catalysts*

Article

Metagenomic Type IV Aminotransferases Active toward (*R*)-Methylbenzylamine

Rokas Statkevičius *, Justas Vaitekūnas, Rūta Stanislauskienė and Rolandas Meškys *

Life Science Center, Vilnius University, Saulėtekio al. 7, 10257 Vilnius, Lithuania
* Correspondence: rokas.statkevicius@gmc.vu.lt (R.S.); rolandas.meskys@bchi.vu.lt (R.M.)

Abstract: Aminotransferases (ATs) are pyridoxal 5′-phosphate-dependent enzymes that catalyze the reversible transfer of an amino group from an amino donor to a keto substrate. ATs are promising biocatalysts that are replacing traditional chemical routes for the production of chiral amines. In this study, an in silico-screening of a metagenomic library isolated from the Curonian Lagoon identified 11 full-length fold type IV aminotransferases that were successfully expressed and used for substrate profiling. Three of them (AT-872, AT-1132, and AT-4421) were active toward (*R*)-methylbenzylamine. Purified proteins showed activity with L- and D-amino acids and various aromatic compounds such as (*R*)-1-aminotetraline. AT-872 and AT-1132 exhibited thermostability and retained about 55% and 80% of their activities, respectively, even after 24 h of incubation at 50 °C. Active site modeling revealed that AT-872 and AT-4421 have an unusual active site environment similar to the AT of *Haliscomenobacter hydrossis*, while AT-1132 appeared to be structurally related to the AT from thermophilic archaea *Geoglobus acetivorans*. Thus, we have identified and characterized PLP fold type IV ATs that were active toward both amino acids and a variety of (*R*)-amines.

Keywords: aminotransferase; (*R*)-selectivity; biocatalysis; thermostability

Citation: Statkevičius, R.; Vaitekūnas, J.; Stanislauskienė, R.; Meškys, R. Metagenomic Type IV Aminotransferases Active toward (*R*)-Methylbenzylamine. *Catalysts* **2023**, *13*, 587. https://doi.org/10.3390/catal13030587

Academic Editors: Chia-Hung Kuo, Chwen-Jen Shieh and Hui-Min David Wang

Received: 20 February 2023
Revised: 9 March 2023
Accepted: 13 March 2023
Published: 15 March 2023

1. Introduction

Chiral amine compounds are widely used as active pharmaceutical ingredients, agricultural chemicals, bioactive natural products, and other value-added compounds [1]. Therefore, the development of widely applicable biocatalytic strategies for their synthesis is essential. Aminotransferases (ATs), also known as transaminases (EC 2.6.1), are increasingly used as an alternative to traditional chemical catalysts. ATs are promising biocatalysts for the development of innovative routes for the production of chiral amines at an industrial scale [2–6]. However, the repertoire of ATs capable of catalyzing the transamination of prochiral ketones into pure chemicals is still limited. An expanded toolbox of available ATs is expected to open new routes for the synthesis of chiral amines.

ATs are pyridoxal 5′-phosphate (PLP)-dependent enzymes that catalyze the reversible transfer of an amino group from an amino donor to a keto substrate. ATs are found in all domains of life: archaea, bacteria, and eukaryotes, where they are involved in the metabolism of amino acids and other nitrogenous compounds [7–9]. The transamination reaction occurs in two steps. First, the amino group of the amino donor is transferred to a pyridoxal phosphate (PLP) covalently bound to a lysine in the active site via a Schiff base. This reaction yields pyridoxamine phosphate (PMP) and the keto product. The second transamination step occurs when the amino group is transferred from the PMP to the keto compound (the amine acceptor), restoring the PLP and forming a new amine compound [10].

PLP-dependent enzymes are classified into different fold types based on their structure and sequence motifs. ATs are contained in fold types I and IV. ATs of fold type IV are further subdivided into branched-chain aminotransferases (BCAATs), D-amino acid aminotransferases (DAATs), and (R)-amine-pyruvate aminotransferases (R-ATs) [11]. In higher eukaryotes, BCATs are responsible for the catabolism of leucine, isoleucine, and valine, and can also deaminate aromatic amino acids. In other organisms, BCATs are used for both the production and degradation of these amino acids [12,13]. Bacteria use DAAT to produce D-amino acids, in particular D-glutamic acid, which is essential for cell wall synthesis [14]. The most recent member of the fold type IV ATs is R-ATs. These ATs are active toward various aliphatic R-amines such as (*R*)-methylbenzylamine, (2R)-aminohexane, and others [15–18]. Höhne et al. [15] revealed an in silico-strategy for the detection of potential R-ATs from metagenomes. Further progress in the identification of different R-ATs was made by Steiner et al. [19] and Jiang et al. [16]. Phylogenetic analysis based on multiple sequence alignments or sequence similarity networks (SSNs) provides reliable data for subgrouping ATs; however, the sequences of potential R-ATs do not form distinct groups. In addition, it is difficult to predict the activity of ATs and the promiscuity of enzyme substrates toward donor or acceptor substrates. ATs are of great importance as potential biocatalysts for the production of pharmaceuticals and other chemical intermediates [2,20–22]; therefore, there is still a need to find suitable R-ATs that can use many different amine acceptors and donors.

In this study, an in silico-screening of a metagenomic library identified 11 full-length fold type IV aminotransferases that were successfully expressed and used for substrate profiling. Three of the tested ATs were active toward *R*-methylbenzylamine.

2. Results and Discussion

2.1. Enzymes Identification and Phylogenetic Analysis

We searched in silico for fold type IV PLP-dependent enzymes in the sequenced metagenomic DNA isolated from the Curonian Lagoon [23]. We have identified eleven genes encoding hypothetical ATs that belong to the fold type IV class. A comparison of the amino acid sequences of the selected ATs (Table S1) (between each other) showed a similarity of 31.6–100%. The homologs of the ATs were identified by a BLASTp search against the nonredundant protein sequences of the NCBI database (Table 1). Eight of the selected ATs were highly similar (91–100%) to their closest homologs and could be accurately assigned to the BCAT group. In contrast, AT-872, AT-1202, and AT-4421 did not show high homology and could not be placed with certainty in any group of fold type IV class ATs.

Table 1. The list of selected ATs.

AT	AT GenBank ID	The Closest Homolog	Identities (%)
AT-872	QVW56595.1	aminotransferase class IV [*Chitinophagaceae bacterium*]	53.28%
AT-1132	ON873786	branched-chain-amino-acid transaminase [*Planctomycetaceae bacterium*]	91.23%
AT-1202	ON873787	aminotransferase class IV [*Candidatus Planktophila vernalis*]	96.80%
AT-1229	ON873788	branched-chain amino acid transaminase [*Burkholderiaceae bacterium*]	99.02%
AT-1638	ON873789	branched-chain amino acid transaminase [*Candidatus Methylopumilus universalis*]	100%
AT-1688	ON873790	branched-chain amino acid transaminase [*Ignavibacteriae bacterium*]	99.34%
AT-4421	ON873791	amino acid aminotransferase [*Sediminibacterium* sp.]	77.45%
AT-7378	ON873792	branched-chain amino acid aminotransferase [*Proteobacteria bacterium*]	97.61%
AT-14307	ON873793	branched-chain-amino-acid transaminase [*Spartobacteria bacterium Tous-C9RFEB*]	95.86%
AT-19987	ON873794	branched-chain amino acid transaminase [*Candidatus Methylopumilus universalis*]	100.00%
AT-35055	ON873795	branched-chain amino acid transaminase [*Rubrivivax* sp.]	99.03%

We attempted to clarify the phylogenetic position of the selected ATs among the characterized BCATs and DAATs capable of deaminating various (*R*)-amines [24–28]. Therefore, we assembled the reviewed sequences of known BCATs and DAATs from UniProtKB, and also added ATs of interest from previous publications [15,19,24,27,29,30]. The constructed phylogenetic tree (Figure S1) revealed that most of the identified BCATs clustered together with known canonical BCATs. AT-1132 and AT-14307 attracted considerable interest, as both of these proteins were grouped with ATs of archaeal origin, which have been shown to deaminate the (*R*)-methylbenzylamine [29] (Figure 1a).

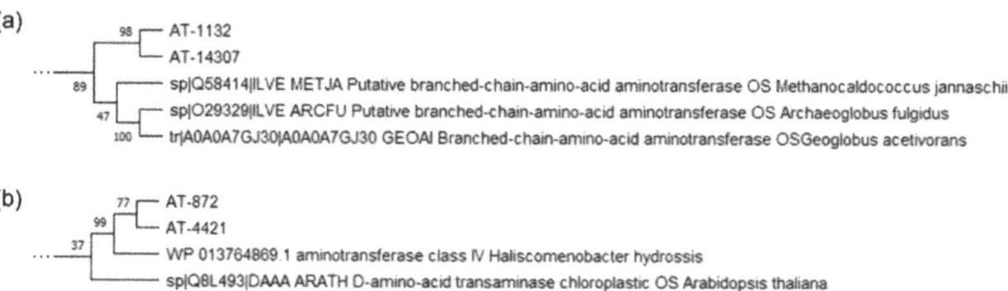

Figure 1. A representative view of the phylogenetic analysis of the selected ATs (**a**) AT-1132 and AT-14307, (**b**) AT-872 and AT-4421 compared to the reviewed sequences of fold type IV aminotransferases. See Figure S1 for a full phylogenetic analysis.

AT-872 and AT-4421 clustered with *Arabidopsis thaliana* chloroplast DAAT and *Haliscomenobacter hydrossis* AT, which, as has been shown previously [30], have an unusual DAAT active site (Figure 1b). According to the phylogenetic analysis, the identified set of ATs was diverse, with a low similarity between its characterized homologs, and, potentially, contained (*R*)-selective counterparts. Therefore, we decided to test all identified AT using (*S*)- and (*R*)-methylbenzylamine (MBA) as substrates.

2.2. Initial Activity Screening

To determine the enzymatic activity of the selected ATs, their genes were cloned into the expression vector pLATE31 and overexpressed *E. coli* strain BL21 (DE3) (Figure S2 and Table S2), as described in Materials and Methods. The measurement of AT activity using cell-free extracts and a universal amino acceptor (pyruvate) showed that none of the ATs tested were active toward *S*-MBA. However, AT-872, AT-1132, and AT-4421 presented low activity in the presence of *R*-MBA. In addition, ATs showed no activity with *o*-xylenediamine, a common chromogenic amino donor used for AT screening and activity measurements [31]. However, some of the tested enzymes showed low but measurable activity in the presence of another chromogenic amino donor, 4-nitrophenylethylamine [32] (Figure S3). AT-14307, although very similar to AT-1132, did not display any appreciable activity toward *R*-MBA. For further investigation, we chose three enzymes (AT-872, AT-1132, and AT-4421) capable of deaminating *R*-MBA. The target proteins were expressed with 6xHis tags and purified by immobilized metal affinity chromatography (Figure S4).

2.3. Temperature Stability and Effect of pH on Activity of the Selected Aminotransferases

We began our characterization by optimizing the pH of the reaction mixture (Figure 2a). The majority of the previously characterized (*R*)-selective ATs were most active at pH 8 or 9 [18,19,33]. AT-1132 exhibited the maximum activity in sodium bicarbonate buffer at pH 9, while AT-4421 showed the highest activity in potassium phosphate buffer at pH 9. Although AT-1132 had a low pH optimum range, AT-4421 remained active (80%) at pH 10 (in sodium carbonate buffer). In contrast, AT-872 preferred higher pH and showed similar activity at pH 10–11.

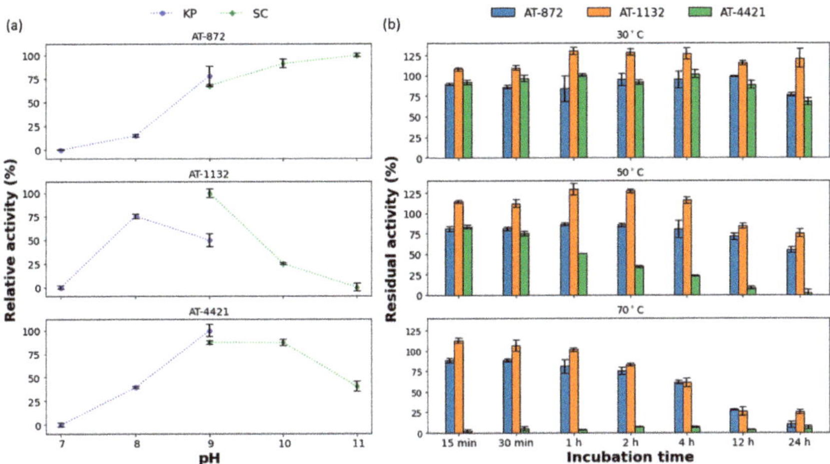

Figure 2. (**a**) The effect of pH on enzyme activity. The highest activity for each enzyme was taken as 100%, KP—potassium phosphate buffer, SC—sodium carbonate buffer; (**b**) Thermostability of ATs. Activity before incubation was taken as 100%. For a detailed description of reaction conditions, see the Methods section. All values are mean ± SD.

The thermostability of the selected ATs was investigated at three temperatures (Figure 2b). At 30 °C, AT-4421 was moderately stable, the activity decreases by 30% after incubation for 24 h. At higher temperatures, a significant decrease with time in AT-4421 activity was observed: 25% of activity was retained after 4 h at 50 °C, while at 70 °C, complete inactivation occurred after 15 min. The other two enzymes were quite thermostable. AT-872 retained about 55% and AT-1132 around 80% of their activities, even after 24 h of incubation at 50 °C. Moreover, these two ATs retained more than 60% of their activity after incubation at 70 °C for 4 h. AT-1132 showed an increase in activity after incubation at all tested temperatures. Such phenomena were observed with certain other thermostable ATs and might be explained by the refolding of a protein that was kept at −20 °C to its natural conformation [34,35]. This increase in activity (above 100%) lasted longer at lower temperatures due to slower protein denaturation rates.

2.4. Substrate Scope of Aminotransferases

We measured ATs' preferences for different amino acceptors (Table 2). AT-872 and AT-4421 were the most active in the presence of pyruvate, while AT-1132 had the highest activity toward α-ketoglutarate.

Table 2. AT activity with different amino acceptors. The highest activity for each enzyme was taken as 100%. All values are mean ± SD.

Amino Acceptor	AT-872	AT-1132	AT-4421
pyruvate	100.0 ± 0.1%	74.5 ± 0.7%	100.0 ± 2.1%
α-ketoglutarate	52.4 ± 1.3%	100.0 ± 1.5%	65.1 ± 4.1%
oxaloacetate	71.9 ± 0.4%	97.3 ± 0.1%	89.3 ± 6.4%
4-methyl-2-oxovaleric acid	61.7 ± 1.3%	95.6 ± 2.6%	38.5 ± 0.1%

The AT-1132 is closely related to BCATs according to protein sequence and its specificity for amino acceptor is similar to BCATs, which are most active with α-ketoglutarate. AT-872 and AT-4421 are homologous to the DAATs, which prefer pyruvate as an amino acceptor [13,36,37]. Thus, the amino acceptor preference of selected ATs correlates with protein sequences.

The conversion rates of the deamination reactions for the different amino donors were also investigated (Figure 3).

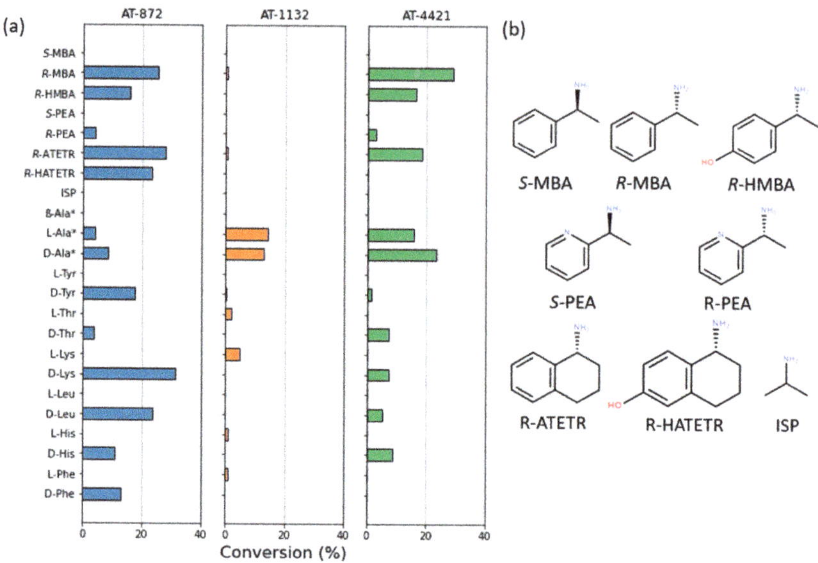

Figure 3. (**a**) The substrate specificity of ATs; (**b**) Structures of the used substrates. Pyruvate was used as an amino acceptor for all reactions except those marked with the "*" symbol where α-ketoglutarate was used. A detailed description of reaction conditions is outlined in the Methods Section. The reaction (100 μL) conditions: 5 mM amino donor and 5 mM pyruvate or α-ketoglutarate, 10 μM PLP, 5 μg of the purified enzyme, and 50 mM sodium carbonate buffer (pH 9). The reactions were performed at 30 °C for 16 h.

AT-1132 was able to deaminate L- and D-alanine and some L-amino acids, although with very low efficiency. AT-872 and AT-4421 were active with D-amino acids and L-alanine. None of the ATs tested were able to use β-alanine or isopropylamine (ISP) as an amino donor. All three ATs were active with amino donor *R*-MBA (AT-872—1.0 U/mg, AT-1132—0.2 U/mg, and AT-4421—0.9 U/mg). These activity values toward *R*-MBA are similar to that of the other described PLP fold type IV ATs [25,29]. AT-872 and AT-4421 were also able to deaminate (*R*)-pyridylethylamine (*R*-PEA). Although a small amount of keto product was detected, the activity of AT-1132 with *R*-PEA was too low to give an accurate estimate of the conversion. Interestingly, AT-4421 was able to deaminate the bulky substrate (*R*)-aminotetralin (*R*-ATETR), while AT-872 was also active with the *R*-ATETR analog (*R*)-hydroxyaminotetralin (*R*-HATETR) and showed more than 20% conversion using equal amounts of amino donor and amino acceptor.

2.5. D-Models of the ATs Active Sites

To better understand the substrate preferences of identified ATs, we used Colab-Fold [38] to create 3D-structure models of the enzymes. All ATs were modeled as dimers. We also used CB-Dock2 [39,40] for template-based docking to dock the ketimine intermediate of MBA bound to PLP into the AT-872 active site.

Although AT-872 and AT-4421 have 40% similarity, their active site environments were almost identical (Figure 4a). The only notable differences were the substitutions at positions 91 and 239 (according to the AT-872 sequence), AT-4421 had Arg instead of Lys91, and Thr was replaced with Ser239. The active sites of AT-872 and AT-4421 corresponded well to the AT of *Haliscomenobacter hydrossis* [30] (35% overall similarity) (Figure 4b). Like all type IV AT [41], the active sites of AT-872 and AT-4421 were divided into two pockets, O and P.

The O pocket of AT-872 consisted of the side chains of Phe30, Arg28 (provided by another monomer), Lys90, and Arg179, thus forming a positively charged pocket. The difference between the O pocket of AT-4421 and AT-872 was Lys91 replaced by Arg, as mentioned above. Neither ATs had the canonical DAAT "carboxylate trap" and most likely used Arg28 and Lys91 (or Arg) side chains to bind the α-carboxyl group of the substrates in a manner similar to the *H. hydrossis* AT. The P-pocket of AT-872 was confined by side chains of Phe36, Ser238, Ser239, and Thr240. Ser239 was replaced by Thr in the AT-4421 P-pocket. Unlike *H. hydrossis* AT, both AT-872 and AT-4421 were capable of deaminating *R*-MBA, *R*-PEA, or *R*-ATETR. This might be explained by the larger and more hydrophobic P-pocket formed when Tyr35 (in *H. hydrossis* AT) was replaced by Phe, and Ile240 (in *H. hydrossis* AT) by Ser or Thr.

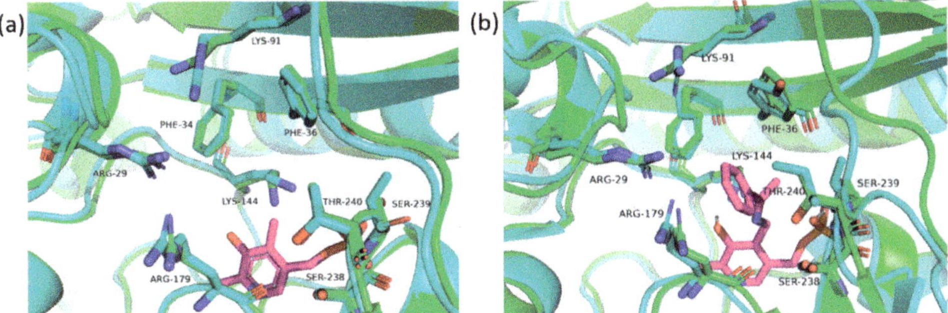

Figure 4. (**a**) Comparison of AT-872 (green) and AT-4421 (cyan) active sites, where AT of *Haliscomenobacter hydrossis* (7p7x) was used for PLP (magenta) placement; and (**b**) active site of the AT-872 (green) superimposed on the AT of *Haliscomenobacter hydrossis* (7p7x) (cyan), with PLP bound to MBA (magenta). In both cases, the amino acid numbering was based on the AT-872.

ColabFold was also used to create a model of the AT-1132 structure. This AT is 50% similar to BCAT from the thermophilic archaea *Geoglobus acetivorans* [29]. Their active sites are almost identical (Figure 5) and, although less so, are similar to the AT from *Thermobaculum terrenum* [42].

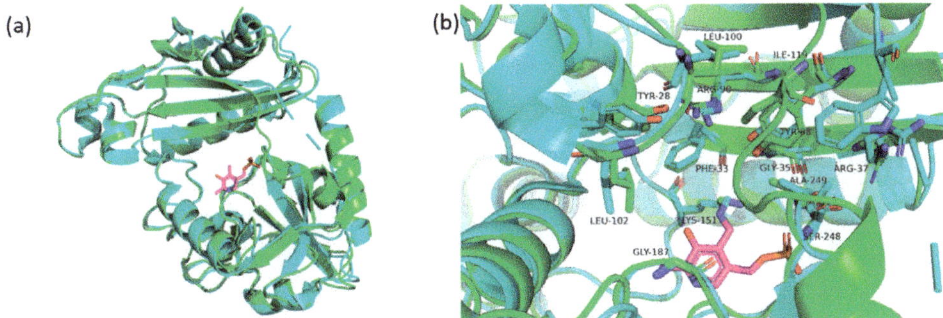

Figure 5. (**a**) The monomer of the AT-1132 (green) superimposed on the monomer of the AT from *Geoglobus acetivorans* (5 cm0) (cyan), PLP (magenta); (**b**) a comparison of AT-1132 (green) and AT from *Geoglobus acetivorans* (cyan) active sites, PLP (magenta). Numbering based on the amino acid of the AT-1132.

The active site of AT-1132 is also divided into O and P pockets. Pocket O consists of Phe33, Arg90, Ile119, and Tyr28, while Leu100 and Leu102 are from the other monomer. The P-pocket of AT-1132 is composed of residues Gly35, Arg38, Tyr89, Gly187, Ser247, and

Ala248. The major difference between AT-1132 and AT from *G. acetivorans* is in the loop region connecting the small subunit to the large domain, where AT-1132 contains Ile119 instead of Trp. It is worth noting that AT-14307, despite an almost identical active site composition, was unable to deaminate *R*-MBA or the activity of this enzyme was too low to be detectable. This might be attributed to subtle differences in the interdomain loop and other flexible regions surrounding the active site.

3. Materials and Methods

3.1. Selection of AT Genes from Metagenome Library and Phylogenetic Analysis

The sequenced metagenome of the Curonian Lagoon [23] was used as a database for the search for potential aminotransferases encoding genes. The experimentally characterized sequences of R-specific (fold type IV) ATs [15,16] were used as a query in BLAST analysis. The hits with the highest scores ($>1 \times 10^{-10}$) were selected for further analysis.

The phylogenetic analysis was performed with the MegaX program [43] version 10.1.8. Multiple sequence alignment was performed using the MUSCLE algorithm. The alignment was manually refined by removing variable regions at the start and the end of the protein. The evolutionary history was inferred by using the maximum likelihood method and JTT matrix-based model, and the bootstrap consensus tree was inferred from 500 replicates.

3.2. Bacterial Strains, Plasmids, Primers, and Standard Cloning Techniques

E. coli strain DH5α was used for cloning experiments. The recombinant proteins were overexpressed in *E. coli* strain BL21 (DE3). The bacterial strains, plasmids, and primers used in this study are listed in Table S3 in the Supplemental Material (SM). Standard molecular biology techniques were performed, as previously described [44]. The genes coding selected ATs were amplified from metagenomic DNA from the Curonian Lagoon using a Phusion Plus PCR Master Mix (Thermo Fisher Scientific, Vilnius, Lithuania) and designed primers (Table S3). The PCR products were extracted from the gel using a GeneJET Gel Extraction Kit (Thermo Fisher Scientific, Vilnius, Lithuania) and were cloned into pLATE31 (C-terminal His-tag) expression vector using aLICator LIC Cloning and Expression Kit 3 (Thermo Fisher Scientific, Vilnius, Lithuania). *E. coli* strain DH5α (Novagen, Darmstadt, Germany) cells were transformed with reaction mixtures and spread on LB agar plates supplemented with 75 µg/mL ampicillin. pDNA (plasmid DNA) from selected bacterial clones was extracted and purified using the ZR Plasmid Miniprep kit (Zymo Research, Irvine, CA, USA). Later, it was sequenced using LIC sequencing primers (Table S3) by the Sanger method (Macrogen Europe, Amsterdam, The Netherlands).

3.3. Expression and Purification of Recombinant Proteins

The *E. coli* BL21 (DE3) (Novagen, Darmstadt, Germany) bacteria harboring the appropriate plasmid were grown with shaking (180 rpm) at 37 °C in 200 mL of LB medium supplemented with 100 µg/mL ampicillin. When OD600 reached 0.6–0.8, protein expression was induced by adding IPTG (final concentration 1.0 mM). After the induction for 12 h at 20 °C temperature, the cells were collected by centrifugation for 20 min, $4000 \times g$ at 4 °C. The cell pellet was resuspended in 25 mL of 50 mM PBS buffer (pH 7.2) and disrupted using ultrasound. The cell debris was removed by centrifugation for 20 min at $15,000 \times g$ and 4 °C. The supernatant was loaded onto a 1 mL of HiTrap Nickel Chelating HP column (GE Healthcare, Uppsala, Sweden) that had been pre-equilibrated with PBS pH 7.2. The column was washed with five column volumes of PBS pH 7.2 containing 30 mM imidazole. The recombinant protein was eluted in PBS pH 7.2 containing 400 mM imidazole. The collected recombinant protein fraction was desalted on a 5 mL Sephadex G25 column (GE Healthcare, Uppsala, Sweden) using PBS pH 7.2. The protein solution was diluted with glycerol (50% final concentration) and supplemented with 50 µM PLP and stored at −20 °C for further use. The purity of the proteins was determined by the SDS-PAGE [45] standard procedure using 12% gels. The protein bands were visualized using Coomassie blue G-250 (Fluka, Buchs, Switzerland) staining. The concentration of purified proteins was determined using a Pierce Coomassie (Bradford) Protein Assay Kit (Thermo Fisher, Rockford, IL, USA) [46].

3.4. Enzyme Assays

The ATs capable of deaminating methylbenzylamine were screened by measuring the amount of formed acetophenone. The reaction (volume 100 µL) conditions were as follows: 10 mM (*S*)- or (*R*)-methylbenzylamine, 10 mM pyruvate as amino acceptor, 10 µM PLP, 50 µL of cell-free lysate (in PBS pH 7.2 buffer), and 50 mM potassium phosphate buffer (pH 9). The reactions were incubated at 30 °C for 16 h. Then, 100 µL of acetonitrile was added to stop the reactions. The reaction mixture was centrifugated for 10 min at $15,000 \times g$ and transferred to the 96-well plates. The formation of acetophenone was determined by the change in UV absorbance at 240 nm using a PowerWave XS plate reader (BioTek, Santa Clara, CA, USA).

To study the effect of pH on enzyme activity, reactions were carried out in buffers of different pH values (50 mM potassium phosphate pH 7–8; 50 mM sodium bicarbonate pH 9–11). The relative enzyme activity was determined by the amount of formed acetophenone. The reaction mixture (100 µL) contained 5 mM (*R*)-methylbenzylamine, 5 mM pyruvate, 10 µM PLP, and 5 µg of the purified enzyme. The reaction was performed at 30 °C for 20 min. The measurement of the acetophenone formed was carried out, as mentioned above. The experiments were conducted in triplicate.

The thermostability of the enzyme was determined by incubating the enzyme at set temperatures (30, 50, or 70 °C) for a specific amount of time (up to 24 h). The remaining enzyme activity after incubation was determined by the amount of formed acetophenone. The reaction mixture (100 µL) contained 5 mM (*R*)-methylbenzylamine, 5 mM pyruvate, 10 µM PLP, 5 µg of the incubated enzyme, and 50 mM sodium carbonate buffer (pH 9). Reactions were performed at 30 °C for 20 min. The measurement of the acetophenone formed was carried out as mentioned above. The experiments were conducted in triplicate.

3.5. Scope of Amino Acceptors and Donors

The optimal amino acceptor was determined by measuring the amount of formed acetophenone. The reactions were carried out using different amino acceptors under the following conditions: 10 mM (*R*)-methylbenzylamine, 5 mM amino acceptor, 10 µM PLP, 5 µg of the purified enzyme, and 50 mM sodium carbonate buffer (pH 9), with a total reaction volume of 100 µL. The reactions were performed at 30 °C for 1 h. The measurement of the acetophenone formed was carried out as mentioned above. The experiments were conducted in triplicate.

Amino donor specificity was determined by measuring the amount of formed alanine or glutamic acid. The reaction (100 µL) conditions: 5 mM amino donor, 5 mM pyruvate or α-ketoglutarate, 10 µM PLP, 5 µg of the purified enzyme, and 50 mM sodium carbonate buffer (pH 9). The reactions were performed at 30 °C for 16 h. The reaction was stopped by adding 500 µL of acetonitrile. The reaction mixture was centrifugated for 20 min at $15,000 \times g$. The amount of formed alanine or glutamic acid was determined by HPLC–MS.

3.6. Analytical Methods

The HPLC–MS analysis was performed using a high-performance liquid chromatography system (Shimadzu, Kyoto, Japan) equipped with a photodiode array (PDA; Shimadzu, Kyoto, Japan) detector and a mass spectrometer (LCMS 2020; Shimadzu, Kyoto, Japan). The samples were mixed with an equal volume of acetonitrile, vortexed for 1 min, clarified by centrifugation at $10,000 \times g$ for 10 min, and subjected to HPLC–MS. The chromatographic separation was conducted using a 150 × 3 mm YMC-Pack Pro C18 column (YMC, Kyoto, Japan) at 40 °C and a mobile phase that consisted of water with 0.1% formic acid (solvent A) and acetonitrile (solvent B) delivered in gradient elution mode at a flow rate of 0.45 mL/min. The elution program was used as follows: isocratic 5% B for 1 min, from 5 to

95% B over 5 min, isocratic 95% B for 2 min, from 95 to 5% B over 1 min, isocratic 5% B for 4 min. The data were analyzed using LabSolutions LCMS software version 5.42.

4. Conclusions

Several PLP fold type IV ATs were identified and characterized in this study. These enzymes are not only active toward D-amino acids but can also transaminate various (R)-amines, which makes them potential biocatalysts. Based on the high thermostability of AT-872 and AT-1132, and the broad substrate spectrum of AT-872 and AT-4421, the engineering of these ATs for application in the production of high-value amino compounds seems feasible. Our results suggest that BCAT and DAAT have biocatalytic potential and that these groups of fold type IV aminotransferases should not be overlooked in the search for R-selective ATs.

Supplementary Materials: The following supporting information can be downloaded at: https://www.mdpi.com/article/10.3390/catal13030587/s1, Figure S1: Evolutionary analysis by Maximum Likelihood method; Figure S2: Protein expression of ATs in *E. coli* BL-21 (DE3). Analysis of cleared lysates using SDS-PAGE; Figure S3: Activities with: (*S*)-methylbenzylamine (S-MBA), (*R*)-methylbenzylamine (R-MBA) and 4-nitrophenylethylamine (4NPEA); Figure S4: Protein purification SDS-PAGE; Table S1: Percent similarity matrix of the chosen ATs; Table S2: AT protein theoretical mass and observed activities and Table S3: Bacterial strains, plasmids, and primers used in this study.

Author Contributions: Conceptualization, R.M., J.V. and R.S. (Rokas Statkevičius); methodology, R.S. (Rokas Statkevičius), R.S. (Rūta Stanislauskienė) and J.V.; software, J.V. and R.S. (Rokas Statkevičius); validation, J.V. and R.M.; formal analysis, R.S. (Rokas Statkevičius); investigation, R.S. (Rokas Statkevičius); resources, J.V. and R.S. (Rūta Stanislauskienė); data curation, R.S. (Rokas Statkevičius) and J.V.; writing—original draft preparation, R.S. (Rokas Statkevičius) and J.V.; writing—review and editing, R.M. and R.S. (Rūta Stanislauskienė); visualization, R.S. (Rokas Statkevičius); supervision, R.M.; project administration, R.M. and J.V.; funding acquisition, R.M. All authors have read and agreed to the published version of the manuscript.

Funding: This project received funding from the European Social Fund (project No. 09.3.3.-LMT-K-712-22-0229) under a grant agreement with the Research Council of Lithuania (LMTLT).

Data Availability Statement: Data is contained within the article or the Supplementary Materials.

Acknowledgments: We thank J. Stankevičiūtė for her technical assistance and help with draft preparation and review.

Conflicts of Interest: The authors declare no conflict of interest.

References

1. Rossino, G.; Robescu, M.S.; Licastro, E.; Tedesco, C.; Martello, I.; Maffei, L.; Vincenti, G.; Bavaro, T.; Collina, S. Biocatalysis: A Smart and Green Tool for the Preparation of Chiral Drugs. *Chirality* **2022**, *34*, 1403–1418. [CrossRef]
2. Savile, C.K.; Janey, J.M.; Mundorff, E.C.; Moore, J.C.; Tam, S.; Jarvis, W.R.; Colbeck, J.C.; Krebber, A.; Fleitz, F.J.; Brands, J.; et al. Biocatalytic Asymmetric Synthesis of Chiral Amines from Ketones Applied to Sitagliptin Manufacture. *Science* **2010**, *329*, 305–309. [CrossRef] [PubMed]
3. Koszelewski, D.; Tauber, K.; Faber, K.; Kroutil, W. ω-Transaminases for the Synthesis of Non-Racemic α-Chiral Primary Amines. *Trends Biotechnol.* **2010**, *28*, 324–332. [CrossRef] [PubMed]
4. Malik, M.S.; Park, E.S.; Shin, J.S. Features and Technical Applications of ω-Transaminases. *Appl. Microbiol. Biotechnol.* **2012**, *94*, 1163–1171. [CrossRef]
5. Slabu, I.; Galman, J.L.; Lloyd, R.C.; Turner, N.J. Discovery, Engineering, and Synthetic Application of Transaminase Biocatalysts. *ACS Catal.* **2017**, *7*, 8263–8284. [CrossRef]
6. Truppo, M.D.; David Rozzell, J.; Turner, N.J. Efficient Production of Enantiomerically Pure Chiral Amines at Concentrations of 50 g/L Using Transaminases. *Org. Process Res. Dev.* **2010**, *14*, 234–237. [CrossRef]
7. Cooper, A.J.L.; Meister, A. An Appreciation of Professor Alexander E. Braunstein. The Discovery and Scope of Enzymatic Transamination. *Biochimie* **1989**, *71*, 387–404. [CrossRef]
8. Braunstein, A.E.; Kritzmann, M.G. Formation and Breakdown of Amino-Acids by Inter-Molecular Transfer of the Amino Group. *Nature* **1937**, *140*, 503–504. [CrossRef]
9. Hayashi, H. Pyridoxal Enzymes: Mechanistic Diversity and Uniformity. *J. Biochem.* **1995**, *118*, 463–473. [CrossRef]

10. Eliot, A.C.; Kirsch, J.F. Pyridoxal Phosphate Enzymes: Mechanistic, Structural, and Evolutionary Considerations. *Annu. Rev. Biochem.* **2004**, *73*, 383–415. [CrossRef]
11. Percudani, R.; Peracchi, A. The B6 Database: A Tool for the Description and Classification of Vitamin B6-Dependent Enzymatic Activities and of the Corresponding Protein Families. *BMC Bioinform.* **2009**, *10*, 273. [CrossRef] [PubMed]
12. Goto, M.; Miyahara, I.; Hayashi, H.; Kagamiyama, H.; Hirotsu, K. Crystal Structures of Branched-Chain Amino Acid Aminotransferase Complexed with Glutamate and Glutarate: True Reaction Intermediate and Double Substrate Recognition of the Enzyme. *Biochemistry* **2003**, *42*, 3725–3733. [CrossRef] [PubMed]
13. Hutson, S. Structure and Function of Branched Chain Aminotransferases. *Prog. Nucleic Acid Res. Mol. Biol.* **2001**, *70*, 175–206. [CrossRef] [PubMed]
14. Radkov, A.D.; Moe, L.A. Bacterial Synthesis of D-Amino Acids. *Appl. Microbiol. Biotechnol.* **2014**, *98*, 5363–5374. [CrossRef] [PubMed]
15. Höhne, M.; Schätzle, S.; Jochens, H.; Robins, K.; Bornscheuer, U.T. Rational Assignment of Key Motifs for Function Guides in Silico Enzyme Identification. *Nat. Chem. Biol.* **2010**, *6*, 807–813. [CrossRef]
16. Jiang, J.; Chen, X.; Zhang, D.; Wu, Q.; Zhu, D. Characterization of (*R*)-Selective Amine Transaminases Identified by in Silico Motif Sequence Blast. *Appl. Microbiol. Biotechnol.* **2015**, *99*, 2613–2621. [CrossRef] [PubMed]
17. Sayer, C.; Martinez-Torres, R.J.; Richter, N.; Isupov, M.N.; Hailes, H.C.; Littlechild, J.A.; Ward, J.M. The Substrate Specificity, Enantioselectivity and Structure of the (*R*)-Selective Amine: Pyruvate Transaminase from Nectria Haematococca. *FEBS J.* **2014**, *281*, 2240–2253. [CrossRef]
18. Pavkov-Keller, T.; Strohmeier, G.A.; Diepold, M.; Peeters, W.; Smeets, N.; Schürmann, M.; Gruber, K.; Schwab, H.; Steiner, K. Discovery and Structural Characterisation of New Fold Type IV-Transaminases Exemplify the Diversity of This Enzyme Fold. *Sci. Rep.* **2016**, *6*, 38183. [CrossRef]
19. Telzerow, A.; Paris, J.; Håkansson, M.; González-Sabín, J.; Ríos-Lombardía, N.; Gröger, H.; Morís, F.; Schürmann, M.; Schwab, H.; Steiner, K. Expanding the Toolbox of R-Selective Amine Transaminases by Identification and Characterization of New Members. *ChemBioChem* **2020**, *22*, 1232–1242. [CrossRef]
20. Fuchs, M.; Farnberger, J.E.; Kroutil, W. The Industrial Age of Biocatalytic Transamination. *Eur. J. Org. Chem.* **2015**, *2015*, 6965–6982. [CrossRef]
21. Kohls, H.; Steffen-Munsberg, F.; Höhne, M. Recent Achievements in Developing the Biocatalytic Toolbox for Chiral Amine Synthesis. *Curr. Opin. Chem. Biol.* **2014**, *19*, 180–192. [CrossRef] [PubMed]
22. Ghislieri, D.; Turner, N.J. Biocatalytic Approaches to the Synthesis of Enantiomerically Pure Chiral Amines. *Top. Catal.* **2014**, *57*, 284–300. [CrossRef]
23. Zilius, M.; Samuiloviene, A.; Stanislauskienė, R.; Broman, E.; Bonaglia, S.; Meškys, R.; Zaiko, A. Depicting Temporal, Functional, and Phylogenetic Patterns in Estuarine Diazotrophic Communities from Environmental DNA and RNA. *Microb. Ecol.* **2021**, *81*, 36–51. [CrossRef]
24. Bezsudnova, E.Y.; Dibrova, D.V.; Nikolaeva, A.Y.; Rakitina, T.V.; Popov, V.O. Identification of Branched-Chain Amino Acid Aminotransferases Active towards (*R*)-(+)-1-Phenylethylamine among PLP Fold Type IV Transaminases. *J. Biotechnol.* **2018**, *271*, 26–28. [CrossRef]
25. Zhai, L.; Xie, Z.; Tian, Q.; Guan, Z.; Cai, Y.; Liao, X. Structural and Functional Analysis of the Only Two Pyridoxal 5′-Phosphate-Dependent Fold Type IV Transaminases in Bacillus altitudinis W3. *Catalysts* **2020**, *10*, 1308. [CrossRef]
26. Bezsudnova, E.Y.; Boyko, K.M.; Nikolaeva, A.Y.; Zeifman, Y.S.; Rakitina, T.V.; Suplatov, D.A.; Popov, V.O. Biochemical and Structural Insights into PLP Fold Type IV Transaminase from Thermobaculum Terrenum. *Biochimie* **2019**, *158*, 130–138. [CrossRef]
27. Zeifman, Y.S.; Boyko, K.M.; Nikolaeva, A.Y.; Timofeev, V.I.; Rakitina, T.V.; Popov, V.O.; Bezsudnova, E.Y. Functional Characterization of PLP Fold Type IV Transaminase with a Mixed Type of Activity from Haliangium Ochraceum. *Biochim. Biophys. Acta-Proteins Proteom.* **2019**, *1867*, 575–585. [CrossRef] [PubMed]
28. Stekhanova, T.N.; Rakitin, A.L.; Mardanov, A.V.; Bezsudnova, E.Y.; Popov, V.O. A Novel Highly Thermostable Branched-Chain Amino Acid Aminotransferase from the Crenarchaeon Vulcanisaeta Moutnovskia. *Enzym. Microb. Technol.* **2017**, *96*, 127–134. [CrossRef]
29. Isupov, M.N.; Boyko, K.M.; Sutter, J.-M.; James, P.; Sayer, C.; Schmidt, M.; Schönheit, P.; Nikolaeva, A.Y.; Stekhanova, T.N.; Mardanov, A.V.; et al. Thermostable Branched-Chain Amino Acid Transaminases From the Archaea Geoglobus Acetivorans and Archaeoglobus Fulgidus: Biochemical and Structural Characterization. *Front. Bioeng. Biotechnol.* **2019**, *7*, 7. [CrossRef]
30. Bakunova, A.K.; Nikolaeva, A.Y.; Rakitina, T.V.; Isaikina, T.Y.; Khrenova, M.G.; Boyko, K.M.; Popov, V.O.; Bezsudnova, E.Y. The Uncommon Active Site of D-amino Acid Transaminase from Haliscomenobacter Hydrossis: Biochemical and Structural Insights into the New Enzyme. *Molecules* **2021**, *26*, 5053. [CrossRef]
31. Green, A.P.; Turner, N.J.; O'Reilly, E. Chiral Amine Synthesis Using ω-Transaminases: An Amine Donor That Displaces Equilibria and Enables High-Throughput Screening. *Angew. Chem. Int. Ed.* **2014**, *53*, 10714–10717. [CrossRef] [PubMed]
32. Baud, D.; Ladkau, N.; Moody, T.S.; Ward, J.M.; Hailes, H.C. A Rapid, Sensitive Colorimetric Assay for the High-Throughput Screening of Transaminases in Liquid or Solid-Phase. *Chem. Commun.* **2015**, *51*, 17225–17228. [CrossRef] [PubMed]
33. Tang, X.L.; Zhang, N.N.; Ye, G.Y.; Zheng, Y.G. Efficient Biosynthesis of (*R*)-3-Amino-1-Butanol by a Novel (*R*)-Selective Transaminase from *Actinobacteria* sp. *J. Biotechnol.* **2019**, *295*, 49–54. [CrossRef] [PubMed]

34. Mathew, S.; Deepankumar, K.; Shin, G.; Hong, E.Y.; Kim, B.G.; Chung, T.; Yun, H. Identification of Novel Thermostable ω-Transaminase and Its Application for Enzymatic Synthesis of Chiral Amines at High Temperature. *RSC Adv.* **2016**, *6*, 69257–69260. [CrossRef]

35. Wang, C.; Tang, K.; Dai, Y.; Jia, H.; Li, Y.; Gao, Z.; Wu, B. Identification, Characterization, and Site-Specific Mutagenesis of a Thermostable ω-Transaminase from Chloroflexi Bacterium. *ACS Omega* **2021**, *6*, 17058–17070. [CrossRef]

36. Inoue, K.; Kuramitsu, S.; Aki, K.; Watanabe, Y.; Takagi, T.; Nishigai, M.; Ikai, A.; Kagamiyama, H. Branched-Chain Amino Acid Aminotransferase of Escherichia Coli: Overproduction and Properties. *J. Biochem.* **1988**, *104*, 777–784. [CrossRef]

37. Tanizawa, K.; Masu, Y.; Asano, S.; Tanaka, H.; Soda, K. Thermostable D-Amino Acid Aminotransferase from a Thermophilic Bacillus Species: Purification, Characterization, and Active Site Sequence Determination. *J. Biol. Chem.* **1989**, *264*, 2445–2449. [CrossRef]

38. Mirdita, M.; Schütze, K.; Moriwaki, Y.; Heo, L.; Ovchinnikov, S.; Steinegger, M. ColabFold: Making Protein Folding Accessible to All. *Nat. Methods* **2022**, *19*, 679–682. [CrossRef]

39. Liu, Y.; Yang, X.; Gan, J.; Chen, S.; Xiao, Z.-X.; Cao, Y. CB-Dock2: Improved Protein-Ligand Blind Docking by Integrating Cavity Detection, Docking and Homologous Template Fitting. *Nucleic Acids Res.* **2022**, *50*, 159–164. [CrossRef]

40. Yang, X.; Liu, Y.; Gan, J.; Xiao, Z.-X.; Cao, Y. Problem Solving Protocol FitDock: Protein-Ligand Docking by Template Fitting. *Brief. Bioinform.* **2022**, *23*, 1–11. [CrossRef]

41. Bezsudnova, E.Y.; Popov, V.O.; Boyko, K.M. Structural Insight into the Substrate Specificity of PLP Fold Type IV Transaminases. *Appl. Microbiol. Biotechnol.* **2020**, *104*, 2343–2357. [CrossRef] [PubMed]

42. Bezsudnova, E.Y.; Nikolaeva, A.Y.; Bakunova, A.K.; Rakitina, T.V.; Suplatov, D.A.; Popov, V.O.; Boyko, K.M. Probing the Role of the Residues in the Active Site of the Transaminase from Thermobaculum Terrenum. *PLoS ONE* **2021**, *16*, e0255098. [CrossRef]

43. Kumar, S.; Stecher, G.; Li, M.; Knyaz, C.; Tamura, K. MEGA X: Molecular Evolutionary Genetics Analysis across Computing Platforms. *Mol. Biol. Evol.* **2018**, *35*, 1547. [CrossRef] [PubMed]

44. Sambrook, J. *Molecular Cloning: A Laboratory Manual/Joseph Sambrook, David W. Russell*; Russell 1957-; David, W., Cold Spring Harbor Laboratory, Eds.; Cold Spring Harbor Laboratory: Cold Spring Harbor, NY, USA, 2001; ISBN 0879695773.

45. Laemmli, U.K. Cleavage of Structural Proteins during the Assembly of the Head of Bacteriophage T4. *Nature* **1970**, *227*, 680–685. [CrossRef] [PubMed]

46. Bradford, M. A Rapid and Sensitive Method for the Quantitation of Microgram Quantities of Protein Utilizing the Principle of Protein-Dye Binding. *Anal. Biochem.* **1976**, *72*, 248–254. [CrossRef] [PubMed]

 *catalysts*

Article

Identification of Cytochrome P450 Enzymes Responsible for Oxidative Metabolism of Synthetic Cannabinoid (1-Hexyl-1*H*-Indol-3-yl)-1-naphthalenyl-methanone (JWH-019)

Ngoc Tran [1], William E. Fantegrossi [2], Keith R. McCain [3], Xinwen Wang [4] and Ryoichi Fujiwara [1,4,*]

[1] Department of Pharmaceutical Sciences, College of Pharmacy, University of Arkansas for Medical Sciences, Little Rock, AR 72205, USA
[2] Department of Pharmacology & Toxicology, College of Medicine, University of Arkansas for Medical Sciences, Little Rock, AR 72205, USA
[3] Department of Pharmacy Practice, College of Pharmacy, University of Arkansas for Medical Sciences, Little Rock, AR 72205, USA
[4] Department of Pharmaceutical Sciences, College of Pharmacy, Northeast Ohio Medical University, Rootstown, OH 44272, USA
* Correspondence: rfujiwara@neomed.edu; Tel.:+1-330-325-6122; Fax: +1-501-325-5936

Abstract: (1-Hexyl-1*H*-indol-3-yl)-1-naphthalenyl-methanone (JWH-019) is one of the second-generation synthetic cannabinoids which as a group have been associated with severe adverse reactions in humans. Although metabolic activation can be involved in the mechanism of action, the metabolic pathway of JWH-019 has not been fully investigated. In the present study, we aimed to identify the enzymes involved in the metabolism of JWH-019. JWH-019 was incubated with human liver microsomes (HLMs) and recombinant cytochrome P450s (P450s or CYPs). An animal study was also conducted to determine the contribution of the metabolic reaction to the onset of action. Using an ultra-performance liquid chromatography system connected to a single-quadrupole mass detector, we identified 6-OH JWH-019 as the main oxidative metabolite in HLMs supplemented with NADPH. JWH-019 was extensively metabolized to 6-OH JWH-019 in HLMs with the K_M and V_{max} values of 31.5 µM and 432.0 pmol/min/mg. The relative activity factor method estimated that CYP1A2 is the primary contributor to the metabolic reaction in the human liver. The animal study revealed that JWH-019 had a slower onset of action compared to natural and other synthetic cannabinoids. CYP1A2 mediates the metabolic activation of JWH-019, contributing to the slower onset of its pharmacological action.

Keywords: synthetic cannabinoids; xenobiotic metabolism; cytochrome P450; oxidation; metabolic activation

Citation: Tran, N.; Fantegrossi, W.E.; McCain, K.R.; Wang, X.; Fujiwara, R. Identification of Cytochrome P450 Enzymes Responsible for Oxidative Metabolism of Synthetic Cannabinoid (1-Hexyl-1*H*-Indol-3-yl)-1-naphthalenyl-methanone (JWH-019). *Catalysts* **2023**, *13*, 1008. https://doi.org/10.3390/catal13061008

Academic Editor: Eun Yeol Lee

Received: 14 April 2023
Revised: 8 June 2023
Accepted: 13 June 2023
Published: 15 June 2023

1. Introduction

Marijuana, *Cannabis sativa*, is one of the most pervasive drugs in the world due to its popularity in both medicinal and recreational use, with a significant increase in consumption being reported in the United States during the 21st century [1]. Synthetic cannabinoids (SCBs) are a large and chemically diverse group of substances. SCBs were initially created as investigatory cannabinoid receptor agonists, which activate central cannabinoid (CB1) and peripheral cannabinoid (CB2) receptors, the same receptors on which the naturally occurring cannabinoids found in marijuana exert their effects [2]. While researchers study the therapeutic potential of SCBs, the use of SCBs has deviated from its original intent, leading to abuse by the general public due to their initial uncontrolled legal status, low cost, ease of availability, difficulty of detection, and pronounced psychoactive effects [3–5]. SCBs are known as being natural and safe to consumers; however, they are neither natural nor safe, causing serious adverse reactions, including seizures, nephrotoxicity, and, in some cases, death [6–8]. With growing awareness and popularity, the number of SC users reporting adverse events dramatically increased over the last decade [9,10]. Case reports and user experiences indicate there is extreme variability in the pattern, duration, and severity of individual clinical courses following SC use. The continued limited understanding of the basic pharmacokinetic and dynamic specifics of individual SC compounds significantly complicates the clinical management of patients seeking medical attention. Care for these patients is largely symptomatic and supportive as, to this point, no specific antidote or therapy is available that targets SCBs in intoxicated users.

The substance (1-Hexyl-1*H*-indol-3-yl)-1-naphthalenyl-methanone (JWH-019) is a member of the second generation of SCBs with the molecular formula $C_{25}H_{25}NO$ and a molecular mass of 355.5 [11]. It is one of the cannabinoid receptor agonists from the napthoylindole family that were synthesized during the 1990s and has seen widespread human recreational use in the U.S. since 2010 [12]. This synthetic compound is more potent and addictive than the natural cannabinoids and is associated with concerns of causing severe complications for human health [13]. While significant limitations in clinical testing have hampered clear delineation and complete characterization of toxicity among SCBs, the medical literature does contain several reports of analytically confirmed involvement of JWH-019 in patients experiencing adverse medical consequence. In 2013, there were 29 subjects reporting to the ER with restlessness, hallucination, vertigo, and anxiousness after ingesting SCBs including JWH-019 [14]. In a 2014 report, a twenty-year-old male is described as having driving skill impairment, vestibular disorder, enlarged pupils, and blunt mood during a traffic stop. In this case a toxicology blood sample was taken after 80 min and the result was positive for four SCs including JWH-019 (1.7 ng/mL) [15]. In 2015, a fifty-year-old man is described as losing consciousness while working and died after being transferred to the hospital. In his initial toxicology findings there was no detection of common psychoactive substances or drug of abuse but, three years later, using the same sample on gas chromatography–mass spectrometry urinalyses the reanalysis found the presence of JWH-019 (278 ng/mL) [16]. Additionally, calls to the United States Poison Control Centers involving exposure to SCBs increased from 14 in 2009 to 6968 by year end of 2011. As the significant harms from SCBs became apparent, an initial attempt in the United States to curb the open availability associated with the sale of many SCBs in the early portion of the last decade was the passage of the Synthetic Drug Abuse Prevention Act in June of 2012. This legislation categorized several chemical structure classes of cannabinoids as schedule I substances. Fifteen SCBs were named in the Act and JWH-019 was included among them [17].

Xenobiotics are transformed in the body by phase I and II drug-metabolizing enzymes (DMEs), such as cytochrome P450s (CYPs) and UDP-glucuronosyltransferases (UGTs), respectively [18,19]. CYPs and UGTs are the most important enzymes that catalyze oxidation and glucuronidation of xenobiotics; therefore, it was hypothesized that CYPs and UGTs are the main enzymes contributing to the metabolism of SCBs. They metabolize lipophilic xenobiotics by introducing a hydrophilic functional group [18,19]. The metabolites formed by DMEs are more readily excreted from the body due to increased water solubility. These detoxification enzymes are abundantly expressed in the liver and extrahepatic tissues, playing an important role in systemic and local metabolism of xenobiotics [19–21]. Oxidized and glucuronidated SCB metabolites have been identified in the blood and urine of SCB users [22,23], suggesting an involvement of CYPs and UGTs in the clearance of SCBs.

The JWH-019 homolog JWH-018 was identified as one of the earliest compounds detected early in the development of increased SCB recreational use [24]. As such, there has been significantly more investigation into this compound and there are many studies on its metabolic process and enzymes involved in its metabolism. JWH-018 shares a similar structure to JWH-019, with the only difference being the length of the alkyl chain. Currently, the process of metabolism of JWH-019 remains unclear. However, based on the similarity in the structure [25], it was hypothesized that JWH-019 would follow the same metabolic pathway as JWH-018 without supporting evidence. Based on the hypothesis, JWH-019 could undergo oxidation to form hydroxyindole (5-hydroxyindole JWH-019), *N*-hexanoic acid (JWH-019 COOH) and two monohydroxylated at the *N*-alkyl chain (5-OH JWH-019 and 6-OH JWH-019) metabolites (Figure 1). According to a recent study that analyzed urine samples from drug users that were positive for JWH-019, three out of four metabolites, 5-OH JWH-019, 6-OH JWH-019, and JWH-019 COOH, have been confirmed [26]. However, it is still unclear which is the primary metabolite or which drug metabolizing enzyme is responsible for the oxidation reaction.

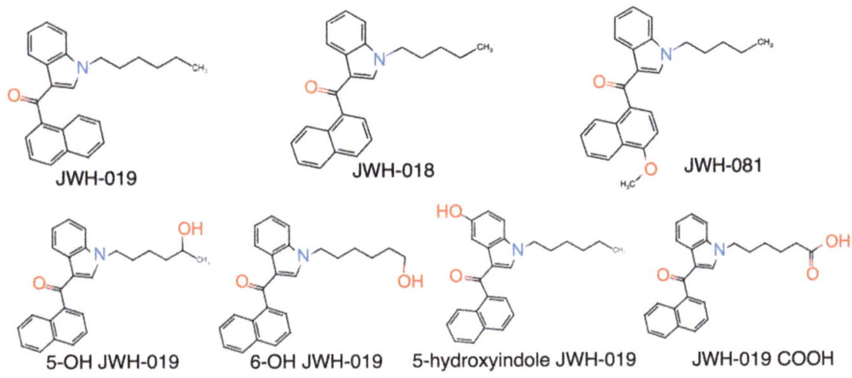

Figure 1. Structures of JWH-019 and its potential metabolites. Chemical structures of SCB JWH-019, -018, and -081 (**top**) and predicted metabolites of JWH-019 (**bottom**), 5-OH JWH-019, 6-OH JWH-019, 5-hydroxyindole JWH-019, and JWH-019 COOH, are shown.

The aim of this study is to identify the main metabolite of JWH-019 and the P450 isoforms that are responsible for the metabolite formation in vitro. We also conducted an animal study to determine the contribution of the metabolic reaction to the onset of action.

2. Results

2.1. Development of a UPLC Method for JWH-019 and Its Metabolites

Currently, four oxidative metabolites of JWH-019 have been suggested, which are 5-OH JWH-019, 6-OH JWH-019, JWH-019 COOH, and 5-hydroxyindole JWH-019. Out of the four postulated metabolites, three, 5-OH JWH-019, 6-OH JWH-019, and JWH-019 COOH, have been previously detected in human urine samples [26]. However, the study was not quantitative. Furthermore, the enzymes that mediate the metabolic reactions of JWH-019 have not been determined to date. The liver is the most important tissue for the metabolism of xenobiotics. Among a number of hepatic detoxification enzymes, CYP is the most abundant and functional enzyme responsible for oxidative metabolism. Therefore, the first aim of this investigation was to identify metabolites of JWH-019 produced in HLMs where CYPs are located.

As a preliminary study, we tested the detection of JWH-019 and its metabolites with positive and negative modes of ionization using incubation mixtures. While the negative mode did not produce any significant peak of any compounds, the positive mode detected JWH-019 and metabolites. The data indicated that the highest sensitivity would be obtained at $m/z = 356.47$ (potentially JWH-019), 386.45 (potentially JWH-019 COOH), and 372.47 (potentially 5-hydroxyindole JWH-019, 5-OH JWH-019, and 6-OH JWH-019). Using the chemical standard, it was confirmed that the peak of $m/z = 356.47$ was JWH-019 with the retention time of 7.6 min. Similarly, we confirmed that a peak at the retention time of 5.5 min was 6-OH JWH-019. The retention times of hydroxyindole JWH-019 and 5-OH JWH-019 were 6.5 min and 6.6 min, respectively (Figure 2). Although we believe that JWH-019 COOH will be detected at $m/z = 386.45$, we were not able to determine the retention time of JWH-019 COOH due to the unavailability of the standard molecule on the market.

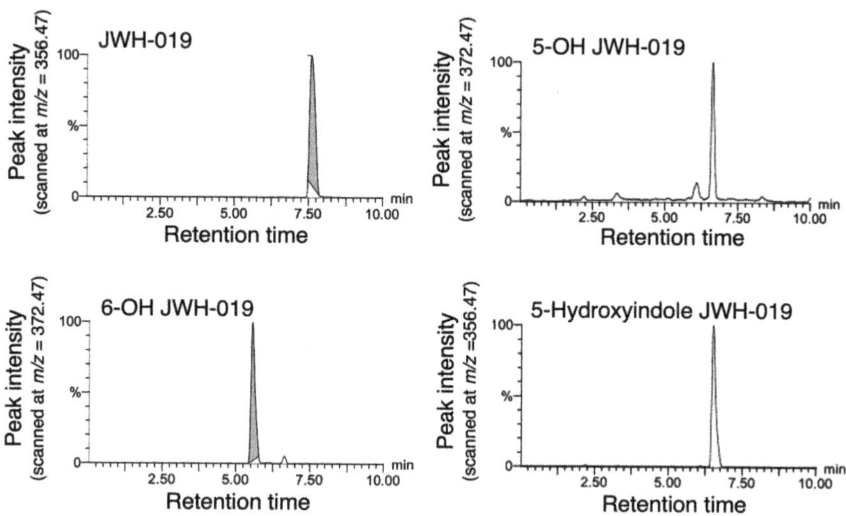

Figure 2. Representative chromatogram of JWH-019 and its main metabolites. Standard solution of JWH-019 and its metabolites was injected to the UPLC-MS system. JWH-019 was detected at m/z = 356.47. 5-OH JWH-019, 6-OH JWH-019, and 5-hydroxyindole JWH-019 were detected at m/z = 372.47. The retention times were 7.6 min for JWH-019, 5.5 min for 6-OH JWH-019, 6.5 min for hydroxyindole JWH-019, and 6.6 min for 5-OH JWH-019, respectively.

To investigate which form of metabolites can be produced by CYPs, JWH-019 (400 μM) was incubated with HLMs for 15 min in the presence and absence of the NADPH regenerating system. A significant peak was observed at 7.6 min regardless of the NADPH regenerating system when the incubated mixture was analyzed at m/z = 356.47, indicating that the substrate was not depleted at all with the 15 min incubation with HLMs (Figure 3A,B). A unique peak was observed at 5.5 min only in a sample that was incubated with NADPH when it was analyzed at m/z = 372.47 (Figure 3C,D), indicating that CYPs are responsible for producing 6-OH JWH-019. The peak size was incubation time dependent, supporting the involvement of enzyme reaction in the production of 6-OH JWH-019 in HLMs. In the same detection setting (m/z = 372.47), another peak was detected at 6.6 min in both incubated mixtures. The peak size, which was much smaller than that of 6-OH JWH-019, was not incubation time dependent, suggesting that a small amount of 5-OH JWH-019 was non-enzymatically produced in HLM regardless of the NADPH regenerating system. There was no peak at 6.5 min when incubation mixtures were analyzed at m/z = 372.47. In addition, there was no significant peak when samples were analyzed at m/z = 386.45 (Figure 3E,F). These data indicate that hydroxyindole JWH-019 or JWH-019 COOH was not produced, or the produced amount was below the limits of detection, in HLMs even with the presence of NADPH.

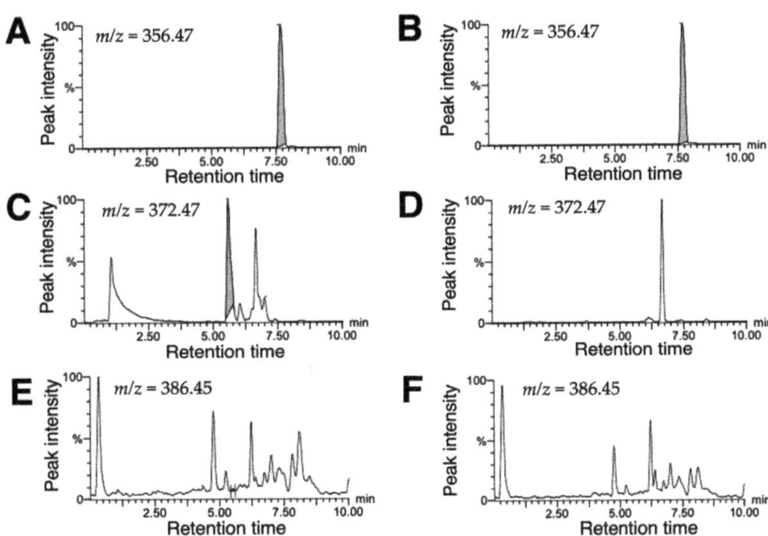

Figure 3. Incubation of JWH-019 in HLMs with and without NADPH. After the incubation of JWH-019 with (**A**) and without (**B**) NADPH, the sample was separated and monitored at m/z = 356.47 to detect the presence of JWH-019. The incubated sample was also monitored at m/z = 372.47 and 386.45 to detect the presence of JWH-019 mono-hydroxy and carboxylated metabolites. A unique peak (6-OH JWH-019) was detected when the parent compound was incubated with NADPH (**C**) but not when incubated without NADPH (**D**). JWH-019 COOH was not detected when JWH-019 was incubated with (**E**) or without (**F**) NADPH.

2.2. Kinetic Analysis of 6-OH JWH-019 Formation in HLMs

After the initial screening, it was determined that HLMs produced 6-OH JWH-019 as a main metabolite of JWH-019 when supplemented with NADPH. To further investigate the enzyme–substrate binding affinity and reaction rate in HLMs, we conducted the enzyme assay at a wide range of substrate concentrations (1–400 μM). The metabolic rate-substrate concentration plots showed a Michaelis–Menten curve (Figure 4). By fitting the data into the Michaelis–Menten equation, the K_M and V_{max} values were estimated to be 31.5 ± 3.3 μM and 432 ± 40 pmol/min/mg protein.

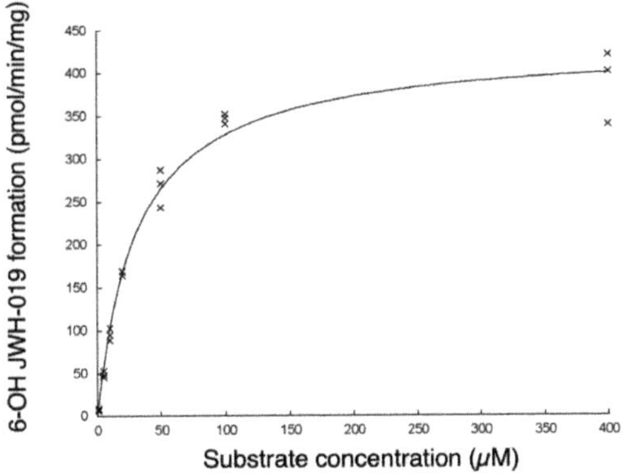

Figure 4. Steady-state kinetic analysis in HLMs. Kinetic constant of 6-OH JWH-019 formation in HLMs was estimated. The kinetic profile followed classical Michaelis–Menten kinetics with a V_{max} value of 432 ± 40 pmol/min/mg and a K_M value of 31.5 ± 3.3 μM.

2.3. Identification of JWH-019-Oxidating Enzymes

CYPs are super-family enzymes. CYP1A2, -2A6, -2B6, -2C8, -2C9, -2C19, -2D6, -2E1, -2J2, -3A4, and -3A5 have been reported as main CYP isoforms expressed and functional in the human liver [27,28]. To identify the CYP isoforms responsible for the formation of 6-OH JWH-019 in HLMs, we used recombinant CYPs and assessed their respective abilities to metabolize JWH-019 using a substrate concentration of 400 μM. Of all the recombinant proteins that were screened, CYP2J2 showed the greatest activity for the formation of 6-OH JWH-019 at a reaction rate of 8.0 $\pm$ 0.3 nmol/min/nmol P450 (Table 1). CYP1A2, -2C19, -2D6, and -3A5 showed moderate activities with their reaction rates of 2.4 $\pm$ 0.1, 4.2 $\pm$ 0.04, 1.6 $\pm$ 0.1, and 2.0 $\pm$ 0.1 nmol/min/nmol P450, respectively. While CYP2B6, -2C8, -2C9, and -3A4 showed slight activities of the 6-OH JWN-019 formation, CYP2A6 or CYP2E1 barely showed the activity.

Table 1. Estimated contribution of CYPs to the 6-OH JWH-019 formation in HLMs.

	Activity (pmol/min/nmol P450)	Abundance (%)	Contribution (%)
CYP1A2	2359 $\pm$ 68	12.1	33.0 $\pm$ 0.9
CYP2A6	25 $\pm$ 7	8.3	0.25 $\pm$ 0.07
CYP2B6	395 $\pm$ 48	2.5	1.14 $\pm$ 0.14
CYP2C8	1238 $\pm$ 17	5.5	7.86 $\pm$ 0.11
CYP2C9	559 $\pm$ 17	16.8	10.8 $\pm$ 0.3
CYP2C19	4195 $\pm$ 43	3.2	15.5 $\pm$ 0.2
CYP2D6	1637 $\pm$ 65	1.8	3.40 $\pm$ 0.14
CYP2E1	2.8 $\pm$ 0.6	14.1	0.05 $\pm$ 0.01
CYP2J2	8000 $\pm$ 334	1.16	10.7 $\pm$ 0.4
CYP3A4	366 $\pm$ 8	30.5	12.9 $\pm$ 0.3
CYP3A5	2014 $\pm$ 104	1.9	4.41 $\pm$ 0.23

Each CYP isoform is differently expressed in the human liver. While CYP3A4 has been known as the isoform that is dominantly expressed in the liver, other isoforms such as CYP2J2 are hardly expressed in hepatic tissue [27,28]. To quantitatively determine the contribution of each CYP isoform to the formation of 6-OH JWH-019 in HLMs, we employed a relative activity factor (RAF) method. The contribution rate (%) was estimated by considering the enzyme activity of the recombinant enzyme and the relative expression level of each CYP isoform in the human liver. Even though CYP2J2 showed the highest activity in the earlier investigation, it was estimated to be the fifth contributor to the overall metabolic reaction in the liver (Table 1). The enzyme that contributes to the metabolism the most was estimated to be CYP1A2 with the contribution rate of 33.6%. CYP2C19, -2C9, and -3A4 were the second, third, and fourth contributors to the formation of 6-OH JWH-019, respectively, with their contribution rates ranging from 10.9 to 15.8%. While CYP2B6, -2C8, -2D6, and -3A5 made a slight contribution (1.1–6.1%), CYP2A6 and CYP2E1 made almost no contribution due to their lowest enzyme activity.

2.4. Pharmacological Activity of JWH-019

Metabolic reaction usually results in reduction of the pharmacological activity of the substances. However, in rare cases, metabolic activation occurs where metabolites show stronger pharmacological activity, which is evidenced by an enhanced pharmacological activity with increased pretreatment time.

Sprague Dawley rats were administered THC (0.3–3.0 mg/kg, I.P., N = 8), JWH-019 (1.0–30 mg/kg, I.P., N = 8), or JWH-081 (1.0–30 mg/kg, I.P., N = 8). All injections were made 30 min prior to sessions. THC showed the strongest activity with its ED_{50} values of 0.85 mg/kg (Figure 5). JWH-081 fully substituted for the discriminative stimulus effects produced by 3 mg/kg of THC (ED_{50} = 4.89 mg/kg). In contrast, JWH-019 only partially substituted for THC when tested 30 min after injection (E_{max} = 46% at 3 mg/kg). We also tested the pharmacological activity of JWH-019 with a pretreatment time of 60 min. This time, interestingly, JWH-019 fully substituted for the discriminative stimulus effects produced by 3 mg/kg of THC (E_{max} = 97% and ED_{50} = 2.7 mg/kg). Prolonged pretreatment time enhanced the pharmacological activity of JWH-019, indicating that an active metabolite was produced in the JWH-019-treated rats.

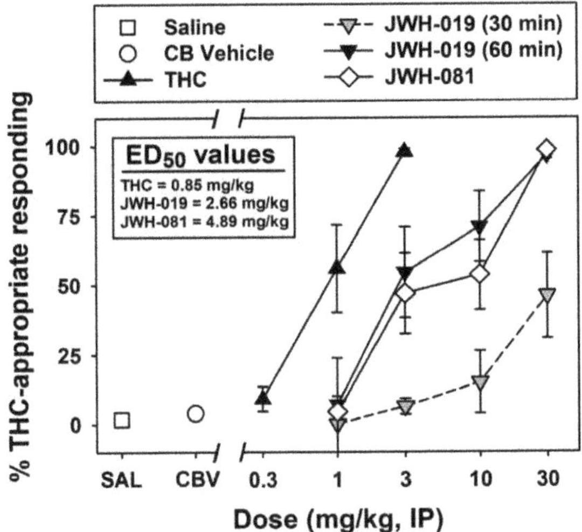

Figure 5. Dose–effect curves for substitution of JWH-019 (1.0–30 mg/kg, I.P., N = 8), THC (0.3–3.0 mg/kg, I.P., N = 8), JWH-081 (1.0–30.0 mg/kg, I.P., N = 8), and cannabinoid vehicle (0.9% saline containing 5% ethanol and 5% emulphor, I.P., N = 8) for THC (3 mg/kg, I.P.). All injections were made 30 min prior to sessions, except for JWH-019, which was also tested following 60 min pretreatment (black inverted triangles). JWH-019 dose dependently produced THC-like discriminative-stimulus effects as reflected in dose-dependent increases in % drug responding when administered 30 or 60 min before sessions, but was more potent and effective when injected 60 min before testing. Vertical lines represent S.E.M. unless the S.E.M. is smaller than the size of the symbol.

3. Materials and Methods

3.1. Materials

Cannabinoids used in the study, JWH-019, JWH-081, and Δ^9-THC, were obtained from the NIDA Drug Supply Program (Research Triangle Institute, Durham, NC, USA). Oxidative metabolites of JWH-019 were obtained from Cayman Chemical (Ann Arbor, MI). Pooled human liver microsomes (HLMs), recombinant P450s, and NADPH Regenerating System Solutions A and B were purchased from Corning (Corning, NY, USA). Recombinant P450 systems used in the study included the expression of cytochrome b5. Water, methanol, and acetic acid at LC/MS grade were purchased from Thermo Fisher Scientific (Hampton, NH, USA). All other chemicals and reagents used for this study were of at least reagent grade and were purchased from Sigma-Aldrich (St. Louis, MO, USA) or Thermo Fisher Scientific, unless specified otherwise.

3.2. Separation and Detection of JWH-019 and Its Metabolites

The parent compound and its oxidized metabolites were separated and identified by the ACQUITY UPLC System with a QDa single-quadrupole mass detector (Waters, Milford, MA, USA). The mobile phases were 0.1% acetic acid (A) and 100% methanol (B), and the flow rate was 0.5 mL/min with an elution gradient of 100% A (0–0.1 min) and a linear gradient from 100% A to 10% A–90% B (0.1–8.5 min). An ACQUITY UPLC BEH C18 column (130 Å, 1.7 μm, 2.1 mm × 50 mm) was re-equilibrated at initial conditions for 1.5 min between runs. The elution was monitored using the MassLynx software (Waters). Injected standards and samples (10 μL) were analyzed with the QDa detector at a positive mode. The optimized parameters of the QDa interface were: source temperature, 120 °C; vaporizer temperature, 600 °C; drying gas (nitrogen) temperature, 600 °C; cone voltage, 15 V; capillary voltage, 800 mV; Gain 1. JWH-019 and its metabolites 5-hydroxyindole JWH-019, JWH-019 COOH, 5-OH JWH-019, and 6-OH JWH-019 were monitored at their $m/z = 356.47$ (JWH-019), 386.45 (JWH-019 COOH), and 372.47 (5-hydroxyindole JWH-019, 5-OH JWH-019, and 6-OH JWH-019). The retention times were 7.6 min for JWH-019, 5.5 min for 6-OH JWH-019, 6.5 min for hydroxyindole JWH-019, and 6.6 min for 5-OH JWH-019, respectively.

3.3. Metabolism of JWH-019 in HLMs and Recombinant P450s

The metabolism of JWH-019 was examined by analyzing the activity of HLMs and recombinant human P450 enzymes (CYP1A2, -2A6, -2B6, -2C8, -2C9, -2C19, -2D6, -2E1, -2J2, -3A4, and -3A5). The substrate (final concentration 1–400 μM) was added to each tube (50 μL) along with protein, water, and buffer (final concentration 0.1 M KPO$_4$, pH 7.4); the reactions were started with the addition of an NADPH-regenerating system (1 mM NADP$^+$, 3 mM glucose 6-phosphate, 3 mM MgCl$_2$; 1 U/mL glucose 6-phosphate dehydrogenase) to ensure the saturation of NADPH, thus enabling cytochrome P450-mediated reactions. As described previously in Jones et al. (2020) [29], we used up to 50 μg of enzyme sources for each incubation tube. We confirmed that we observed the enzyme activity in a protein amount-dependent manner. Controls omitting the substrate, protein, and NADPH were included with each assay. Reactions were incubated at 37 °C for 0–60 min and were terminated by the addition of three-time volume of methanol. Protein and other particulates were precipitated by centrifugation at 12,000× g for 5 min, and the supernatant was subsequently analyzed by the UPLC-QDa system as described above. All reactions were performed in triplicate.

3.4. Kinetic Assay and Data Analysis

Incubation conditions were optimized for time and protein concentration, which is typically 15–30 min of incubation time and 1.0 mg/mL of protein concentration. All reactions were performed within the linear range of metabolite formation. Other than substrate concentrations and incubation times, the reaction mixture composition and analytical methods were identical to those described for the above screening assays. Incubations were carried out with HLMs in the presence of various concentrations of the substrate (1–400 μM) for 15 min at 37 °C.

Kinetic parameters were estimated from fitted curves using a program (http://www.ic50.tk, accessed on 1 July 2021) designed for non-linear regression analysis. The Michaelis–Menten equation,

$$V = V_{\max} \cdot [S]/(K_M + [S]),$$

was used to calculate the K_M and $V_{\max}$ values, where V is the velocity of the reaction, S is the substrate concentration, K_M is the Michaelis–Menten constant, and $V_{\max}$ is the maximum velocity. Kinetic constants are reported as the mean ± SD of triplicate experiments. Quantification of 6-OH JWH-019 was performed by comparing the peak height to that of the authentic standard.

3.5. Assessment of the Pharmacological Activity

All procedures were identical, with some modifications, to those in previous methods [30,31]. All procedures were carried out in accordance with the Guide for Care and Use of Laboratory Animals as adopted and promulgated by the National Institutes of Health (NIH). The experimental protocol was approved by the Institutional Animal Care and Use Committee (IACUC) at the University of Arkansas for Medical Sciences. The detailed method for the animal study can be found as a supplemental material.

The percentage of drug-appropriate responding was calculated by dividing the total number of responses on the drug-appropriate lever by the total number of responses. Rate of responding was calculated by dividing the total number of responses by the session time (excluding the post-response TO periods) and expressed as percentage of saline (control) rate of responding. These data were shown as mean values (±S.E.M.) for groups of subjects at each drug dose. If subjects were tested repeatedly under a single condition, the data were pooled for an individual subject and then averaged into a group mean.

Full substitution was operationally defined as: (1) 85% or more of the group responses on the drug-appropriate lever and (2) the group mean was significantly different from that of saline. For all analyses, $p < 0.05$ was considered statistically significant. The statistical significance was assessed by appropriate one-way repeated-measures analyses of variance (ANOVA). A post hoc Bonferroni t test was used for all pairwise comparisons. Dose–effect curves (DECs) for percent drug-appropriate responding were analyzed using standard linear regression techniques, from which ED50 values (50% effective dose for the substitution) with 95% confidence limit (95% Cl) values [32] will be calculated. Only points on the linear part of the ascending portions of the DECs were used. For test substances that did not fully substitute for the training drug, standard ANOVA was conducted to determine whether drug substitution differed from saline controls. The maximal substitution (E_{max}) was also compared.

4. Discussion

While a few SCB based drug products (dronabinol and nabilone) have obtained U.S. Food and Drug Administration approval for human medicinal use, the number of unapproved SCBs used recreationally has rapidly expanded in popularity, due to their low cost, ease of availability, and difficulty of detection [4,5,33]. Importantly, a significant portion of patients admitted to emergency rooms after use of SCBs are adolescents [34]. However, currently there are limited data to explain the high risk of SCB toxicity in adolescents. Moreover, there is no emergency therapy that can be used to target SCBs in intoxicated users. In order to develop therapeutic treatments for detoxification of SCBs, it is essential to determine whether the metabolic process results in detoxification of SCBs or production of their active metabolite., Additionally, the understanding of SCBs' toxicokinetic properties is unknown due to the uncontrolled and rapid emergence of SCBs, These knowledge deficits make investigations into their metabolic profiles vital for developing improved risk assessment and prognosis of clinical course following recreational use misadventure.

A previous study extensively investigated the metabolic pathway of JWH-018, finding that at least 13 metabolites were formed in NADPH-supplemented HLMs [35]. One of the major metabolic reactions of JWH-018 was mono-hydroxylation at the naphthalene ring system, the indole moiety, or the alkyl side chain. Carboxylation of the alkyl chain produced carboxylated JWH-018 in HLMs. Meanwhile, in a case of JWH-019, the chemical structure of which is almost identical to that of JWH-018 except for the length of the alkyl chain, 6-OH JWH-019 was identified as the main oxidative metabolite in HLMs. This observation might appear inconsistent with the earlier investigation that detected 5-OH JWH-019 and JWH-019 COOH in urine samples from drug users that were positive for JWH-019 [26]. One explanation for this is a potential involvement of cytosolic enzymes or extrahepatic tissues in the production of 5-OH JWH-019 and JWH-019 COOH. In fact, DMEs including CYPs and carboxylases are widely expressed and functional in various extrahepatic tissues [36,37].

One of the most significant findings in our study was the identification of CYP1A2 as the major contributor to the mono-hydroxylation of JWH-019. The expression of CYP1A2 is highly regulated by an aryl hydrocarbon receptor (AhR), which is a ligand-activated transcription factor. Environmental contaminants such as dioxin, polycyclic aromatic hydrocarbons, and a component of ambient air pollution and cigarette smoke activate the function of AhR [38], causing induction of the CYP1A2 expression. Moreover, CYP1A2 is a polymorphic enzyme. While many of the gene polymorphisms on human *CYP1A2* result in reducing its enzyme activity, some genetic mutation can induce the activity [39]. Although increased metabolism might seem helpful, that is not the case if the metabolic reaction produces an active metabolite. As evidenced by our investigation, 6-OH JWH-019 seems to be an active metabolite (Figure 5). Therefore, genetic mutation and/or exposure to AhR-activating environmental contaminants might be an important contributing factor associated with development of JWH-019 toxic reactions in some individuals.

Another interesting observation in our study was the great ability of CYP2J2 to form 6-OH JWH-019 (Figure 4). Due to the lowest expression in the human liver, its contribution to the overall metabolism of JWH-019 in the body was lower (Table 1). However, CYP2J2 is primarily expressed in the cardiovascular system, especially cardiomyocytes and endothelial cells [40,41]. Due to its abundance in the heart, along with the ability to form the active metabolite, CYP2J2 might be involved in an increased susceptibility to cardiac toxicity of JWH-019.

In conclusion, we investigated the CYP-mediated oxidation metabolism of JWH-019 in NADPH-supplemented HLMs. We identified 6-OH JWH-019 as the main oxidative metabolite in contrast to the case of JWH-018, which is metabolized to at least 13 oxidative metabolites in HLMs. The RAF method concluded that CYP1A2 was the highest contributor to the metabolism of JWH-019 in the liver. Prolonged metabolic process resulted in increased pharmacological activity of JWH-019 in our animal study, suggesting that JWH-019 is metabolized to an active metabolite, which is most likely 6-OH JWH-019. Smoking and genetic polymorphisms in the *CYP1A2* gene might increase the risk of JWH-019-inducing toxic reactions in individuals.

Author Contributions: Conceptualization, W.E.F. and R.F.; methodology, N.T., W.E.F. and R.F.; validation, N.T., W.E.F. and R.F.; formal analysis, N.T.; investigation, N.T., W.E.F. and R.F.; resources, W.E.F.; data curation, N.T.; writing—original draft preparation, N.T., W.E.F. and R.F.; writing—review and editing, N.T., W.E.F., K.R.M., X.W. and R.F.; visualization, N.T.; supervision, W.E.F. and R.F.; project administration, N.T., W.E.F. and R.F.; funding acquisition, W.E.F. and R.F. All authors have read and agreed to the published version of the manuscript.

Funding: This work was supported in part by the National Institutes of Health National Institute on Drug Abuse [Grants DA039143_W.E.F.]. The study was also supported by the University of Arkansas for Medical Sciences (UAMS) COBRE program, with a grant from the National Institute of General Medical Sciences (NIGMS) [2P20GM109005_R.F.].

Data Availability Statement: All data supporting the reported results can be found in the published article.

Acknowledgments: We thank (UAMS) Analytical and Bioanalytical Core (CORE B) for supporting quantification of small molecules.

Conflicts of Interest: The authors declare that the research was conducted in the absence of any commercial or financial relationships that could be construed as a potential conflict of interest.

References

1. Hasin, D.S.; Saha, T.D.; Kerridge, B.T.; Goldstein, R.B.; Chou, S.P.; Zhang, H.; Jung, J.; Pickering, R.P.; Ruan, W.J.; Smith, S.M.; et al. Prevalence of marijuana use disorders in the United States between 2001–2002 and 2012–2013. *JAMA Psychiatry* **2015**, *72*, 1235–1242. [CrossRef]
2. Howlett, A.C.; Abood, M.E. CB₁ and CB₂ Receptor Pharmacology. *Adv. Pharmacol.* **2017**, *80*, 169–206.
3. Schneir, A.B.; Cullen, J.; Ly, B.T. "Spice" girls: Synthetic cannabinoid intoxication. *J. Emerg. Med.* **2011**, *40*, 296–299. [CrossRef]
4. Palamar, J.J.; Acosta, P. Synthetic cannabinoid use in a nationally representative sample of US high school seniors. *Drug. Alcohol. Depend.* **2015**, *149*, 194–202. [CrossRef] [PubMed]
5. Winstock, A.R.; Barratt, M.J. Synthetic cannabis: A comparison of patterns of use and effect profile with natural cannabis in a large global sample. *Drug. Alcohol. Depend.* **2013**, *131*, 106–111. [CrossRef]
6. Lapoint, J.; James, L.P.; Moran, C.L.; Nelson, L.S.; Hoffman, R.S.; Moran, J.H. Severe toxicity following synthetic cannabinoid ingestion. *Clin. Toxicol.* **2011**, *49*, 760–764. [CrossRef]
7. Law, R.; Schier, J.; Martin, C.; Chang, A.; Wolkin, A. Increase in reported adverse health effects related to synthetic cannabinoid use—United States, January–May 2015. *MMWR Morb. Mortal. Wkly. Rep.* **2015**, *64*, 618.
8. Kasper, A.M.; Ridpath, A.D.; Gerona, R.R.; Cox, R.; Galli, R.; Kyle, P.B.; Parker, C.; Arnold, J.K.; Chatham-Stephens, K.; Morrison, M.A.; et al. Severe illness associated with reported use of synthetic cannabinoids: A public health investigation (Mississippi, 2015). *Clin. Toxicol.* **2019**, *57*, 10–18. [CrossRef]

9.   American Association of Poison Control Centers. Synthetic Cannabinoid Data. 23 September 2015. Available online: http://www.aapcc.org/ (accessed on 1 March 2022).

10. Connors, J.M. Hemorrhagic highs from synthetic cannabinoids–A new epidemic. *N. Engl. J. Med.* **2018**, *379*, 1275–1277. [CrossRef] [PubMed]

11. Shanks, K.G.; Dahn, T.; Behonick, G.; Terrell, A. Analysis of First and Second Generation Legal Highs for Synthetic Cannabinoids and Synthetic Stimulants by Ultra-Performance Liquid Chromatography and Time of Flight Mass Spectrometry. *J. Anal. Toxicol.* **2012**, *36*, 360–371. [CrossRef] [PubMed]

12. Logan, B.K.; Reinhold, L.E.; Xu, A.; Diamond, F.X. Identification of synthetic cannabinoids in herbal incense blends in the United States. *J. Forensic Sci.* **2012**, *57*, 1168–1180. [CrossRef]

13. Müller, H.H.; Kornhuber, J.; Sperling, W. The behavioral profile of spice and synthetic cannabinoids in humans. *Brain Res. Bull.* **2016**, *126*, 3–7. [CrossRef]

14. Hermanns-Clausen, M.; Kneisel, S.; Szabo, B.; Auwärter, V. Acute toxicity due to the confirmed consumption of synthetic cannabinoids: Clinical and laboratory findings. *Addiction* **2013**, *108*, 534–544. [CrossRef]

15. Musshoff, F.; Madea, B.; Kernbach-Wighton, G.; Bicker, W.; Kneisel, S.; Hutter, M.; Auwärter, V. Driving under the influence of synthetic cannabinoids ("Spice"): A case series. *Int. J. Legal Med.* **2014**, *128*, 59–64. [CrossRef] [PubMed]

16. Anzillotti, L.; Marezza, F.; Calò, L.; Banchini, A.; Cecchi, R. A case report positive for synthetic cannabinoids: Are cardiovascular effects related to their protracted use? *Leg. Med.* **2019**, *41*, 101637. [CrossRef] [PubMed]

17. Brents, L.K.; Prather, P.L. The K2/Spice phenomenon: Emergence, identification, legislation and metabolic characterization of synthetic cannabinoids in herbal incense products. *Drug. Metab. Rev.* **2014**, *46*, 72–85. [CrossRef]

18. Rendic, S. Summary of information on human CYP enzymes: Human P450 metabolism data. *Drug. Metab. Rev.* **2002**, *34*, 83–448. [CrossRef]

19. Fisher, M.B.; Paine, M.F.; Strelevitz, T.J.; Wrighton, S.A. The role of hepatic and extrahepatic UDP-glucuronosyltransferases in human drug metabolism. *Drug. Metab. Rev.* **2001**, *33*, 273–297. [CrossRef]

20. Hirth, J.; Watkins, P.B.; Strawderman, M.; Schott, A.; Bruno, R.; Baker, L.H. The effect of an individual's cytochrome CYP3A4 activity on docetaxel clearance. *Clin. Cancer Res.* **2000**, *6*, 1255–1258. [PubMed]

21. Ding, X.; Kaminsky, L.S. Human extrahepatic cytochromes P450: Function in xenobiotic metabolism and tissue-selective chemical toxicity in the respiratory and gastrointestinal tracts. *Annu. Rev. Pharmacol. Toxicol.* **2003**, *43*, 149–173. [CrossRef]

22. Vikingsson, S.; Josefsson, M.; Gréen, H. Identification of AKB-48 and 5F-AKB-48 Metabolites in Authentic Human Urine Samples Using Human Liver Microsomes and Time of Flight Mass Spectrometry. *J. Anal. Toxicol.* **2015**, *39*, 426–435. [CrossRef] [PubMed]

23. Diao, X.; Scheidweiler, K.B.; Wohlfarth, A.; Pang, S.; Kronstrand, R.; Huestis, M.A. In Vitro and In Vivo Human Metabolism of Synthetic Cannabinoids FDU-PB-22 and FUB-PB-22. *AAPS J.* **2016**, *18*, 455–464. [CrossRef]

24. Patton, A.L.; Seely, K.A.; Yarbrough, A.L.; Fantegrossi, W.; James, L.P.; McCain, K.R.; Fujiwara, R.; Prather, P.L.; Moran, J.H.; Radominska-Pandya, A. Altered metabolism of synthetic cannabinoid JWH-018 by human cytochrome P450 2C9 and variants. *Biochem. Biophys. Res. Commun.* **2018**, *498*, 597–602. [CrossRef]

25. Aung, M.M.; Griffin, G.; Huffman, J.W.; Wu, M.; Keel, C.; Yang, B.; Showalter, V.M.; Abood, M.E.; Martin, B.R. Influence of the N-1 alkyl chain length of cannabimimetic indoles upon CB(1) and CB(2) receptor binding. *Drug. Alcohol. Depend.* **2000**, *60*, 133–140. [CrossRef] [PubMed]

26. Hutter, M.; Broecker, S.; Kneisel, S.; Franz, F.; Brandt, S.D.; Auwärter, V. Metabolism of synthetic cannabinoid receptor agonists encountered in clinical casework: Major in vivo phase I metabolites of JWH-007, JWH-019, JWH-203, JWH-307, UR-144, XLR-11, AM-2201, MAM-2201 and AM-694 in human urine using LC-MS/MS. *Curr. Pharm. Biotechnol.* **2018**, *19*, 144–162. [CrossRef] [PubMed]

27. Jonsson-Schmunk, K.; Schafer, S.C.; Croyle, M.A. Impact of nanomedicine on hepatic cytochrome P450 3A4 activity: Things to consider during pre-clinical and clinical studies. *J. Pharm. Investig.* **2018**, *48*, 113–134. [CrossRef]

28. Wu, Z.; Lee, D.; Joo, J.; Shin, J.H.; Kang, W.; Oh, S.; Lee, D.Y.; Lee, S.J.; Yea, S.S.; Lee, H.S.; et al. CYP2J2 and CYP2C19 are the major enzymes responsible for metabolism of albendazole and fenbendazole in human liver microsomes and recombinant P450 assay systems. *Antimicrob. Agents Chemother.* **2013**, *57*, 5448–5456. [CrossRef]

29. Jones, S.; Yarbrough, A.L.; Fantegrossi, W.E.; Prather, P.L.; Bush, J.M.; Radominska-Pandya, A.; Fujiwara, R. Identifying cytochrome P450s involved in oxidative metabolism of synthetic cannabinoid N-(adamantan-1-yl)-1-(5-fluoropentyl)-1H-indole-3-carboxamide (STS-135). *Pharmacol. Res. Perspect.* **2020**, *8*, e00561. [CrossRef]

30. Gannon, B.M.; Williamson, A.; Suzuki, M.; Rice, K.C.; Fantegrossi, W.E. Stereoselective Effects of Abused "Bath Salt" Constituent 3,4-Methylenedioxypyrovalerone in Mice: Drug Discrimination, Locomotor Activity, and Thermoregulation. *J. Pharmacol. Exp. Ther.* **2016**, *356*, 615–623. [CrossRef]

31. Hiranita, T.; Soto, P.L.; Tanda, G.; Katz, J.L. Lack of cocaine-like discriminative-stimulus effects of σ-receptor agonists in rats. *Behav. Pharmacol.* **2011**, *22*, 525–530. [CrossRef]

32. Snedecor, G.W.; Cochran, W.G. *Statistical Methods*, 6th ed.; Iowa State University Press: Ames, IA, USA, 1967.

33. Keyes, K.M.; Rutherford, C.; Hamilton, A.; Palamar, J.J. Age, period, and cohort effects in synthetic cannabinoid use among US adolescents, 2011–2015. *Drug. Alcohol. Depend.* **2016**, *166*, 159–167. [CrossRef] [PubMed]

34. Bush, D.M.; Woodwell, D.A. Update: Drug-related emergency department visits involving synthetic cannabinoids. In *The CBHSQ Report*; Substance Abuse and Mental Health Services Administration (US): Rockville, MD, USA, 2014.

35. Wintermeyer, A.; Moller, I.; Thevis, M.; Jubner, M.; Beike, J.; Rothschild, M.A.; Bender, K. In vitro phase I metabolism of the synthetic cannabimimetic JWH-018. *Anal. Bioanal. Chem.* **2010**, *398*, 2141–2153. [CrossRef] [PubMed]
36. Zhang, Q.Y.; Dunbar, D.; Ostrowska, A.; Zeisloft, S.; Yang, J.; Kaminsky, L.S. Characterization of human small intestinal cytochromes P-450. *Drug Metab. Dispos.* **1999**, *27*, 804–809. [PubMed]
37. Di, L. The Impact of Carboxylesterases in Drug Metabolism and Pharmacokinetics. *Curr. Drug. Metab.* **2019**, *20*, 91–102. [CrossRef]
38. Rogers, S.; de Souza, A.R.; Zago, M.; Iu, M.; Guerrina, N.; Gomez, A.; Matthews, J.; Baglole, C.J. Aryl hydrocarbon receptor (AhR)-dependent regulation of pulmonary miRNA by chronic cigarette smoke exposure. *Sci. Rep.* **2017**, *7*, 40539. [CrossRef] [PubMed]
39. Laika, B.; Leucht, S.; Heres, S.; Schneider, H.; Steimer, W. Pharmacogenetics and olanzapine treatment: CYP1A2*1F and serotonergic polymorphisms influence therapeutic outcome. *Pharmacogenomics J.* **2010**, *10*, 20–29. [CrossRef] [PubMed]
40. Delozier, T.C.; Kissling, G.E.; Coulter, S.J.; Dai, D.; Foley, J.F.; Bradbury, J.A.; Murphy, E.; Steenbergen, C.; Zeldin, D.C.; Goldstein, J.A. Detection of human CYP2C8, CYP2C9, and CYP2J2 in cardiovascular tissues. *Drug. Metab. Dispos.* **2007**, *35*, 682–688. [CrossRef]
41. Evangelista, E.A.; Kaspera, R.; Mokadam, N.A.; Jones, J.P., 3rd; Totah, R.A. Activity, inhibition, and induction of cytochrome P450 2J2 in adult human primary cardiomyocytes. *Drug. Metab. Dispos.* **2013**, *41*, 2087–2094. [CrossRef]

Article

Thermodynamic and Kinetic Investigation on *Aspergillus ficuum* Tannase Immobilized in Calcium Alginate Beads and Magnetic Nanoparticles

Jônatas de Carvalho-Silva [1,†], Milena Fernandes da Silva [2,†], Juliana Silva de Lima [3,†], Tatiana Souza Porto [4], Luiz Bezerra de Carvalho, Jr. [3] and Attilio Converti [5,*]

[1] Northeast Biotechnology Network, Federal University of the Agreste of Pernambuco (UFAPE), Av. Bom Pastor, s/n–Boa Vista, Garanhuns 55292-270, PE, Brazil; jonatas_carvalho.jcs@hotmail.com
[2] Northeast Strategic Technologies Center (CETENE), Ministry of Science, Technology and Innovation (MCTI), Av. Prof. Luis Freire, 01, Cidade Universitária, Recife 50740-545, PE, Brazil; milena.silva@cetene.gov.br
[3] Laboratory of Immunopathology Keizo Asami (LIKA), Department of Biochemistry, Federal University of Pernambuco, Av. Prof. Moraes Rego, Recife 50670-901, PE, Brazil; ju_db_08@hotmail.com (J.S.d.L.); lbcj.br@gmail.com (L.B.d.C.J.)
[4] Department of Animal Morphology and Physiology, Federal Rural University of Pernambuco (UFRPE), Rua Dom Manuel de Medeiros, s/n, Dois Irmãos, Recife 52171-900, PE, Brazil; tatiana.porto@ufrpe.br
[5] Department of Civil, Chemical and Environmental Engineering, University of Genoa (UNIGE), Pole of Chemical Engineering, via Opera Pia 15, 16145 Genoa, Italy
* Correspondence: converti@unige.it
† These authors contributed equally to this work.

Citation: de Carvalho-Silva, J.; da Silva, M.F.; de Lima, J.S.; Porto, T.S.; de Carvalho, L.B., Jr.; Converti, A. Thermodynamic and Kinetic Investigation on *Aspergillus ficuum* Tannase Immobilized in Calcium Alginate Beads and Magnetic Nanoparticles. *Catalysts* 2023, 13, 1304. https://doi.org/10.3390/catal13091304

Academic Editors: Chia-Hung Kuo, Chwen-Jen Shieh and Hui-Min David Wang

Received: 24 August 2023
Revised: 14 September 2023
Accepted: 16 September 2023
Published: 18 September 2023

Abstract: Tannase from *Aspergillus ficuum* was immobilized by two different techniques for comparison of kinetic and thermodynamic parameters. Tannase was either entrapped in calcium alginate beads or covalently-immobilized onto magnetic diatomaceous earth nanoparticles. When immobilized on nanoparticles, tannase exhibited lower activation energy (15.1 kJ/mol) than when immobilized in alginate beads (31.3 kJ/mol). Surprisingly, the thermal treatment had a positive effect on tannase entrapped in alginate beads since the enzyme became more solvent exposed due to matrix leaching. Accordingly, the proposed mathematical model revealed a two-step inactivation process. In the former step the activity increased leading to activation energies of additional activity of 3.1 and 26.8 kJ/mol at 20–50 °C and 50–70 °C, respectively, while a slight decay occurred in the latter, resulting in the following thermodynamic parameters of denaturation: 14.3 kJ/mol activation energy as well as 5.6–9.7 kJ/mol standard Gibbs free energy, 15.6 kJ/mol standard enthalpy and 18.3–29.0 J/(K·mol) standard entropy variations. Conversely, tannase immobilized on nanoparticles displayed a typical linear decay trend with 43.8 kJ/mol activation energy, 99.2–103.1 kJ/mol Gibbs free energy, 41.1–41.3 kJ/mol enthalpy and −191.6/−191.0 J/(K·mol) entropy of denaturation. A 90-day shelf-life investigation revealed that tannase immobilized on nanoparticles was approximately twice more stable than the one immobilized in calcium alginate beads, which suggests its use and recycling in food industry clarification operations. To the best of our knowledge, this is the first comparative study on kinetic and thermodynamic parameters of a tannase produced by *A. ficuum* in its free and immobilized forms.

Keywords: tannase; *Aspergillus*; immobilization; calcium alginate beads; magnetic nanoparticles; kinetics; thermodynamics

1. Introduction

The immobilization of enzymes, when combined with the fundamental principles of thermodynamics and kinetics, has proven to be a highly useful tool for the successful application of biocatalysis in industrial processes [1,2]. If, on the one hand, immobilization can enhance the stability and reusability of a biocatalyst, on the other hand, knowledge of kinetic and thermodynamic parameters—such as decimal reduction time, half-life, activation energy, Gibbs free energy, enthalpy and entropy—is essential to predict the reaction mechanism and behavior of any process using immobilized enzymes as well as biocatalyst thermostability at a given operating temperature [1–3]. These aspects taken together are key issues that must be considered to make the application of enzyme-based industrial processes even more economic [1,4].

Tannase (or tannin acyl hydrolase, E.C. 3.1.1.20) is a biotechnologically important enzyme that catalyzes the hydrolysis of ester and depside bonds present in tannins, such as tannic acid among others, and releases gallic acid and glucose [5]. For this reason, many industrial processes require tannase to improve their efficiency [6–8]. Such an enzyme can be extracted from vegetable and animal sources. However, for being an extracellular enzyme, microorganisms are its main producers for industrial applications [6]. Among them, fungi (e.g., *Aspergillus* species) [7], bacteria [8] and yeasts [9] are generally used for large scale production of tannases, since their enzymes are very stable in wide pH and temperature ranges [7,9,10].

Being versatile enzymes, tannases have a lot of different applications in a number of industrial sectors, such as food, beverage, animal feed, chemical and pharmaceutical industries, as well as in wastewater treatment and other bioprocesses [6,7]. In particular, thanks to their capability of increasing tannin solubility, tannases are broadly used to improve the clarification of beer, juices and wine, thus reducing turbidity, improving color appearance, and softening the costumer's undesired astringent taste of tannins [8,11].

Despite its undisputed potential, the drawbacks connected with the use of free enzymes like tannases in industrial applications (e.g., insufficient stability and inability to recover/recycle the biocatalyst) make the process expensive [7,12]. Enzyme immobilization is one of the strategies developed to overcome these limitations [7,12,13]. After immobilization, some features of the enzymes can change, being able to improve their catalytic capacity and thermal stability, vary the pH range where they can be used, as well as facilitate their separation/recovery by simple physical operations [12,14–16].

According to the required application, research efforts have focused on the development of more effective enzyme immobilization techniques with better catalytic features, seeking suitable/attractive carrier materials to be used as biocompatible support matrices to immobilize biomolecules. However, according to Silva et al. [14], upon immobilization the enzyme activity may be altered by some factors such as crosslink agents, binding mode, microenvironment, diffusion, protein aggregation, molecular polarization, partition, conformational changes, induction and structural flexibility.

Larosa et al. [17] have reported that, among the various available immobilization methods, entrapment in calcium alginate beads is an effective method to immobilize tannase and to preserve its catalytic activity because it does not involve chemical modification, thereby ensuring easiness of use, mild conditions and low cost as its main advantages. Recently, Intisar et al. [18] have reported that alginate biopolymer, an anionic polysaccharide derived from brown seaweeds [19], has gained noteworthy consideration from researchers and industrialists as a suitable material for developing entrapped enzyme preparations with enhanced activities because of its versatile characteristics, such as no toxicity, relatively high availability, low cost, simple preparation, good biocompatibility and reusability.

On the other hand, according to Cabrera et al. [20], diatomaceous earth, a lightweight mineral clay composed especially of amorphous hydrated silica, has gained attention thanks to its favorable properties, such as low cost, ready availability, high cation exchange capacity, chemical inertness, porous structure, large surface area, and low thermal conductivity. In previous studies, *Aspergillus ficuum* tannase was immobilized by two different techniques, namely entrapment in calcium alginate beads [19] and covalent immobilization onto magnetic nanoparticles composed of diatomaceous earth coated with polyaniline [21]. The results of these works suggest that immobilized tannase systems are promising in food applications to improve tea quality and to remove tannins from aromatic beverages.

Nowadays, enzyme immobilization and kinetic/thermodynamic studies have attracted much interest [1,22–24]. Indeed, an overview of the literature published over the last two years up to the beginning of August 2023 has revealed that there has been an increase in the contributions related to "enzyme immobilization and thermodynamic and kinetic", with no less than 815 research articles (based on the Science Direct search engine: http://www.sciencedirect.com, accessed on 10 August 2023). However, there are a few investigations on the kinetic and thermodynamic parameters of immobilized or free tannases, especially those of the comparative type like this.

Based on this background, the present work aimed to determine the kinetic and thermodynamic parameters of activity and thermal stability of *A. ficuum* tannase immobilized either in calcium alginate beads or on magnetic nanoparticles. For this purpose, the performance of both preparations was compared for the first time to that of the free enzyme, providing valuable information for a better understanding of reaction mechanism and performance prediction in possible enzyme-based industrial applications.

2. Results and Discussion

The *Aspergillus ficuum* tannase was previously immobilized on two supports with different techniques, namely entrapment in calcium alginate beads [19] and covalent immobilization on magnetic nanoparticles composed of polyaniline-coated diatomaceous earth (mDE-PANI-tannase) [21]. In this work, both immobilized-enzyme preparations were subjected to kinetic and thermodynamic modeling to determine the parameters involved in both catalysis and protein thermal denaturation, whose related equations and definitions have been mostly reported for two different enzyme preparations [25,26] and partially summarized in Section 3.3. In addition, a comparative study of such parameters with those of the free enzyme was also performed.

2.1. Thermodynamic Parameters of Tannase-Catalyzed Reaction

Figure 1A illustrates the straight lines obtained by plotting, according to Arrhenius, the experimental results of tannin hydrolysis using free tannase, tannase immobilized in calcium alginate beads and mDE-Pani-tannase. In particular, the straight lines on the right refer to the increase in activity, described by the Arrhenius equation, resulting from a temperature rise from 20 to 40 °C for both immobilized enzyme preparations and from 20 to 30 °C for the free enzyme. This 10 °C difference in the optimal temperature of catalysis (T_{opt}) can be ascribed to conformational changes that the enzyme underwent due to immobilization in different supports [21]. Free tannase exhibited the highest activation energy of reaction (E) (51 kJ/mol, $R^2 = 0.98$) followed by tannase immobilized in calcium alginate beads (31.3 kJ/mol, $R^2 = 0.96$) and by mDE-PANI-tannase (15.1 kJ/mol, $R^2 = 0.99$). The lowest value of E observed for the enzyme immobilized on magnetic nanoparticles could have been due to the covalent nature of bonds involved in such a type of immobilization, which may have allowed the enzyme to be present only on the support surface and therefore more available for contact with the substrate [14]. As reported by Fernandes et al. [27], lower activation energies are desirable for commercial enzymes, as they imply lower expenses during the process.

Unlike what occurred for E, the standard enthalpy variation of the enzyme unfolding equilibrium ($\Delta H°_u$) [25,26], which was estimated from the slopes of straight lines obtained at temperatures higher than T_{opt} (on the left in Figure 1A), showed an opposite behavior, with the highest value being observed for tannase immobilized in calcium alginate beads (29.6 kJ/mol, $R^2 = 0.97$), followed by mDE-PANI-tannase (17.6 kJ/mol, $R^2 = 0.90$) and free tannase (17.6 kJ/mol, $R^2 = 0.99$). In this case, the higher this value, the greater the stability of the biocatalyst at high temperatures [4]. This outcome corroborates with what was expected from the technique used to immobilize tannase. The entrapment technique confined the enzyme within the matrix, which gave it greater protection against sudden variations or prolonged exposure to a given temperature. As explained better by Boudrant et al. [28], a proper immobilization protocol is important to achieve higher performance of the biocatalist and mainly to enhance its stability.

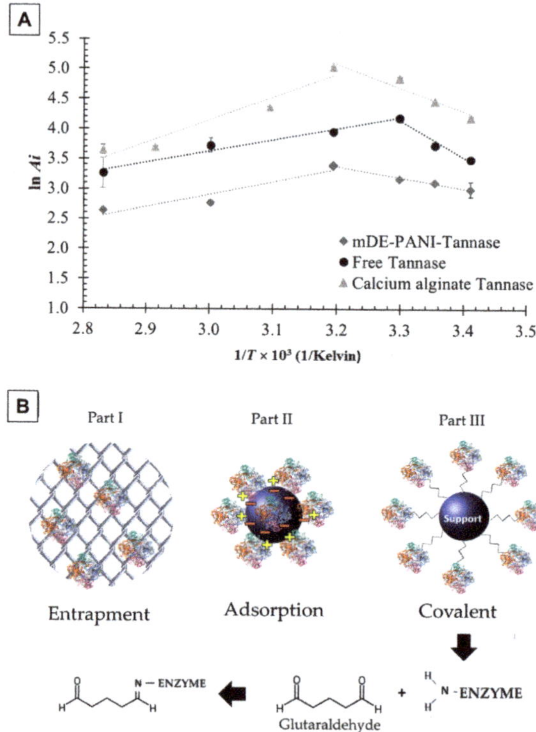

Figure 1. (**A**) Arrhenius-type plots for determination of the activation energy of the reaction catalyzed by free tannase, tannase entrapped in calcium alginate beads and tannase covalently immobilized on magnetic nanoparticles (mDE-PANI-tannase). (**B**) Schematic representations of the main immobilization techniques.

The entrapment technique imprisons the enzyme molecules inside the support (Figure 1B, part I), and the pore size influences directly the enzyme activity. Differently from adsorption (Figure 1B, part II) and covalent immobilization, it does not use chemical bonds or electrostatic forces, and the enzyme is physically trapped in the matrix. On other hand, covalent immobilization (Figure 1B, part III) resorts to chemical bonds. However, the enzyme structure cannot make a bond with polymer supports like alginate and polyaniline, for which a powerful crosslinking agent such as glutaraldehyde is needed. Glutaraldehyde, a bifunctional compound with an aldehyde group in each extremity, is able to make a chemical bond with amino groups from the side chains (Figure 1B, part III); so, the enzyme is immobilized on the support surface. Both techniques have their advantages depending mainly on the application [14].

It is noteworthy that the parameter $\Delta H^{\circ}{}_{u}$ refers to the enthalpic energy related to enzyme unfolding, which interferes with the formation of enzyme-substrate complex. As previously described by Abellanas-Perez et al. [29], the enzyme has a specific affinity for a particular substrate, which depends on its tridimensional structure; so, any change in structure leads to a loss in biological activity. As explained by Da Silva et al. [4], the tridimensional structures of enzymes are supported by non-covalent bonds such as hydrogen bonds, Van der Waals forces, dipole-dipole interactions, ion exchange, etc., which are easily broken by motion due to a temperature increase. At low temperatures there is little vibration, and enzyme molecules are present in rigid forms. At higher temperatures, due to some breaks of these bonds, enzyme molecules undergo structural modifications and acquire more malleable conformations, which allows the substrate to more easily fit in the active site and, as a consequence, the affinity to increase. However, when too many bonds are broken simultaneously, the enzyme cannot form its complex with the substrate, and thus a decline in the reaction rate is observed [30].

2.2. Thermodynamics of Tannase Thermal Inactivation

As explained earlier, thermodynamics of thermal inactivation of the enzyme is related to energy involved in the loss of protein structure. This occurs when the enzyme is subjected to prolonged exposure to a given temperature.

2.2.1. Free Tannase Thermal Inactivation

Figure 2A shows the straight lines obtained by plotting the natural logarithm of the free enzyme residual activity (Ψ) as a function of time at different temperatures in order to determine the related constants of thermal inactivation (denaturation) (k_d). These k_d values were later used to estimate the activation energy of free enzyme thermal inactivation (E_d = 53.5 kJ/mol) through the Arrhenius equation (Figure 2B). The straight lines showed high determination coefficients (R^2), and the values of kinetic and thermodynamic parameters of free tannase denaturation are listed in Table 1.

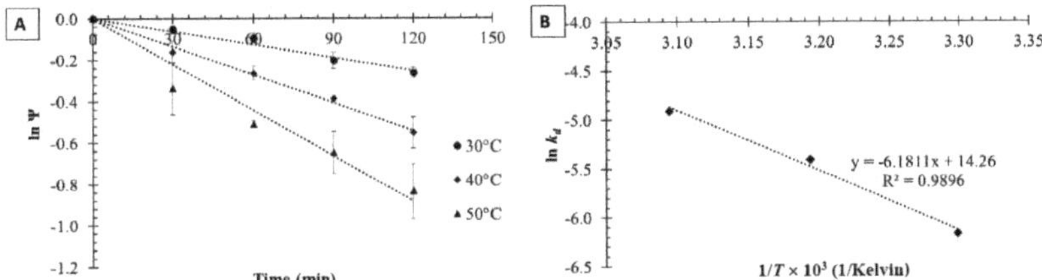

Figure 2. Semi-log plots of residual activity (Ψ) of free tannase vs. time at different temperatures (**A**) and Arrhenius-type semi-log plot of the first-order denaturation constant (k_d) vs. reciprocal temperature ($1/T$) used to estimate the activation energy of free tannase thermal inactivation (**B**).

Table 1. Kinetic and thermodynamic parameters of free tannase denaturation.

Temperature (°C)	k_d [1] (min⁻¹)	R^2	ΔH_d [2] (kJ/mol)	ΔG_d [3] (kJ/mol)	ΔS_d [4] (J/(K·mol))	$t_{1/2}$ [5] (min)	D-Value [6] (min)
30	0.0021	0.98	50.99	100.15	−162.15	330.07	1096.67
40	0.0045	0.99	50.91	101.55	−161.72	154.03	511.78
50	0.0074	0.98	50.82	103.54	−163.13	93.67	311.22

[1] k_d = first-order denaturation constant. [2] ΔH_d = activation enthalpy of denaturation. [3] ΔG_d = activation Gibbs free energy of denaturation. [4] ΔS_d = activation entropy of denaturation. [5] $t_{1/2}$ = half-life. [6] D-value = decimal reduction time.

The activation Gibbs free energy of enzyme denaturation (ΔG_d), which reflects the amount of remaining energy in the protein structure after exposure to a certain temperature [31], increased from 100.15 kJ/mol at 30 °C to 103.54 kJ/mol at 50 °C, thereby highlighting a certain thermal stability. This result suggests that, within the tested temperature range and exposure time (120 min), free tannase denaturation was likely to be reversible, as better explained in Section 3.3. Such an assumption is corroborated by the negative values of activation entropy of free tannase denaturation listed in Table 1 (ΔS_d). In fact, according to Wahba et al. [23], negative entropy values suggest greater enzyme stability and that the irreversible denaturation process is a further step forward involving higher temperatures and longer times.

Consistently, the enthalpy of denaturation (ΔH_d), which reflects the amount of broken non-covalent bonds, showed only a slight variation. In fact, considering that the energy to break a non-covalent bond can be estimated at about 5.4 kJ/mol [30], in the temperature range under investigation only 9 non-covalent bonds were likely to be broken in free tannase, a number that does not seem to have been sufficient for the denaturation process to become irreversible. In this respect, it is important to notice that enzymes with thermostability characteristics in their free form are more suitable to be immobilized in supports, as they can be recovered for later applications.

Other kinetic parameters evaluated for free tannase were the half-life ($t_{1/2}$) and the decimal reduction time (*D*-value), which characterize the thermal inactivation behavior at each temperature to which an enzyme is exposed [24]. Both decreased by increasing temperature, showing the temperature influence in reducing the overall activity due to thermal inactivation. The thermal resistance constant (*Z*-value), a parameter of sensitivity to temperature variation, was then estimated from the slope of thermal– death–time plot of log*D*-value versus temperature. It revealed that an increase of about 19 °C is required for *D*-value to be reduced by one log cycle, i.e., by 90% [14].

2.2.2. Thermal Inactivation of Tannase Immobilized on Magnetic Nanoparticles

The kinetic and thermodynamic parameters of thermal inactivation of mDE-PANI-tannase were close to those found for the free enzyme. As can be seen in Figure 3, the semi-log plots used to determine the values of k_d (Figure 3A) and E_d (Figure 3B) showed high R^2 values (Table 2), demonstrating that mDE-PANI-tannase behavior can be described by the equations outlined in Section 3.3. However, this tannase immobilization technique reduced the value of E_d to 43.9 kJ/mol, i.e., by approximately 10 kJ/mol compared to the free enzyme. Considering that the higher this parameter, the greater the stability of a biocatalyst, such a reduction suggests that the enzyme was slightly more sensitive to temperature variations [14].

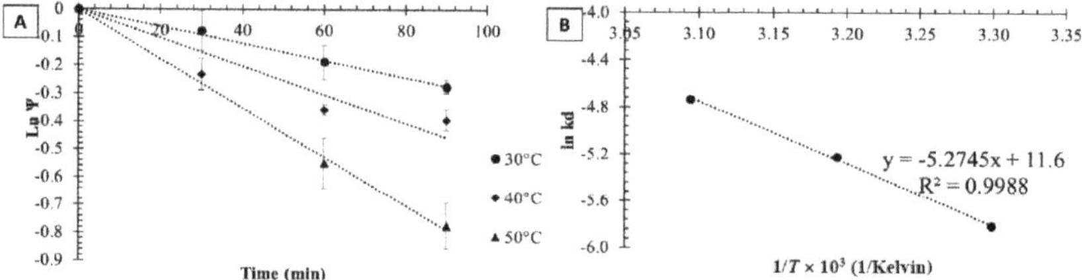

Figure 3. Semi-log plots of residual activity (Ψ) of mDE-PANI-tannase (starting hydrolysis specific activity of 342 U/mg) vs. time at different temperatures (**A**) and Arrhenius-type semi-log plot of the first-order denaturation constant (k_d) vs. reciprocal temperature ($1/T$) used to estimate the activation energy of mDE-PANI-tannase thermal inactivation (E_d) (**B**).

Table 2. Kinetic and thermodynamic parameters of mDE-PANI-tannase denaturation.

Temperature (°C)	k_d [1] (min^{-1})	R^2	ΔH_d [2] (kJ/mol)	ΔG_d [3] (kJ/mol)	ΔS_d [4] (J/(K·mol))	$t_{1/2}$ [5] (min)	*D*-Value [6] (min)
30	0.0030	0.99	41.33	99.25	−191.04	231.05	767.67
40	0.0054	0.96	41.25	101.08	−191.05	128.36	426.48
50	0.0088	0.99	41.17	103.08	−191.58	78.77	261.70

[1] k_d = first-order denaturation constant. [2] ΔH_d = activation enthalpy of denaturation. [3] ΔG_d = activation Gibbs free energy of denaturation. [4] ΔS_d = activation entropy of denaturation. [5] $t_{1/2}$ = half-life. [6] *D*-value = decimal reduction time.

This behavior is corroborated by the decrease of *Z*-value from 19 to 16 °C. The values of the other kinetic denaturation parameters, namely, *D*-value and $t_{1/2}$ (Table 2), were also smaller than those of free tannase (Table 1), confirming the lower thermostability of this enzyme preparation.

ΔG_d increased from 99.25 to 103.08 kJ/mol with increasing temperature from 30 to 50 °C (Table 2), suggesting that tannase maintained its reversible profile of thermal denaturation after immobilization on magnetic nanoparticles. This occurrence is consistent with the results of de Lima et al. [21], who reported that tannase immobilized on magnetic diatomaceous earth nanoparticles coated with polyaniline could be reused up to 10 times losing only 34% of its initial activity. ΔS_d also had negative values ($-191.58/-191.04$ J/(K·mol)), which are characteristic of reversible inactivation. Finally, ΔH_d (41.17–41.33 kJ/mol) was reduced by almost 10 kJ/mol compared to the free enzyme, which corresponds to the breakdown of approximately eight non-covalent bonds.

2.2.3. Thermal Inactivation of Tannase Immobilized in Calcium Alginate Beads

Unlike what was expected from a typical thermal stability study, *A. ficuum* tannase entrapped in calcium alginate beads exhibited an unusual behavior throughout the whole tested temperature range (20–70 °C). While the free form of tannase and the mDE-PANI-tannase showed the characteristic decay profile of protein denaturation (Figures 2A and 3A), the activity of tannase entrapped in alginate beads, after an initial increase until a maximum value after about 30 min, progressively decreased still keeping above its starting value (Figure 4). This behavior is not commonly observed. Only Rodríguez-Lopez et al. [32] reported the same behavior for mushroom polyphenol oxidase upon inactivation using an 80 °C hot water bath and 22.6 W/cm^3 microwave power irradiation. However, there is no mathematical model available in the literature on other biocatalysts that acted in the same or similar way.

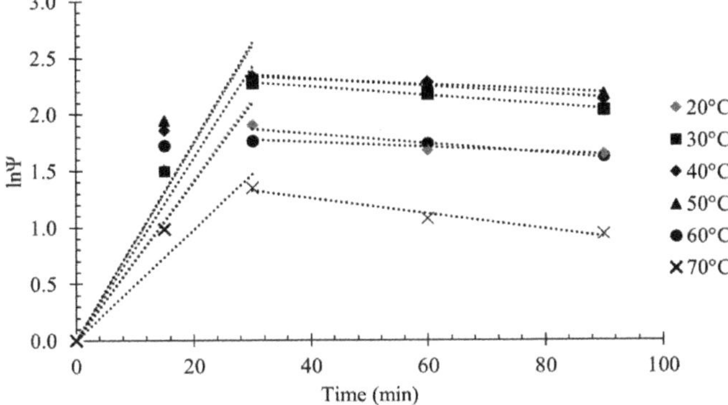

Figure 4. Semi-log plots of residual activity (Ψ) of tannase entrapped in calcium alginate beads (starting hydrolysis activity of 127.5 U/mL) vs. time at different temperatures.

Therefore, a new mathematical approach to the kinetic and thermodynamic parameters of the biocatalyst entrapped in calcium alginate beads was proposed in the present study. Since the calcium alginate beads were porous polymeric spheres where tannase had been entrapped, they may have suffered leaching due to exposition to temperatures higher than the optimum one, which may have partially degraded their polymeric structure and increased pore size (Figure 5, part II). Obviously, a similar mechanism can be proposed for the action of other external agents such as suboptimal pH, chemical agents and so on. Therefore, the contact between enzyme and outer environment, i.e., the solvent in the case of residual activity tests performed in this study at temperatures higher than that optimal one or reaction medium in industrial applications, became progressively more effective along the starting 30 minutes. However, when the enzyme was exposed for a longer period, its structure began to be denatured, and there was a decrease in its activity (Figure 5, Part III). Nonetheless, the residual activity still remained above the initial one. Calcium alginate probably conferred extra protection to tannase, so that an activity lower than the initial one would have required exposition longer than 100 min.

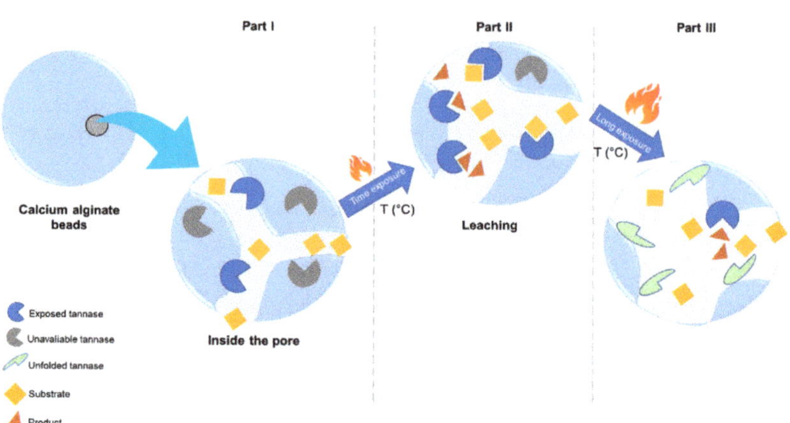

Figure 5. Mechanism proposed in this study for the degradation of calcium alginate beads exposed to temperatures higher than the optimum one.

As mathematically described by Equation (5) onwards in Section 3.3, the kinetic constant of the resultant of these two phenomena is the sum of both positive (k_L) and negative (k_d) contributions in terms of residual activity. The values of these constants are gathered in Table 3 along with the respective determination coefficients.

Table 3. Kinetic constants describing the activity increase due to matrix leaching (k_L) and activity decrease due to subsequent denaturation (k_d) of tannase entrapped in calcium alginate beads.

Temperature (°C)	k_L (min^{-1})	R^2	k_d (min^{-1})	R^2
20	0.079	0.96	0.0043	0.85
30	0.0808	0.99	0.0040	0.99
40	0.087	0.96	0.0035	0.93
50	0.0879	0.95	0.0024	0.95
60	0.070	0.91	0.0023	0.87
70	0.0491	0.97	0.0068	0.96

It is possible to observe in Figure 4 that the straight lines describing the variations of $\ln\Psi$ over time did not obey the traditional one-step decreasing profile as the temperature was raised from 20 to 70 °C. Since these constants are related to the activation energies involved in the two proposed phenomena, it is possible to identify in Figure 6A two different regions, the former between 20 and 50 °C (straight line on the right) and the latter between 50 and 70 °C (straight line on the left). Thus, it was possible to estimate two activation energies linked to the increase in tannase activity resulting from matrix leaching (E_{L1} = 3.1 kJ/mol and E_{L2} = 26.8 kJ/mol, respectively) (Figure 6A) and an activation energy linked to the decrease in activity due to denaturation under prolonged exposure (E_d = 14.3 kJ/mol) (Figure 6B).

As is well known, the Gibbs free energy variation is a parameter that measures the spontaneity of any process, in that, positive values are characteristic of a spontaneous process, negative values of a non-spontaneous process and a null value of a process in thermodynamic equilibrium. As can be observed in Table 4, the standard Gibbs energy variations related to both the activity increase due to leaching of calcium alginate beads ($\Delta G°_L$) and to the activity decrease due to denaturation of the entrapped enzyme after prolonged exposure ($\Delta G°_d$) were negative, demonstrating that the increase in biocatalyst activity was a spontaneous phenomenon characteristic of the support outwear. However, the values of $\Delta G°$ that referred to the combination of the two phenomena ($\Delta G°_{RT}$) were positive, suggesting that the increase in activity was a non-spontaneous phenomenon forced by the leaching process.

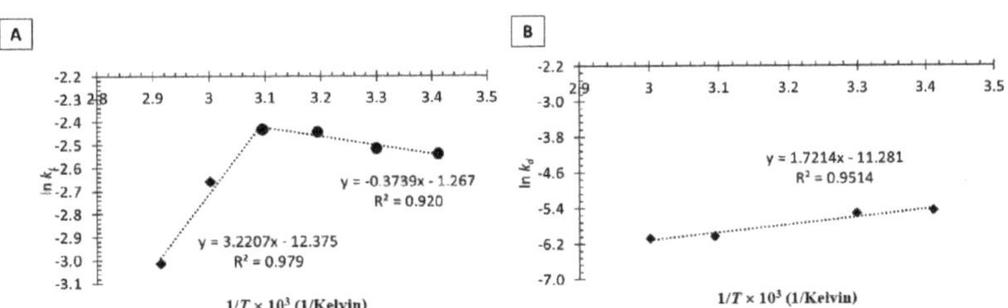

Figure 6. Semi-log plots used to estimate the activation energies linked to the increase in activity resulting from matrix leaching (E_{L1} and E_{L2}) (**A**) and to the decrease in activity due to enzyme denaturation under prolonged exposure (E_d) (**B**). Enzyme system: tannase entrapped in calcium alginate beads.

Table 4. Thermodynamic parameters referred to the activity of tannase entrapped in calcium alginate beads. Parameters referred to the increase in activity resulting from matrix leaching are denoted by the subscript "*L*", those referred to the decrease in activity due to enzyme denaturation under prolonged exposure by the subscript "*d*" and those referred to combination of both phenomena by the subscript "*RT*". Standard variations of Gibbs free energy ($\Delta G°$), enthalpy ($\Delta H°$) and entropy ($\Delta S°$).

Temperature (°C)	$\Delta G°_L$ (kJ/mol)	$\Delta G°_d$ (kJ/mol)	$\Delta G°_{RT}$ (kJ/mol)	$\Delta H°_L$ [1] (kJ/mol)	$\Delta H°_d$ (kJ/mol)	$\Delta H°_{RT}$ (kJ/mol)	$\Delta S°_L$ (J/(K·mol))	$\Delta S°_d$ (J/(K·mol))	$\Delta S°_{RT}$ (J/(K·mol))
20	−87.9	−95.0	7.1	29.9	14.3	15.6	401.9	372.9	29.0
30	−90.9	−98.5	7.6	29.9	14.3	15.6	398.6	372.2	26.4
40	−93.8	−102.2	8.4	29.9	14.3	15.6	395.1	372.1	23.0
50	−96.9	−106.6	9.7	29.9	14.3	15.6	392.3	374.1	18.2
60	−100.6	−110.1	9.5	29.9	14.3	15.6	391.7	373.3	18.4
70	−104.7	−110.4	5.6	29.9	14.3	15.6	392.3	363.3	29.0

[1] $\Delta H°_L$ was calculated as the sum (E_L) of both activation energies linked to the activity increase resulting from matrix leaching (E_{L1}) and activity decrease due to denaturation after prolonged exposure (E_{L2}) (Figure 6A and Equation (21)).

Standard enthalpy variations related to increased activity due to support leaching ($\Delta H°_L$) and decreased activity due to denaturation after prolonged exposure ($\Delta H°_d$) did not vary with rising temperature, being 29.9 and 14.3 kJ/mol, respectively, while the one that referred to the combination of the two phenomena ($\Delta H°_{RT}$) was 15.6 kJ/mol in the entire temperature range studied. This value corresponds approximately to the breaking of only three non-covalent bonds after 90 min of biocatalyst exposure. Finally, the fact that the related entropy variations were positive and very small ($18.2 \leq \Delta S°_{RT} \leq 29.0$ J/(K·mol)) confirms that, despite representing an irreversible denaturation process, such thermal inactivation was poorly significant and that the biocatalyst had excellent thermostability.

2.2.4. Inactivation of Immobilized Enzyme under Storage Conditions

Tannase immobilized either on magnetic nanoparticles or in calcium alginate beads was finally subjected to a shelf-life study to determine the kinetic parameters often used to characterize the stability of enzyme preparations to be used in industrial applications, i.e., the half-life ($t_{1/2}$) and the decimal reaction time (*D*-value).

During experiments for 90 days at 4 °C (Figure 7) mDE-PANI-tannase showed significantly greater stability than tannase entrapped in calcium alginate beads. Indeed, $t_{1/2}$ (217 days) and *D*-value (720 days) of the former preparation were almost twice those of the latter ($t_{1/2} = 116$ days; *D*-value = 384 days).

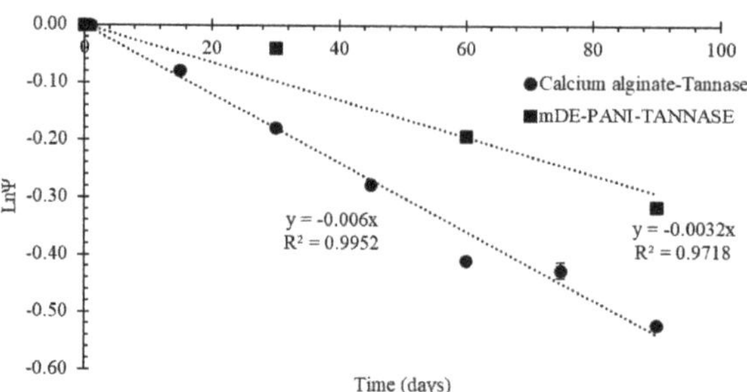

Figure 7. Semi-log plots of the residual activity coefficient (Ψ) vs. time during shelf-life experiments at 4 °C of *A. ficuum* tannase immobilized in calcium alginate beads and on magnetic nanoparticles (mDE-PANI-tannase).

3. Materials and Methods

Commercial tannase (300 U/g) from *Aspergillus ficuum* used in this study was purchased from Sangherb Bio-Tech (Xi'an, China). It was immobilized either in calcium alginate beads or on magnetic nanoparticles.

3.1. Tannase Immobilization

3.1.1. Immobilization in Calcium Alginate Beads

Tannase was entrapped in calcium alginate beads as described by de Lima et al. [19]. Briefly, 10 mL of *A. ficuum* tannase solution (6 mg/mL, corresponding to enzyme activity of 127.5 U/mL) in 0.2 M acetate buffer, pH 5.0, were mixed with 30 mL of 3.0% (w/v) sodium alginate solution and dropped in 60 mL of 2.0% (w/v) $CaCl_2$ solution at 4 °C under stirring for beads formation. The resulting tannase-loaded beads, with approximately 0.4 mm mean diameter, were left to cure in the same $CaCl_2$ solution for 4.0 h, collected, washed twice with distilled water, suspended in the above working buffer and finally stored at 4 °C for further use.

3.1.2. Immobilization on mDE-PANI Nanoparticles

Diatomaceous earth (DE) particles were magnetized (mDE) according to Cabrera et al. [20]. A black precipitate was collected and coated with polyaniline (PANI) as follows. The precipitate was submitted to treatment with 0.1 M $KMnO_4$ solution at 25 °C for 60 min, and nanoparticles were washed with distilled water and immersed into 0.25 M aniline solution prepared in 2.0 M HCl at 4 °C for 30 min. Then, they were washed with distilled water and 1.0 M HCl to remove residual aniline using an external 0.6-T magnetic field (Ciba Corning Medical Diagnostics, Walpole, MA, USA). Finally, mDE-PANI nanoparticles were washed several times, dried at 50 °C and stored at 25 °C for later use.

mDE-PANI nanoparticles were activated with glutaraldehyde (2.5% (v/v)) by stirring at 25 °C for 2 h and washed several times with distilled water for eliminating unreacted glutaraldehyde. Tannase from *A. ficuum* prepared in 0.2 M sodium acetate buffer (1 mL of 3 mg/mL solution, corresponding to 342 U/mg specific activity), pH 5.0, was incubated with mDE-PANI nanoparticles (0.05 g) at 4 °C under mild stirring for 20 h. Then, tannase immobilized on mDE-PANI nanoparticles (mDE-PANI-tannase) was collected by an external 0.6-T magnetic field, and the supernatants were used for protein determination according to the Bradford method [33] using bovine serum albumin as a standard. The immobilized derivatives were stored in sodium acetate buffer at 4 °C until use.

3.2. Determination of Free and Immobilized Tannase Activity

Tannase activity was measured according to Pinto et al. [34]. Briefly, 0.1 mL of the free enzyme solution in 0.2 M acetate buffer, pH 5.0, or 15 calcium alginate beads with immobilized tannase or 100 µg of tannase immobilized on 0.05 g mDE-PANI were added to

2.0 mL of 0.05% (*w/v*) tannic acid solution. After homogenization for 7 min at 30 °C, 100 μL of this reaction mixture were added to 150 μL of ethanolic rhodanine solution (0.667% *w/v*) and allowed to react for 5 min. Then, 100 μL of 0.5 M KCl and, after 2.5 min, 2.15 mL of distilled water were added to the mixture. The formation of a complex with maximum absorbance at 520 nm was followed by means of a UV–Vis spectrophotometer, model Lambda 25 (Perkin Elmer, Wellesley, MA, USA). One unit of tannase was defined as the amount of enzyme necessary to obtain 1 μmol of gallic acid per minute under the assay conditions (30 °C for 7 min). Control experiments were also performed using immobilized beads without tannase [19].

3.3. Thermodynamic Modeling

Tannase, either free or immobilized in calcium alginate beads or magnetic nanoparticles, was incubated for 2 h in 0.2 M sodium acetate buffer, pH 5.0, at various temperatures (from 20 to 90 °C), and the residual enzyme activities were determined as described in Section 3.2.

Activation energy of the enzyme-catalyzed reaction (E) and standard enthalpy variation of the enzyme unfolding equilibrium ($\Delta H°_u$) were estimated from semi-log plots of the starting enzyme activity (A_i) versus the reciprocal temperature ($1/T$) in the temperature range of 20–70 °C. In particular, as described elsewhere [25,26], the linearized log form of the Arrhenius equation was used to estimate E in the temperature ranges of 20–40 °C for immobilized enzyme preparations and 20–30 °C for free tannase:

$$A_i = A_o e^{-\frac{E}{RT}} \tag{1}$$

while $\Delta H°_u$ was estimated from the slope of the left straight line of the Arrhenius-type plot of $\ln A_i$ at temperatures higher than the optimal ones (30 and 40 °C, respectively).

Thermal inactivation of most free or covalently immobilized enzymes can be described by the following general deactivation model [35]:

$$N \underset{k'_2}{\overset{k'_1}{\rightleftarrows}} U \overset{k'_3}{\rightarrow} D \tag{2}$$

where N is the native biomolecule in its totally active state, while U and D the biomolecule in its reversibly unfolded and ultimate irreversibly denatured states, while k'_1, k'_2 and k'_3 are the rate constants of enzyme unfolding, folding and denaturation, respectively.

This was the case of both free tannase and tannase immobilized on magnetic nanoparticles, for which the traditional kinetic and thermodynamic approach proposed for other enzyme systems was successful [25,26].

However, after immobilization some enzymes may become more thermostable due to variations in their tertiary structure induced by crosslinking agents or the matrix itself. Since the entrapment of enzymes in calcium alginate beads implies their immobilization within the matrix during the gelation process (Figure 5, part I), the access of substrate takes place through the polymer pores. Organic polymers like this may be subject to degradation when exposed to chemical, physical or microbiological agents. In this case, their pores may increase in size and number, leaving the immobilized enzyme, i.e., tannase in this study, more exposed to the reaction medium and allowing the substrate to enter the pores more easily (Figure 5, part II). Equation (2) can then be rewritten in order to take into account such an increase in enzyme activity due to exposure of the polymeric matrix to these agents:

$$I \overset{k_0}{\rightarrow} A \underset{k_2}{\overset{k_1}{\rightleftarrows}} U \overset{k_3}{\rightarrow} D \tag{3}$$

where I and A are the immobilized-enzyme preparations at the beginning and after exposure, k_0 is the rate constant of this phenomenon, while k_1, k_2 and k_3 are the new unfolding, folding and denaturation constants.

Assuming that polymer degradation is an irreversible and first-order process, enzyme activity tends to increase over time:

$$\frac{dA}{dt} = kA \tag{4}$$

where k is the overall kinetic constant and A the enzyme activity.

Because enzymes are molecules with biological activity, they progressively lose their activity due to the denaturation process. In fact, when exposed for long time to agents such as high temperature in the present case, non-covalent bonds can be broken, and their three-dimensional structure becomes predominantly unfolded (Figure 5, part III). Normally, as is the case of free tannase and tannase immobilized on magnetic nanoparticles, this decay shows a linear trend and can be treated as an irreversible first-order reaction. On the other hand, in the case of tannase entrapped in calcium alginate beads, k is influenced by these opposite contributions, i.e., the activity increase resulting from polymer degradation, described by the degradation constant (k_L), and the activity reduction due to enzyme denaturation described by the denaturation constant (k_d):

$$k = k_L - k_d \tag{5}$$

The contribution of the former phenomenon is more pronounced than that of the latter when $k > 0$, and vice versa when $k < 0$. To better understand the meaning of this constant, it is important to remember that the enzyme is in a thermodynamic folding/unfolding equilibrium, which is governed by k_1 and k_2, while the formation of final product is governed by k_3. The overall kinetic constant can then be written as follows [36]:

$$k = \frac{k_1 k_3}{k_2 + k_3} \tag{6}$$

By equaling Equations (5) and (6), we obtain the Equation:

$$k_L - k_d = \frac{k_1 k_3}{k_2 + k_3} \tag{7}$$

that simplifies to Equation (8) when $k_3 \ll k_2$:

$$k_L - k_d = k_3 K_{eq} \tag{8}$$

where K_{eq} is the constant of the folding/unfolding equilibrium.

Considering that $k_3 = \frac{k_B T}{h}$ [37], one can write:

$$K_{eq} = \frac{k_L h}{k_B T} - \frac{k_d h}{k_B T} \tag{9}$$

where T is the absolute temperature, h the Planck's constant, and k_B the Boltzmann's constant.

To describe the thermal inactivation kinetics, Ortega et al. [38] proposed a multi-fraction inactivation model, which supposes the existence of multiple enzyme fractions, each of which can be independently analyzed with first-order kinetics. Considering the well-known relationship between the standard variation of Gibbs free energy ($\Delta G°$) and K_{eq}:

$$\Delta G^o = -RT \ln K_{eq} \tag{10}$$

and applying the approach of Ortega et al. [38] to the above two contributions governed by k_L and k_d, we can calculate this parameter for the tannase system under investigation ($\Delta G°_{TR}$) as:

$$\Delta G^o{}_{TR} = -\left(RT \ln \frac{k_L h}{k_B T} - RT \ln \frac{k_d h}{k_B T} \right) \tag{11}$$

Considering that:

$$\Delta G^o = \Delta H^o - T\Delta S^o \tag{12}$$

under the equilibrium conditions described above, ΔG°_{TR} can also be described by the equation:

$$\Delta G^o_{TR} = (\Delta H^o_{\,L} - T\Delta S^o_{\,L}) - (\Delta H^o_{\,d} - T\Delta S^o_{\,d}) \tag{13}$$

By equaling the right members of Equations (11) and (13), we obtain:

$$RT \ln\left(\frac{k_d}{k_L}\right) = \Delta H^o_{\,L} - T\Delta S^o_{\,L} - \Delta H^o_{\,d} + T\Delta S^o_{\,d} \tag{14}$$

$$RT \ln\left(\frac{k_d}{k_L}\right) = \Delta H^o_{\,L} - \Delta H^o_{\,d} + T(\Delta S^o_{\,d} - \Delta S^o_{\,L}) \tag{15}$$

$$\ln\left(\frac{k_d}{k_L}\right) = \frac{\Delta H^o_{\,L}}{RT} - \frac{\Delta H^o_{\,d}}{RT} + \frac{(\Delta S^o_{\,d} - \Delta S^o_{\,L})}{R} \tag{16}$$

By differentiating Equation (16) with respect to T, we can write:

$$\frac{d\left(\ln \frac{k_d}{k_L}\right)}{dT} = -\frac{\Delta H^o_{\,L}}{RT^2} + \frac{\Delta H^o_{\,d}}{RT^2} \tag{17}$$

Applying the Arrhenius equation to the two above-mentioned contributions governed by k_L and k_d, we can write the equation:

$$\frac{k_d}{k_L} = \frac{A_o\, e^{-E_L/RT}}{B_o\, e^{-E_d/RT}} \tag{18}$$

where E_L and E_d are their respective activation energies, while A_o and B_o are the corresponding pre-exponential factors.

The linearized version of this equation:

$$\ln \frac{k_d}{k_L} = -\frac{(-E_L + E_d)}{RT} + \ln \frac{A_o}{B_o} \tag{19}$$

can be differentiated with respect to T to omit the constant term, thus leading to the equation:

$$\frac{d\left(\ln \frac{k_d}{k_L}\right)}{dT} = \frac{(-E_L + E_d)}{RT^2} \tag{20}$$

Equaling the right terms of Equations (17) and (20), the enthalpy of the reaction catalyzed by tannase entrapped in calcium alginate beads can be calculated as the difference of E_d and E_L:

$$\Delta H^o_{\,RT} = \Delta H^\circ_{\,d} - \Delta H^\circ_{\,L} = E_d - E_L \tag{21}$$

Considering Equation (12), one can calculate the entropy of tannase-catalyzed reaction for this enzyme preparation by the equation:

$$\Delta S^\circ_{\,RT} = \frac{\Delta H^\circ_{\,RT} - \Delta G^\circ_{\,RT}}{T} \tag{22}$$

3.4. Shelf Life Stability of the Immobilized Enzyme Preparations

Storage stability of both free and immobilized tannase, with starting activity of about 170 U/mL, was checked by determining the residual activity after long-time storage (90 days) at 4 °C. Free tannase was stored as solution in 0.2 M acetate buffer, pH 5.0, while the immobilized one was stored in wet form. Enzyme activity was determined at regular time intervals (15 days). The kinetic shelf-life parameters of tannase immobilized either in mDE-PANI or in calcium alginate beads were evaluated for 90 days to 4 °C. The enzyme activity was performed every 15 days using tannic acid as a substrate as described in Section 3.2. k_d, $t_{1/2}$ and D-value were determined as described by Silva et al. [14].

4. Conclusions

The *Aspergillus ficuum* tannase, which was immobilized either by entrapment in alginate calcium beads or covalently on magnetic nanoparticles, showed good stability when subjected to different temperatures and exposed to the solvent for a long period. The thermodynamic analysis of the reaction revealed that tannase immobilized on nanoparticles had the lowest activation energy and, therefore, would be the most appropriate enzyme preparation to conduct low-cost industrial tannin degradation treatments. However, thermodynamics and kinetics of biocatalyst denaturation showed that, although the enzyme either in its free or immobilized form is subject to a reversible denaturation mechanism, calcium alginate immobilization ensured greater stability for longer. Using this entrapment technique, tannase hydrolytic activity was increased due to leaching of support accompanied by pore enlargement, which, in addition to allowing greater biocatalyst contact with the substrate, provided greater protection against thermal inactivation after 90 min of incubation. Finally, shelf-life tests performed on immobilized biocatalysts at 4 °C for 90 days revealed that the enzyme immobilized on magnetic nanoparticles kept its activity for almost twice the time as the enzyme entrapped in calcium alginate beads. Such findings suggest that this tannase in both immobilized forms, which showed great potential for tannin degradation and thermal stability, could be profitably exploited for applications in the food industry.

Author Contributions: Conceptualization, A.C., J.d.C.-S., M.F.d.S. and T.S.P.; methodology, J.d.C.-S., A.C., M.F.d.S. and J.S.d.L.; software, J.d.C.-S., A.C., M.F.d.S. and T.S.P.; validation, A.C., J.d.C.-S., M.F.d.S. and T.S.P.; formal analysis, J.d.C.-S., J.S.d.L. and M.F.d.S.; investigation, J.d.C.-S., A.C., J.S.d.L. and M.F.d.S.; resources, L.B.d.C.J., A.C. and T.S.P.; data curation, J.d.C.-S., A.C. and M.F.d.S.; writing—original draft preparation, J.d.C.-S. and M.F.d.S.; writing—review and editing, J.d.C.-S., M.F.d.S., T.S.P. and A.C.; visualization, J.d.C.-S., M.F.d.S., A.C. and T.S.P.; supervision, A.C., L.B.d.C.J. and T.S.P.; project administration, A.C., L.B.d.C.J. and T.S.P.; funding acquisition, L.B.d.C.J., A.C. and T.S.P. All authors have read and agreed to the published version of the manuscript.

Funding: This study was funded in part by the CAPES (Coordenação de Aperfeiçoamento de Pessoal de Nível Superior–Brazil, grant number 88881.131854/2016-01).

Data Availability Statement: Not applicable.

Acknowledgments: J.D.C.S. would like to thank FACEPE (Foundation for Science and Technology support of the Pernambuco State, Recife, Brazil) for complementary scholarship DCR-004-3.06/21 and CNPq (National Council for Scientific and Technological Development, Brazil) for scholarship 300883/2021-8. This article is in part a result of the projects APQ-0210-3.06/19 and APQ-0726-5.07/21 from FACEPE. Furthermore, M.F.D.S. would also like to thank CETENE/MCTI (Centro de Tecnologias Estratégicas do Nordeste/Ministério da Ciência, Tecnologia e Inovação) and CNPq, grant number 301123/2023-3.

Conflicts of Interest: The authors declare no conflict of interest.

References

1. Eixenberger, D.; Kumar, A.; Klinger, S.; Scharnagl, N.; Dawood, A.W.H.; Liese, A. Polymer-grafted 3D-printed material for enzyme immobilization—designing a smart enzyme carrier. *Catalysts* **2023**, *13*, 1130. [CrossRef]
2. de Oliveira, R.L.; dos Santos, V.L.V.; da Silva, M.F.; Porto, T.S. Kinetic/thermodynamic study of immobilized β-fructofuranosidase from *Aspergillus tamarii* URM4634 in chitosan beads and application on invert sugar production in packed bed reactor. *Food Res. Int.* **2020**, *137*, 109730. [CrossRef] [PubMed]
3. Silva, J.D.C.; de França, P.R.L.; Converti, A.; Porto, T.S. Pectin hydrolysis in cashew apple juice by *Aspergillus aculeatus* URM4953 polygalacturonase covalently-immobilized on calcium alginate beads: A kinetic and thermodynamic study. *Int. J. Biol. Macromol.* **2019**, *126*, 820–827. [CrossRef] [PubMed]

4.   Da Silva, O.S.; de Oliveira, R.L.; Silva, J.D.C.; Converti, A.; Porto, T.S. Thermodynamic investigation of an alkaline protease from *Aspergillus tamarii* URM4634: A comparative approach between crude extract and purified enzyme. *Int. J. Biol. Macromol.* **2018**, *109*, 1039–1044. [CrossRef]

5.   Sayed Mostafa, H. Production of low-tannin *Hibiscus sabdariffa* tea through D-optimal design optimization of the preparation conditions and the catalytic action of new tannase. *Food Chem. X.* **2023**, *17*, 100562. [CrossRef] [PubMed]

6.   Mansor, A.; Samat, N.; Mat Amin, N.; Syaril Ramli, M.; Siva, R. Chapter. Microbial tannase production from agro-industrial byproducts for industrial applications. In *Microbial Bioprocessing of Agri-Food Wastes. Industrial Enzymes*, 1st ed.; Molina, G., Usmani, Z., Sharma, M., Benhida, R., Kuhad, R.C., Gupta, V.K., Eds.; CRC Press: Boca Raton, FL, USA, 2023; p. 40, ISBN 9781003341017.

7.   Benucci, I.; Lombardelli, C.; Esti, M. Innovative continuous biocatalytic system based on immobilized tannase: Possible prospects for the haze-active phenols hydrolysis in brewing industry. *Eur. Food Res. Technol.* **2023**. [CrossRef]

8.   Kumar, M.; Mehra, R.; Yogi, R.; Singh, N.; Salar, R.K.; Saxena, G.; Rustagi, S. A novel tannase from *Klebsiella pneumoniae* KP715242 reduces haze and improves the quality of fruit juice and beverages through detannification. *Front. Sustain. Food Syst.* **2023**, *7*, 1173611. [CrossRef]

9.   Song, L.; Wang, X.-C.; Feng, Z.-Q.; Guo, Y.-F.; Meng, G.-Q.; Wang, H.-Y. Biotransformation of gallate esters by a pH-stable tannase of mangrove-derived yeast *Debaryomyces hansenii*. *Front. Mol. Biosci.* **2023**, *10*, 1211621. [CrossRef] [PubMed]

10. Ristinmaa, A.S.; Coleman, T.; Cesar, L.; Weinmann, A.L.; Mazurkewich, S.; Branden, G.; Hasani, M.; Larsbrink, J. Structural diversity and substrate preferences of three tannase enzymes encoded by the anaerobic bacterium *Clostridium butyricum*. *J. Biol. Chem.* **2022**, *298*, 101758. [CrossRef] [PubMed]

11. Abdel-Naby, M.A.; El-Tanash, A.B.; Sherief, A.D.A. Structural characterization, catalytic, kinetic and thermodynamic properties of *Aspergillus oryzae* tannase. *Int. J. Biol. Macromol.* **2016**, *92*, 803–811. [CrossRef] [PubMed]

12. Xie, J.; Zhang, Y.; Simpson, B. Food enzymes immobilization: Novel carriers, techniques and applications. *Curr. Opin. Food Sci.* **2022**, *43*, 27–35. [CrossRef]

13. Maghraby, Y.R.; El-Shabasy, R.M.; Ibrahim, A.H.; El-Said Azzazy, H.M. Enzyme immobilization technologies and industrial applications. *ACS Omega* **2023**, *8*, 5184–5196. [CrossRef] [PubMed]

14. Silva, J.D.C.; de França, P.R.L.; Converti, A.; Porto, T.S. Kinetic and thermodynamic characterization of a novel *Aspergillus aculeatus* URM4953 polygalacturonase. Comparison of free and calcium alginate-immobilized enzyme. *Process Biochem.* **2018**, *74*, 61–70. [CrossRef]

15. Miguel Júnior, J.; Mattos, F.R.; Costa, G.R.; Zurlo, A.B.R.; Fernandez-Lafuente, R.; Mendes, A.A. Improved catalytic performance of Lipase Eversa® Transform 2.0 via immobilization for the sustainable production of flavor esters—Adsorption process and environmental assessment studies. *Catalysts* **2022**, *12*, 1412. [CrossRef]

16. Valls-Chivas, A.; Gómez, J.; Garcia-Peiro, J.I.; Hornos, F.; Hueso, J.L. Enzyme–iron oxide nanoassemblies: A review of immobilization and biocatalytic applications. *Catalysts* **2023**, *13*, 980. [CrossRef]

17. Larosa, C.; Salerno, M.; de Lima, J.S.; Meri, R.M.; da Silva, M.F.; de Carvalho, L.B.; Converti, A. Characterisation of bare and tannase-loaded calcium alginate beads by microscopic, thermogravimetric, FTIR and XRD analyses. *Int. J. Biol. Macromol.* **2018**, *115*, 900–906. [CrossRef]

18. Intisar, A.; Hedar, H.; Sharif, A.; Ahmed, E.; Hussain, N.; Hadibarata, H.; Ali Shariati, M.; Smaoui, S. Chapter 21–Enzyme immobilization on alginate biopolymer for biotechnological applications. In *Microbial Biomolecules. Emerging Approach in Agriculture, Pharmaceuticals and Environment Management. Developments in Applied Microbiology and Biotechnology*; Elsevier: Amsterdam, The Netherlands, 2023; pp. 471–488, ISBN 978-0-323-99476-7. [CrossRef]

19. de Lima, J.S.; Cabrera, M.P.; Casazza, A.A.; da Silva, M.F.; Perego, P.; de Carvalho, L.B., Jr.; Converti, A. Immobilization of *Aspergillus ficuum* tannase in calcium alginate beads and its application in the treatment of boldo (*Peumus boldus*) tea. *Int. J. Biol. Macromol.* **2018**, *118*, 1989–1994. [CrossRef]

20. Cabrera, M.P.; da Fonseca, T.F.; de Souza, R.V.B.; de Assis, C.R.D.; Marcatoma, J.Q.; Maciel, J.D.C.; Neri, D.F.M.; Soria, F.; de Carvalho, L.B., Jr. Polyaniline-coated magnetic diatomite nanoparticles as a matrix for immobilizing enzymes. *Appl. Surf. Sci.* **2018**, *457*, 21–29. [CrossRef]

21. de Lima, J.S.; Cabrera, M.P.; Motta, C.M.D.S.; Converti, A.; Carvalho, L.B., Jr. Hydrolysis of tannins by tannase immobilized onto magnetic diatomaceous earth nanoparticles coated with polyaniline. *Food Res. Int.* **2018**, *107*, 470–476. [CrossRef]

22. Gao, X.; Pan, H.; Tian, S.; Su, L.; Hu, Z.; Qiao, C.; Liu, Q.; Zhou, C. Co-immobilization of bienzyme HRP/GO$_x$ on highly stable hierarchically porous MOF with enhanced catalytic activity and stability: Kinetic and thermodynamic studies. *J. Environ. Chem. Eng.* **2023**, *11*, 110684. [CrossRef]

23. Wahba, M.I.; Saleh, S.A.A.; Mostafa, F.A.; Abdel Wahab, W.A. Immobilization impact of GEG-Alg-SPI as a carrier for *Aspergillus niger* MK981235 inulinase: Kinetics, thermodynamics, and application. *Bioresour. Technol. Rep.* **2022**, *18*, 101099. [CrossRef]

24. De Oliveira, R.L.; da Silva, M.F.; da Silva, S.P.; Cavalcanti, J.V.F.L.; Converti, A.; Porto, T.S. Immobilization of a commercial *Aspergillus aculeatus* enzyme preparation with fructosyltransferase activity in chitosan beads: A kinetic/thermodynamic study and fructo-oligosaccharides continuous production in enzymatic reactor. *Food Bioprod. Process.* **2020**, *122*, 169–182. [CrossRef]

25. De Oliveira, R.L.; da Silva, S.P.; Converti, A.; Porto, T.S. Production, biochemical characterization, and kinetic/thermodynamic study of inulinase from *Aspergillus terreus* URM4658. *Molecules* **2022**, *27*, 6418. [CrossRef] [PubMed]

26. De Oliveira, R.L.; Claudino, E.S.; Converti, A.; Porto, T.S. Use of a sequential fermentation method for the production of *Aspergillus tamarii* URM4634 protease and a kinetic/thermodynamic study of the enzyme. *Catalysts* **2021**, *11*, 963. [CrossRef]

27. Fernandes, L.M.G.; Carneiro-da-Cunha, M.N.; Silva, J.D.C.; Porto, A.L.F.; Porto, T.S. Purification and characterization of a novel *Aspergillus heteromorphus* URM 0269 protease extracted by aqueous two-phase systems PEG/citrate. *J. Mol. Liq.* **2020**, *317*, 113957. [CrossRef]

28. Boudrant, J.; Woodley, J.M.; Fernandez-Lafuente, R. Parameters necessary to define an immobilized enzyme preparation. *Process Biochem.* **2020**, *90*, 66–80. [CrossRef]

29. Abellanas-Perez, P.; Carballares, D.; Fernandez-Lafuente, R.; Rocha-Martin, J. Glutaraldehyde modification of lipases immobilized on octyl agarose beads: Roles of the support enzyme loading and chemical amination of the enzyme on the final enzyme features. *Int. J. Biol. Macromol.* **2023**, *248*, 125853. [CrossRef] [PubMed]

30. Heidtmann, R.B.; Duarte, S.H.; de Pereira, L.P.; Braga, A.R.C.; Kalil, S.J. Caracterização cinética e termodinâmica de β-galactosidase de *Kluyveromyces marxianus* CCT 7082 fracionada com sulfato de amônio. *Braz. J. Food Technol.* **2012**, *15*, 41–49. [CrossRef]

31. Khatun, S.; Riyazuddeen; Yasmeen, S. Unraveling the thermodynamics, enzyme activity and denaturation studies of triprolidine hydrochloride binding with model transport protein. *J. Mol. Liq.* **2021**, *337*, 116569. [CrossRef]

32. Rodríguez-López, J.N.; Fenoll, L.G.; Tudela, J.; Devece, C.; Sánchez-Hernández, D.; de los Reyes, E.; García-Cánovas, F. Thermal Inactivation of Mushroom Polyphenoloxidase Employing 2450 MHz Microwave Radiation. *J. Agric. Food Chem.* **1999**, *47*, 3028–3035. [CrossRef]

33. Bradford, M.M. A rapid and sensitive method for the quantitation of microgram quantities of protein utilizing the principle of protein-dye binding. *Anal. Biochem.* **1976**, *72*, 248–254. [CrossRef] [PubMed]

34. Pinto, G.A.S.; Couri, S.; Gonçalves, E.B. Replacement of methanol by ethanol on gallic acid determination by rhodanine and its impacts on the tannase assay. *Electron. J. Environ. Agric. Food Chem.* **2006**, *5*, 1560–1568.

35. Da Silva, O.S.; Silva, J.D.C.; de Almeida, E.M.; Sousa, F.; Gonçalves, O.S.L.; Sarmento, B.; Neves-Petersen, M.T.; Porto, T.S. Biophysical, photochemical and biochemical characterization of a protease from *Aspergillus tamarii* URM4634. *Int. J. Biol. Macromol.* **2018**, *118*, 1655–1666. [CrossRef]

36. Michel, D. Simply conceiving the Arrhenius law and absolute kinetic constants using the geometric distribution. *Phys. A Stat. Mech. Appl.* **2013**, *392*, 4258–4264. [CrossRef]

37. Converti, A.; Pessoa, A., Jr.; Silva, J.D.C.; de Oliveira, R.L.; Porto, T.S. Thermodynamics applied to biomolecules. In *Pharmaceutical Biotechnology: A Focus on Industrial Application*; Pessoa, A., Jr., Vitolo, M., Long, P.F., Eds.; CRC Press: Boca Raton, FL, USA, 2021; Volume 1, pp. 29–42, ISBN 9781000399844.

38. Ortega, N.; de Diego, S.; Perez-Mateos, M.; Busto, M.D. Kinetic properties and thermal behaviour of polygalacturonase used in fruit juice clarification. *Food Chem.* **2004**, *88*, 209–217. [CrossRef]

 *catalysts* 

Article

Comparative Evaluation of the Asymmetric Synthesis of (S)-Norlaudanosoline in a Two-Step Biocatalytic Reaction with Whole *Escherichia coli* Cells in Batch and Continuous Flow Catalysis

Adson Hagen Arnold and Kathrin Castiglione *

Institute of Bioprocess Engineering, Friedrich-Alexander-Universität Erlangen-Nürnberg, Paul-Gordan-Str. 3, 91052 Erlangen, Germany
* Correspondence: kathrin.castiglione@fau.de

Abstract: Opioids are important analgesics, and their pharmaceutical application is increasing worldwide. Many opioids are based on benzylisoquinoline alkaloids (BIA) and are still industrially produced from *Papaver somniferum* (opium poppy). (S)-norlaudanosoline ((S)-NLS) is a complex BIA and an advanced intermediate for diverse pharmaceuticals. The efficient synthesis of this scaffold could pave the way for a plant-independent synthesis platform. Although a promising biocatalytic route to (S)-NLS using norcoclaurine synthase (NCS) and ω-transaminase (TAm) has already been explored, the cost-effectiveness of this process still needs much improvement. Therefore, we investigated whether the synthesis could also be performed using whole cells to avoid the use of (partially) purified enzymes. With an optimized mixing ratio of TAm- and NCS-containing cells in batch biotransformations, 50 mM substrate was converted within 3 h with more than 90% yield and a high enantiomeric excess of the product (95%). To further increase the space–time yield, the cells were immobilized to enable their retainment in fixed-bed reactors. A comparison of glass beads, Diaion HP-2MG and alginate revealed that the addition of Diaion during bacterial growth led to the most active immobilisates. To facilitate sustained production of (S)-NLS, a fixed-bed setup was constructed based on lithographically printed columns from biocompatible PRO-BLK 10 plastic. The continuous production at two scales (5 mL and 50 mL columns) revealed insufficient system stability originating from biocatalyst leaching and inactivation. Thus, while the use of whole cells in batch biotransformations represents an immediate process improvement, the transfer to flow catalysis needs further optimization.

Keywords: chiral synthesis; flow-reactor; immobilized cells; dopamine; stereolithography; transaminase; benzylisoquinoline alkaloids

Citation: Arnold, A.H.; Castiglione, K. Comparative Evaluation of the Asymmetric Synthesis of (S)-Norlaudanosoline in a Two-Step Biocatalytic Reaction with Whole *Escherichia coli* Cells in Batch and Continuous Flow Catalysis. *Catalysts* **2023**, *13*, 1347. https://doi.org/10.3390/catal13101347

Academic Editors: Chwen-Jen Shieh, Hui-Min David Wang and Chia-Hung Kuo

Received: 22 August 2023
Revised: 2 October 2023
Accepted: 4 October 2023
Published: 5 October 2023

1. Introduction

Benzylisoquinoline alkaloids (BIAs) are a family of secondary metabolites with a wide variety of pharmaceutical applications. Important examples of this family are the opioid analgesics morphine and codeine, the antitussive noscapine or the antispasmodic drug papaverine [1]. Opioids are classified as essential medicines by the World Health Organization (WHO), due to their usefulness in reducing pain for trauma patients and in palliative care [2], with 10 billion standard daily doses consumed per year [3]. For industrial production, typically a largely mechanized process is used in which fully mature plants are harvested, dried and then pulverized, before extracting the active component [4], which is then refined for use in pharmaceuticals. Environmental factors such as pests, diseases, and extreme weather events make poppy supply chains unstable, threatening the secure availability of the medicines derived from them. Therefore, there is great interest in developing efficient and scalable production platforms that would not only secure supply but also overcome the capacity limitations of the plant-based production.

More than 30 chemical and chemo-enzymatic synthesis pathways of morphine and derivatives are known, but none of them could be developed to the point where industrial production at market prices would be possible, as the synthesis of these chiral molecules is very complex and currently not cost-competitive [5]. The first full biosynthesis of opioids in yeast was published in 2015 but was also not able to deter industrial production away from using the opium poppy as the raw material, due to the very low titers achieved [6]. In the same year, an elegant biocatalytic route to the important BIA scaffold structure (S)-norlaudanosoline ((S)-NLS) was published [7], which involves two enzymatic transformations and the affordable starting materials dopamine (DA) and pyruvate, as shown in Scheme 1. Since (S)-NLS can be diversified into various target structures, including the pharmaceuticals morphine and codeine or the neuromuscular-blocking drug cisatracurium [1,8,9], the efficient industrial-scale production of this molecule would represent a milestone on the way to a plant-independent synthesis platform.

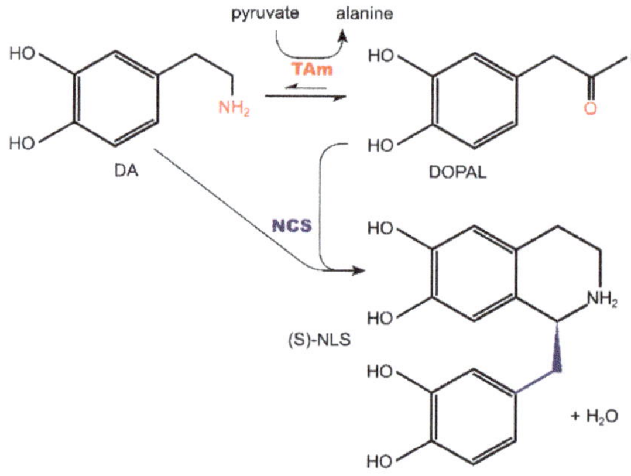

Scheme 1. In a ω-transaminase (TAm)-catalyzed reaction, dopamine (DA) is converted into 3,4-dihydroxyphenylacetaldehyde (DOPAL). Subsequently, norcoclaurine synthase (NCS) synthesizes (S)-norlaudanosoline ((S)-NLS) from DA and DOPAL. The functional groups relevant for the TAm reaction are highlighted in red, while the stereochemical bond generated by the asymmetric NCS reaction is depicted as a blue wedge.

In the two-step enzymatic synthesis of (S)-NLS, dopamine is first converted to 3,4-dihydroxyphenylacetaldehyde (DOPAL) at the expense of pyruvate using transaminase (EC 2.6.1.B16), which catalyzes the transfer of an amine group from an amine donor to either a ketone or aldehyde with 5′-pyridoxalphosphate (PLP) as the amine shuttling cofactor [10]. A transaminase subgroup, the so-called ω-transaminases (TAms), use non-α position amino acids or amine compounds without carboxylic group as donors and have been studied intensively due to their high relevance for the production of chiral molecules, e.g., in fine chemistry and the pharmaceutical industry. The TAm-catalyzed reactions are generally reversible [10].

In the triangular cascade to (S)-norlaudanosoline, the second step is catalyzed by norcoclaurine synthase (NCS, EC 4.2.1.78), which combines dopamine and the formed aldehyde, thereby creating the BIA core structure. NCS, which is found in plants, catalyzes a stereoselective Pictet–Spengler cyclisation, where arylethylamine and aldehyde/ketone condense followed by a ring closure.

One general problem of the reaction is the oxidative degeneration of DA and DOPAL, which can be reduced by the exclusion of oxygen [11–13]. Another challenge is the inherent instability of DOPAL and its tendency to create polymers with the surrounding molecules via a radical reaction [11], often creating black eumelanin [14]. This can be mitigated by adding the antioxidative ascorbate to the reaction solution [15].

A spontaneous reaction of DA and DOPAL to racemic NLS can be observed in the presence of several conventional buffer ions, for example, phosphate. This unwanted spontaneous reaction can be minimized by using only specific buffers, for example, HEPES buffer, and by keeping the DOPAL (or theoretically DA) concentration low [11,16,17]. Here, the triangular cascade synthesis of (S)-NLS using TAm and NCS is beneficial since it generates and utilizes the reactive aldehyde DOPAL in situ and achieved yields of up to 87% using (partially) purified enzymes [7].

To further optimize the biocatalytic synthesis of (S)-NLS from dopamine and pyruvate and its cost effectiveness, the use of whole cells instead of (partially) purified enzymes and the establishment of a continuous process in flow are two options. While the use of whole cells lowers the cost of the catalyst preparation, continuous processing can intensify biocatalytic reactions, thereby reducing the size of the process equipment, improving product quality and lowering energy and solvent consumption, as well as the waste generated [18]. Ultimately, all these aspects reduce the footprint of the process and might also improve its safety [19]. While diverse TAm reactions (e.g., [20–22]) and also the NCS reaction [15] have been successfully performed in flow setups, the triangular synthesis of (S)-NLS has not been investigated in flow so far.

The aim of this work was to elucidate whether the asymmetric two-step synthesis of (S)-NLS can be performed with whole-cell catalysts and whether the joint implementation of the TAm and NCS reactions in a continuous flow system provides access to a more efficient synthesis of the BIA scaffold structure. Possible benefits could be a higher space–time yield, a reduced spontaneous background reaction, and thus, a higher optical purity of the product, as proposed by Lechner et al. [15]. Furthermore, simple 3D-printed columns should be used as fixed-bed reactors. The method of 3D printing has long been used in the manufacturing industry to produce design prototypes and accelerate developments, made possible by the design freedom inherent in additive manufacturing in combination with the low production costs [23], and we wanted to take advantage of these benefits in this study as well.

2. Results

To investigate the asymmetric two-step enzymatic synthesis of the BIA scaffold structure (S)-NLS with whole cell catalysts in batch biotransformations and flow setups, two recombinant *E. coli* strains expressing the genes for the enzymes TAm and NCS were cultivated separately. Subsequently, the cells were washed and lyophilized. Both biocatalysts showed good expression levels of the target proteins (Supporting Figures S1 and S2). Typical specific activities of TAm-containing cells were in the range of 30 U per gram cell dry mass. Due to the high (non-specific) reactivity of the aldehyde DOPAL, the specific activity of NCS cells was not determined directly with the substrate in excess, but as combined activity in batch experiments to identify the optimal cell ratio.

2.1. Batch Biotransformations

To achieve the maximum enzymatic activity and to test the interaction of both enzymes, different mass ratios of lyophilized TAm- and NCS-containing cells were combined into mixtures. The different cell mixtures were suspended in the reaction buffer, and the reactions were performed in miniaturized stirred-tank reactors at the 12 mL scale. To minimize the oxidation and subsequent polymerization of substrates and intermediates, ascorbate was used, and the reactors' headspaces were gassed with nitrogen. The antibiotic gentamicin was added to prevent bacterial growth, e.g., on pyruvate as the carbon source. The enzyme activity of TAm and NCS in the cell mixtures was measured in tandem, as the NCS reaction is dependent on the supply of the intermediate DOPAL via TAm. The TAm (1st cascade reaction) activity is in turn increased by the presence of NCS (2nd cascade reaction), which uses DOPAL in its reaction. To obtain results that are comparable to those of the cell-free enzyme system used by Lichman et al. [7], samples were taken after 3 h. Measurements were done for (S)-NLS and (R)-NLS, to determine the yield (referring to dopamine) and enantiomeric excess (ee), as well as for alanine and pyruvate, to gain a measure of the TAm activity, as summarized in Table 1.

Table 1. Activity screening of varied TAm/NCS-cell ratios in miniaturized stirred-tank reactors at the 12 mL scale.

Cells Containing		Ratio				
TAm	NCS	TAm-Containing Cells/ NCS-Containing Cells	Yield [a]	Ee [b]	Alanine	Pyruvate
[mg]	[mg]	[-]	[%]	[%]	[mM]	[mM]
12	12	1	11	37	22	9
36	12	3	11	55	23	-
60	12	5	12	39	26	-
12	66	0.18	34	87	10	11
36	66	0.55	35 ± 1 [c]	81 ± 1 [c]	23 ± 1 [c]	-
60	66	0.91	36	83	21	-
12	120	0.10	67	94	26	18
36	120	0.30	62	90	23	-
60	120	0.50	54	99	24	-
36	228	0.16	91	95	25	-
36	336	0.11	91	97	22	-
36	444	0.081	92	97	15	-
36	552	0.065	85	97	18	-
36	660	0.055	96	96	29	-

[a] Reaction conditions: 50 mM HEPES (pH 7.4), 50 mM dopamine, 50 mM pyruvate, 4 g L^{-1} ascorbate, 25 µg mL^{-1} gentamicin, 3 h, and 37 °C. [b] ee: enantiomeric excess. [c] The data represent the results of single-replicate experiments, except for one triplicate (n = 3) showing the low standard deviation of the determined values.

The central finding of this experiment was the fact that mainly the mass of NCS-containing cells in the mixture was crucial for the yield, while the TAm-containing cell mass above a minimal amount nearly did not factor in. In the assays with only 12 mg of TAm-containing cells, residual pyruvate concentrations of 9 mM, 11 mM and 18 mM were found. Increased amounts of TAm led to a complete pyruvate depletion during the reaction. Pyruvate is used up much more than would be expected by the reaction scheme. This has been anticipated as a necessary cost of using intact cells, which still contain metabolic enzymes. These can metabolize some of the pyruvate as it is a central part of their carbon metabolism [24]. Although gentamicin inhibits the protein biosynthesis and prevents the growth of bacteria, metabolic enzymes already present in the cells can still degrade the carbon source pyruvate. This was also shown in a previous experiment, where tests with 25 mM pyruvate and without gentamicin led to a complete pyruvate depletion in all samples. The increasing residual pyruvate concentration with the increasing concentration of NCS-containing cells could indicate an inhibiting effect of the final product on the cellular metabolism of *E. coli*, preventing the additional use of pyruvate as a carbon source.

While increased amounts of TAm-containing cells with a given amount of NCS-containing cells led to only an insignificant increase in the total yield in two of the three assays, for 120 mg of NCS-containing cells, higher amounts of TAm led to an even lower yield (from 67 over 62 to 54%). This could indicate a negative effect of DOPAL accumulation created by excess TAm. The synergy between the catechol and aldehyde groups of DOPAL strongly enhances its reactivity, possibly leading to enzyme inactivation by modification of functional protein residues, protein cross-linking and protein aggregation [25].

Increased amounts of NCS-containing cells in the reaction setup resulted in higher yields: 12 mg of catalyst gave 11% yield, while 66 mg resulted in 35%. In the experiments with 36 mg of TAm-containing cells, the yield seemed to linearly increase with the NCS-containing cell mass until reaching 91% at 228 mg. Thereafter, the yield stopped increasing steadily. The observed maximum yield of about 90% could be due to product inhibition of NCS, which could be analogous to the inhibitory effect of (S)-norcoclaurine detected in an earlier study [15]. This trihydroxylated compound represents the BIA scaffold structure occurring in plant BIA syntheses and differs from the tetrahydroxylated (S)-NLS by the absence of a hydroxy group at the isoquinoline moiety.

In general, ee followed the trend of the yield. An increasing amount of NCS in the reaction mixture led to an increase in enantioselective enzyme activity compared to the non-selective spontaneous Pictet–Spengler reaction and thus to a higher optical purity. Due to the single measurements, the effects that were observed with very high cell concentrations are of limited significance. Fluctuations in the yield and ee could also be due to measurement uncertainties. This implies that amounts of the NCS catalyst higher than 228 mg do not reliably improve the output under these conditions. The mixture of 36 mg TAm-containing and 228 mg NCS-containing cell mass, corresponding to a mass ratio of 3:19, seemed optimal, although as shown by the pyruvate depletion, even lower ratios of TAm to NCS cells could also be successful.

Compared to the study performed by Lichman et al. [7], who used *E. coli* cell extract for TAm activity in combination with purified NCS enzymes, slightly higher yields were reached in this work. At substrate concentrations of 50 mM DA and 25 mM pyruvate, an 85% conversion (referring to DA) was reported after 3 h with a product ee of 96% (S). Due to the differences in catalyst preparations, the enzyme loadings are hard to compare. However, this result underlines that cheaply producible whole-cell catalysts are suitable for use in this two-step enzyme reaction and there is no need for enzyme purification. Whole cells were also employed in the study performed by Lechner et al. [15], who investigated the NCS-catalyzed reaction of DA and 4-hydroxyphenylacetaldehyde to (S)-norcoclaurine. At a catalyst concentration of 10 mg mL^{-1}, 27% conversion with >98% ee was obtained. This catalyst loading corresponds well to the experiments performed in this study, with 120 mg NCS-containing cells in a reaction volume of 12 mL, where yields of $\geq$54% with 90–99% ee were obtained, depending on the amount of TAm used. Although the syntheses are not directly comparable because of the different enzymatic reactions, the mass transport across two cell walls, which is necessary due to the use of catalysts with separately expressed TAm and NCS genes, does not seem to be severely rate-limiting since the reaction is faster than the one-step biotransformation to (S)-norcoclaurine, where the substrates must cross only one cell wall. This is supported by the fact that the enzymatic activities of NCS with DOPAL and 4-hydroxyphenylacetaldehyde, respectively, are typically in the same order of magnitude [26].

The comparison with other data from the literature is difficult in general as early reports either (1) used phosphate buffer leading to significant amounts of racemic NLS [16,27,28], (2) aim at slightly different BIAs such as norcoclaurine [15,29] or (S)-reticuline [30–32] or (3) only use a single-enzyme reaction [15,20,33].

2.2. Whole Cell Catalyst Immobilization on Carriers

To establish the two-step enzymatic reaction using cells in a flow setup, the cell mixture must be immobilized in the reactor, e.g., on a suitable carrier material. Immobilization has several advantages pertaining to (1) product purification, as less biomass must be removed, (2) reusability, as enzymes can theoretically be used until fully degraded, and (3) volumetric activity, as higher local cell concentrations can theoretically be created [34–36]. Thus, the immobilization of the cell mixture on carrier particles was investigated.

Five distinct types of catalyst particles were tested. Diaion HP-2MG represents methacrylic ester copolymer particles, which are hydrophilic and allow cells to adsorb to their surface. Two different methods of cell adsorption on Diaion particles were assessed here: (1) The incubation of particles with suspended lyophilized cells (Diaion incubated) or (2) the addition of particles during the growth of cells expressing the enzymes (Diaion overgrown). The dried particles with cells adsorbed on them were subsequently mixed. (3) As an even cheaper adsorber material, glass beads were selected and also incubated with the lyophilized cells [37]. For an established encapsulation option, calcium alginate beads were chosen. The high porosity of cross-linked alginate combined with the potential to use the full particle rather than just the surface implies the possibility of achieving high volumetric reaction rates, especially compared to the adsorption-based particles. The alginate beads were created using two types of preparation: (4) pure calcium alginate beads (uncoated alginate) or (5) calcium alginate beads additionally coated twice, first with polyDADMAC, which binds as cation to the anionic alginate on the bead surface, and subsequently in sodium-alginate, which binds to the polyDADMAC now on the surface, creating layers that are linked together at multiple sites (coated alginate).

For the five different types of particles loaded with the (3:19) mixture of TAm and NCS cells, respectively, 1.5 g of each were added to the miniaturized stirred-tank reactors containing 12 mL reaction medium in a batch setup. Headspace aeration with nitrogen was no longer carried out, as ascorbate was found to be sufficient to prevent oxidation, while contact with oxygen was still minimized for the solutions. The setup and reaction medium were otherwise kept the same as with the screening of cell ratios.

The yields obtained with the different preparations of the immobilized cells are summarized in Table 2. Preparations with glass did not lead to significant product formation (0.9% yield) and thus no ee could be determined. The optical purities obtained with all other catalyst particles were in the range between 89 and 97%. Due to measurement uncertainties, the differences between the ee values cannot be considered statistically significant. The selection of the catalyst preparation for further experiments was therefore based mainly on the yield obtained. The second lowest yield was observed for uncoated alginate particles (19.9%), followed by coated alginate particles (33.5%). The use of incubated Diaion led to a yield of 52.4%, whereas overgrown Diaion particles led to an even higher yield of 64.1%, corresponding to 55 mg of product. The addition of ascorbate and minimization of oxygen contact seemed to prevent the formation of insoluble polydopamine and eumelanin derivatives observed in other studies [20], as no blackening of any solution was observed.

Table 2. Yields, productivity and enantiomeric excess achieved with different catalyst particles. Additionally, the cellular dry mass (CDM) and the cell loading are shown.

Carrier	Yield [a]	Space–Time Yield	Overall Productivity	Cellular Dry Mass	Cell Loading	ee
[-]	[%]	$[\mu mol\ mL^{-1}\ h^{-1}]$	$[mg_{S\text{-}NLS}\ g_{CDM}^{-1}\ h^{-1}]$	$[g_{CDM}]$	$[g_{CDM}\ g_{carrier}^{-1}]$	[%]
Uncoated alginate	19.9 ± 4.5	1.66	1.52	0.06 ± 0.01	0.04	97 ± 4
Coated alginate	33.5 ± 9.0	2.79	1.22	0.14 ± 0.03	0.09	89 ± 11
Glass	0.9 ± 1.1	0.08	-	-[c]	-	-
Diaion HP-2MG incubated	52.4 ± 10.6	4.36	0.72	0.35 ± 0.03	0.23	93 ± 3
Diaion HP-2MG overgrown	64.1 ± 6.5	5.34	-	-[b]	-	95 ± 4

[a] Reaction conditions: 50 mM HEPES (pH 7.4), 50 mM dopamine, 50 mM pyruvate, and 4 g L^{-1} ascorbate. 3 h; 37 °C determined by chiral HPLC. n = 3. [b] Not measured. [c] No dry mass detectable.

While the cellular dry mass for the experiment with glass was below the limit of detection, the value for the overgrown Diaion could not be determined because the proliferation of cells during preparation did not allow for a mass balance. When comparing the cellular dry mass for the other three immobilizations, it was lowest with uncoated alginate at 0.06 ± 0.01 mg per batch of 1.5 g carriers, followed by coated alginate at 0.14 ± 0.03 mg, while incubated Diaion HP2MG fixed 0.35 ± 0.03 mg cell dry mass. A slight turbidity, later identified by microscopy as intact cells, occurred in the solution of alginate particles suggesting that the alginate particles lose cells during the reaction. This could indicate a constant loss of catalyst in the solution of alginate particles through a combination of dissolved alginate matrix and cells slowly washing out through openings in the matrix. Binding *E. coli* on Diaion seemed to be comparatively more stable and no turbidity was observed.

Although glass beads seem to be a bad choice for this type of use, they are still the cheapest option of the investigated particle types and thus should not be easily discounted. To increase adsorption and thus yield in future studies, (1) beads could be overgrown as was done with Diaion in this study and (2) surfaces could be covered by a thin coating of

Fe$_2$O$_3$ or other metal oxides, to increase the adsorption strength [38] or (3) cells could be linked covalently by aminopropyltrimethoxy silane and glutaraldehyde to the carrier to increase cell density [39].

The yield was related to the determined cellular dry mass to make statements about the productivity of the cells in the different batches. Here, uncoated alginate scored highest (1.52 mg$_{S\text{-NLS}}$ g$_{CDM}^{-1}$ h^{-1}), followed by coated alginate (1.22 mg$_{S\text{-NLS}}$ g$_{CDM}^{-1}$ h^{-1}) and incubated Diaion (0.72 mg$_{S\text{-NLS}}$ g$_{CDM}^{-1}$ h^{-1}). While productivities were even higher in alginate, this is to be expected, as the lower amount of biomass in these preparations led to lower product accumulation and thus lower product inhibition, but also a lower space–time yield, which was deemed paramount for a later application.

Though the exact productivity values could not be calculated for overgrown Diaion particles, as it is unknown how much cell mass is deposited during incubation, they were found to reach the highest yield combined with an acceptable ee of 95% and were thus chosen as the biocatalyst for subsequent experiments.

To better interpret the higher yield of overgrown Diaion compared to incubated Diaion, both particle populations as well as untreated Diaion particles and Diaion particles left to hydrate in 0.9% NaCl for one day were measured using a Mastersizer 2000 (Table 3). Sizes of wet Diaion (d (0.5) = 580 ± 2 μm), incubated Diaion (d (0.5) = 598 ± 4 μm) and overgrown Diaion (d (0.5) = 546 ± 7 μm) were different; the overgrown Diaion particles were smaller. The relatively lower diameter could be due to the abrasion of the particles by the shaking during cultivation.

Table 3. Diaion HP-2MG particle sizes as measured by Mastersizer in μm. (n = 5).

Particle Preparation	d (0.1)	d (0.5)	d (0.9)
Untreated	426 ± 23	574 ± 17	737 ± 15
1d in 0.9% NaCl	460 ± 6	580 ± 2	735 ± 13
Incubated with lyophilized cells	486 ± 2	598 ± 4	737 ± 5
Overgrown (added during induction)	462 ± 9	546 ± 7	643 ± 4

2.3. Preliminary Experiments in Packed-Bed Flow

Having established an optimum ratio of TAm-containing to NCS-containing cells and having chosen overgrown Diaion particles as the best catalytic preparation, a preliminary test in a flow setup was carried out. The flow setup could be advantageous over a batch setup as it could allow production over a longer time scale than batch setups, if the enzyme activity can be maintained, while still keeping concentrations of DA and DOPAL relatively low. This in turn minimizes the unselective background reaction of DA and DOPAL to racemic NLS, which can occur spontaneously, and also reduces the risk of substrate polymerization to polydopamine or eumelanin derivatives.

For the preliminary experiment, a 5 mL-reactor (reactor A) with typical cylindrical geometry was designed and additively manufactured. To facilitate manual filling and assembly, the column was printed in two parts and complementary screw threads were added to both parts. For the inlet and outlet, the form of a male Luer-Lock connection was used, to allow an easy connection with the periphery.

Before both reactor parts were screwed together, the inside was carefully filled with the overgrown Diaion particle mixture creating the particle bed. The reactor was then connected to tubes, placed in a temperature-controlled incubator at 37 °C and was filled with HEPES buffer. During the reaction, a HPLC pump was used to continuously feed substrate at a flow rate of 0.5 mL min^{-1}, while a fraction collector was used to collect the outflowing fluid. The first fraction contained the outflow of the first 10 min (5 mL), while each subsequent fraction contained the outflow of 20 min (10 mL). Samples were taken from each fraction and the concentrations of both NLS enantiomers were measured to calculate the yield and ee.

After the reaction, the particle mass was measured again to be able to compute porosity. The bed volume was determined to be 4.26 mL accommodating 0.47 g of Diaion particles, which corresponds to 0.11 g cm^{-3}. The total fluid volume of the filled column A was determined as 1.6 mL resulting in a porosity of 37% (Φ = 0.37). This corresponds well with the literature values for dense random packing (Φ = 0.359–0.375) and thus, this column packing was deemed acceptable [40].

The advantage of additive manufacturing is the fast and inexpensive design of prototypes for preliminary experiments to clarify individual questions. The small column was used to investigate if any adverse effects of the "Figure 4®PRO-BLK 10" plastic were detected (which was not the case) and if the immobilized cell preparation appears to be sufficiently stable in the flow experiments to allow for larger-scale experiments needing more material. From the combination of the flow rate and the hold-up volume of the column, a residence time of 3.2 min was calculated. Compared to the batch experiment with Diaion (overgrown), which was performed with 3-times more catalyst and a 56-times longer reaction time, a much lower yield than 64.1% (Table 2) was expected. As shown in Figure 1, a yield of around 6% was reached over 80 column volumes (CV), with a consistent ee of 97% and an overall productivity of more than 118 mg$_{(S)\text{-NLS}}$ g$_{CDM}$$^{-1}$ h^{-1}, assuming that these particles had at least the same cell loading as the incubated Diaion.

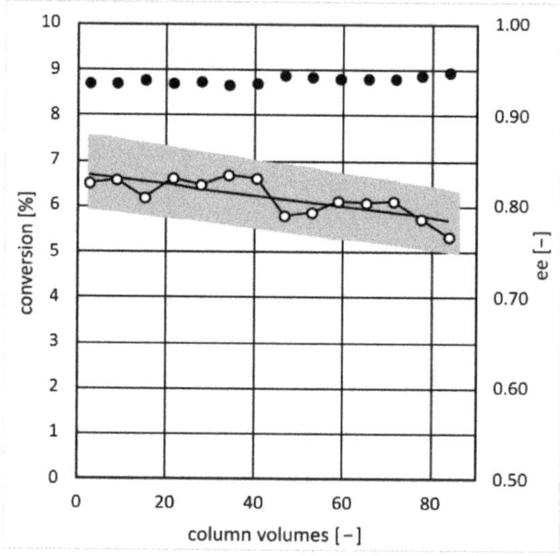

Figure 1. Time-course of the synthesis of (S)-norlaudanosoline using column A with 0.47 g of particles. White circles: data points yield; black line: linear trend; greyed area: method deviation around trend line. Black circles: data points, ee. Reaction conditions: 50 mM HEPES (pH 7.4), 50 mM dopamine, 50 mM pyruvate, 4 g L^{-1} ascorbate, 4 g L^{-1} ascorbate, 25 µg mL^{-1} gentamicin, 37 °C and 0.5 mL min^{-1} flow. ee at 0.97 determined by chiral HPLC, trend equation y = − 0.012x + 6.7; R^2 = 0.59.

The yield dropped from about 6.5% at the start to 5.8% at CV 80, corresponding to a trend of constantly decreasing yield. Over the time course of the experiment of 5 h, the product concentration measured at the outlet decreased by 18%, pointing to the fact that the system stability is not high enough for long-time operation. To further substantiate this finding, a confirmational large scale-experiment was conducted.

2.4. Rescaled Packed Bed Reactor

To create comparability to previous experiments performed in batch mode, a residence time of about 180 min was envisioned, which would correspond to a 60-fold increase in contact time compared to the preliminary experiment with a residence time of 3.2 min. This was achieved by lowering the flow rate from 0.5 mL min^{-1} to 0.1 mL min^{-1}, which was the minimum flow rate compatible with the fraction collector and designing a column with a 12-times increased inner volume of 48 mL. A cone shaped part was added to the inlet and outlet section, to accommodate flow through the wider main cylindrical column part. The total bed volume was determined to be 51.1 mL, which was filled with 15.05 g of a 1:2 mixture of overgrown Diaion particles carrying TAm or NCS and untreated Diaion particles carrying no cells. For an industrial application, of course, the column should be filled with active material only and the flow rate should be increased proportionally, while minding the increased pressure. This was only carried out in this way to cope with the system restriction regarding the minimum flow rate without using too much catalyst. Nevertheless, a 3.7-fold higher catalyst load was used compared to the batch process performed with Diaion (overgrown) (5.5 g versus 1.5 g). The total fluid volume was computed as 18.9 mL, resulting in an average residence time of 189 min at a flow rate of 0.1 mL min^{-1}.

Figure 2 shows the progression of the yield of the reaction and the ee of the product as a function of column volumes. After replacing NaCl in the void volume of the column with a substrate solution during CV 1, the yield rose rapidly to 88.3%. This is higher than the yield obtained in the batch process with Diaion (overgrown), which can be explained by the higher amount of catalysts in the system. The yield stayed stable for about 1–2 CVs (~3–6 h). Afterwards, the yield decreased consistently and with increasing intensity until the end of the experiment to 58%. This means a proportional loss of yield of about 34% from CV 1.1 to 4.3. The achieved ee was consistent at 94 ± 4%. The system achieved a production of 318 mg of (S)-NLS, corresponding to an overall productivity of more than 29 mg$_{\text{(S)-NLS}}$ g$_{\text{CDM}}^{-1}$ h^{-1}, assuming that these particles had at least the same cell loading as the incubated Diaion. The deducted productivity dropped 4-fold compared to the small column, mainly because more biomass was used over a longer run-time of the experiment. Analyzing the data further, the observed trend was best fitted as an increasing exponential decay, which seemed unusual. An extrapolation showed that the yield would approach zero after 5.5 CVs of active production, confirming the finding that this setup is not suitable for long-term operation.

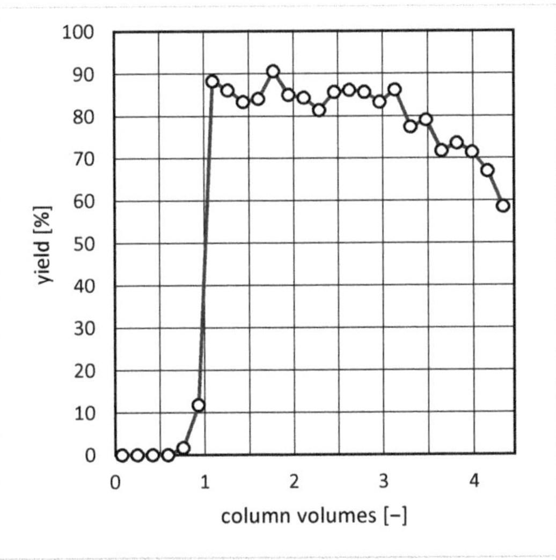

Figure 2. Time-course of the synthesis of (S)-norlaudanosoline using column B. White circles: data points; reaction conditions: 50 mM HEPES (pH 7.4), 50 mM dopamine, 50 mM pyruvate, 4 g L^{-1} ascorbate, 25 µg mL^{-1} gentamicin, 37 °C and 0.1 mL min^{-1} flow. ee determined by chiral HPLC.

MTT-testing in the collected fractions showed a continuous loss of biomass from CV 1 when starting at 0.4 mg mL^{-1} at CV 1 and increasing to 2 mg mL^{-1} at CV 4. This resulted in a loss of 171 mg of biomass over the whole experiment, assuming the same cell loading for these particles as it was determined for incubated Diaion. This corresponds to a loss of at least 16% of biomass over the course of the reaction as shown in Figure 3.

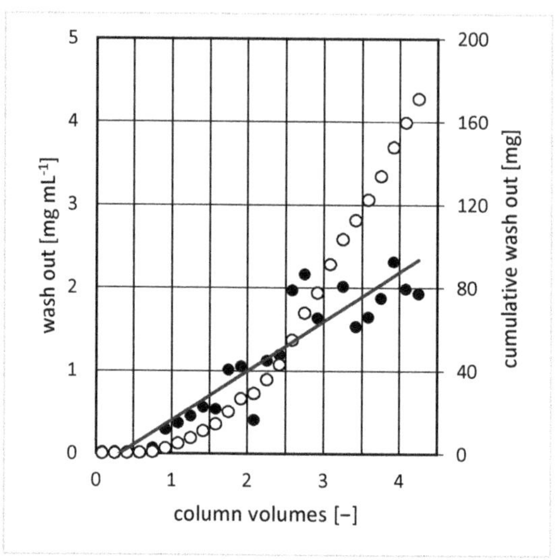

Figure 3. Time-course of wash out (black circles) during the synthesis using column B; black line: linear trend; white circles: cumulative wash out. Reaction conditions as above, trend equation $y = 0.596 \cdot x - 0.194$, $R^2 = 0.85$.

Other factors that were considered as possibly contributing to this loss of yield could be the loss of substrate due to polymerization and catalyst inactivation. Polymerization seems unlikely, since these effects should be prevented by the addition of ascorbate to the substrate solution and the mechanical barrier of the reactor walls, preventing direct contact with the atmosphere. Also, the substrate and product fluid were monitored for the emergence of black precipitates, which would indicate such reactions, but were not found. The stability of the biocatalysts was tested by incubating cells at 37 °C (without substrate) before their use in the reaction. In this experiment, a loss of activity of 16% after 24 h was observed, increasing to a loss of 48% activity after 48 h (Supporting Figure S3). This indicates that the loss of activity is connected to enzyme inactivation. After four CVs (meaning 12 h), this could account for about 10% of the 25% yield loss, but not for all of it. This catalyst inactivation, in conjunction with the wash out of cellular material from the particles, could be the reason for the observed loss of activity. As the wash out was detected via an MTT-test and assuming that the loss of activity correlates directly to the loss of cellular material, a loss of 16% yield could be due to wash out. This seems the most likely explanation at least for the bulk of the activity loss that was observed.

Concerning experimental options to increase the performance, the enzyme degradation occurring over time might be due to the fact that NCS is susceptible to protease attacks originating from the *E. coli* cells. In this case, the identification of the responsible protease(s), for example, by a screening of deletion mutants, and the use of the corresponding knock-out strain for protein expression might increase the performance greatly [41], via a better long-term stability of the enzymes involved in the synthesis.

It would also be of interest to measure the maximum amount of cells washed out of the immobilization particles and to examine how many cells remain bound on the Diaion carriers, as it is possible that only the extra outer layers of bacterial cells that adsorb onto already bound cells and not directly onto the carrier itself are more easily washed away under the pressure of the column flow. If so, a "single layered" particle would create less yield loss during use. The resulting reduction of the cell number per particle could then be compensated for by increasing the amount of enzymatically active particles per column.

To increase the adsorption strength of and to decrease the wash-out from Diaion particles, the *silCoat* technology proposed by Findeisen et al. could be adapted [42]. Solid carriers are coated with a silicon layer containing cells or enzymes, thus reducing the wash-out. Alternatively, thin coatings with metal oxides or covalent linking, as proposed for glass beads, could also be utilized.

While the separation of enzyme-containing cells from Diaion particles and their subsequent wash-out as well as enzyme inactivation were identified as key challenges preventing a long-time use of the column system, it already produced, in its current state, a decent amount of (S)-NLS.

3. Materials and Methods

3.1. Chemicals, Medium and Buffers

All chemicals, Diaion HP-2MG (13601 Supelco Diaion®® HP-2MG), glass beads (18406–500G, diameter ~5 mm) and sodium alginate (W201502) were obtained from Merck KGaA (Darmstadt, Germany). Polydiallyldimethylammonium chloride (short: PolyDADMAC, Polyquat 40 U50 A) was obtained from Katpol-Chemie GmbH (Bitterfeld, Germany). For lysogeny broth (LB)-medium, 10 g L^{-1} tryptone, 10 g L^{-1} NaCl, 10 g L^{-1} yeast extract were dissolved in deionized water (DIW). The pH was adjusted to 6.8. The buffers used in this study are HEPES buffer (50 mM, pH 7.4) and MOPS buffer (10 mM MOPS, pH 7.0). Adjustments of pH were carried out with NaOH and HCl if needed.

3.2. Expression of Proteins

Table 4 gives an overview of the proteins recombinantly produced in *Escherichia coli* in this study.

Table 4. List of proteins, source organisms and encoding genes used in this study.

Source Organism	Uniprot Entry	Gene Name	Abbreviation
Chromobacterium violaceum	Q7NWG4_CHRVO	*CV_2025*	TAm
Papaver somniferum	NCS2_PAPSO	*Ps_NCS2*	NCS
Mus musculus	Q64433	*groES*	-
Mus musculus	P63038	*groEL*	-
Escherichia coli (strain K12)	P0A850	*tig*	-

A plasmid encoding the NCS from *Papaver somniferum* (Ps_NCS2) was kindly donated by the working group of Prof. Wolfgang Kroutil (Institute of Chemistry, University of Graz, Graz, Austria). The gene encoding TAm (CV_2025) was synthesized by Eurofins Genomics (Ebersberg, Germany) and codon optimized for the expression in *E. coli* (using the GENEius tool of Eurofins Genomics). Both coding regions were cloned into pET28a(+) vectors and expressed separately in *E. coli* BL21(DE3) (Novagen, Darmstadt, Germany). For the recombinant production of NCS, the cells were co-transformed with the plasmid pG-Tf2 from the Takara chaperone plasmid set (Takara Bio inc., Shiga, Japan). This plasmid bears the genes *groES-groEL-tig*, which code for the GroES/GroEL chaperonins and the trigger factor. In a previous study, it has been shown that the simultaneous expression of chaperones had a positive effect on the specific NCS activity of the recombinant cells [15].

LB-medium supplemented with 50 µg mL^{-1} kanamycin was used for (pre)culture. For the cells co-transformed with the pG-Tf2 plasmid, additionally, 25 µg mL^{-1} of chloramphenicol was used. The precultures were incubated overnight at 150 rpm at 37 °C in an INFORS HT Multitron incubator (50 mm shake stroke, Infors AG, Bottmingen). The main cultures were inoculated with the preculture to reach an optical density at 600 nm (OD$_{600}$) of 0.1.

The main cultures were grown at 150 rpm and 37 °C in 2 L shake flasks without baffles filled with 400 mL media and 5 ng mL^{-1} tetracycline when NCS was to be expressed, until an OD_{600} of 0.5–0.7 was reached. Expression of NCS and TAm was started by adding 0.5 mM isopropyl-β-D-thiogalactopyranoside (IPTG). Cells were then grown for 16 h at 20 °C. Cells were harvested via centrifugation for 30 min, 4500 g, 4 °C, washed with HEPES buffer with 20% (v/v) glycerol and stored at −20 °C until use. For lyophilization, cells with TAm or NCS were pelleted and frozen in HEPES buffer at −20 °C for one day before freeze-drying at −80 °C for three days or until completely dry.

3.3. Batch Biotransformations

To investigate the whole cell biotransformations in batch and to identify the optimum ratio of TAm- and NCS-containing *E. coli* biomass, different ratios of the lyophilized whole cell preparations were weighed with a precision balance and added to 12 mL of substrate solution containing 50 mM pyruvate, 50 mM dopamine and 4 g L^{-1} ascorbic acid in 50 mM HEPES buffer (pH 7.4). The reactions were performed in miniaturized stirred-tank reactors (bioREACTOR48) from 2mag AG (Munich, Germany) equipped with S-type stirrers driven at 500 rpm [43]. The reactor temperature was set to 37 °C, while headspace cooling (15 °C) reduced evaporation. Prior to the reaction, the headspace was flushed with nitrogen and during the reaction, >0.1 bar of nitrogen was attached to minimize oxidation. Sealing tape was used to seal contact surfaces. The sampling aperture was closed with a fitting PTFE plastic plug except during sampling. Yield was measured 3 h after starting the reaction.

3.4. Immobilization of Lyophilized Cells

Lyophilized cells were mixed in a ratio of 3 g TAm-containing cells to 19 g NCS-containing cells, representing the optimum ratio determined during the batch experiments. Different materials and methods were used for subsequent immobilization. While the immobilization of lyophilized cells on the medium polarity adsorbent resin Diaion HP-2MG and glass beads followed a single protocol, the confinement of cells in alginate was performed with and without additional coating with PolyDADMAC. After immobilization, Diaion and glass-based particles were lyophilized before use, while alginate particles were not.

3.5. Incubated Diaion and Glass Beads

For Diaion (incubated) and glass, the method used by Findeisen et al. was adapted [42]. Thus, 22 g L^{-1} of mixed catalyst was added to 30 mL (final volume) HEPES buffer containing 25 µg mL^{-1} gentamicin and 50 g L^{-1} of carrier. The dispersion was incubated for 16 h at 20 °C. The supernatant was removed, and the particles were washed twice with 50 mL HEPES buffer. Cellular dry mass immobilized on the particles was determined by separate lyophilization of supernatant and incubated particles. Added mass on the particles and reduced dry mass of the supernatant fraction were assessed as biological loading on the particles.

3.6. Uncoated Alginate

For uncoated alginate, 11 g L^{-1} of the mixed cells was dispersed into MOPS buffer. Subsequently, 10 g L^{-1} of sodium alginate was added. Beads were produced by adding droplets of the solution to a stirred (300 rpm, magnetic stirrer) 100 mM $CaCl_2$, 10 mM MOPS (pH 7) solution using the Encapsulator Medical IEM-40 2 (nozzle 300, pump speed 6.8 mL min^{-1}, 677 Hz) (Inotech, Nabburg, Germany). Particles were incubated for 30 min at 25 °C before washing with 50 mL MOPS buffer. Cellular dry mass in the particles was determined by dissolution of particles in a covering solution of 0.5 M citric acid for 8 h. The fluid was centrifuged, resuspended in 50 mL MOPS buffer, and centrifuged again. The pellet was lyophilized and weighed.

3.7. Coated Alginate

Coated alginate particles were created by using alginate particles (created as described above), which subsequently were incubated in 50 mL of 1% PolyDADMAC. Particles were stirred (300 rpm) for 30 min at 25 °C, before washing with 50 mL MOPS buffer. The incubation was repeated in a solution of sodium alginate (1%), then the washing was done again. Finally, the particles were incubated in 1% PolyDADMAC as before and subsequently washed again. Movement of the particles between containers and washing was done with a sieve.

3.8. Immobilization of Growing Cells

The Diaion adsorbent resin was also used in another protocol in which the cells were immobilized by adding the carrier material to the growing cells to achieve biofilm formation [42]. For the preparation of Diaion (overgrown), 0.5 g of Diaion HP-2MG particles was added at the time point of induction to 400 mL of bacterial culture.

3.9. Column Printing

The 3D-printing was done with a stereolithographic "Figure 4®Modular" printer (3D Systems, Moerfelden-Walldorf, Germany) using "Figure 4®PRO-BLK 10" biocompatible plastic (ISO10993-5 [44], ISO10993-10 [45]). Columns were precleaned manually with 70% EtOH before use. Figure 4 shows a technical drawing of column A, while Figure 5 shows column B.

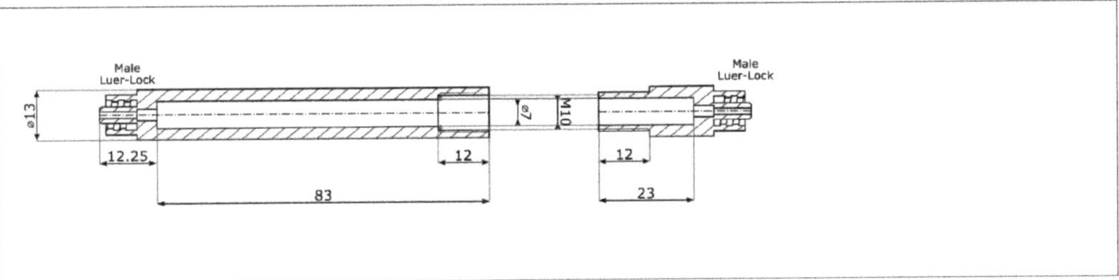

Figure 4. Dimensioning of column A (screwed column inner length 94 mm, inner diameter 9 mm and volume 3.6 mL).

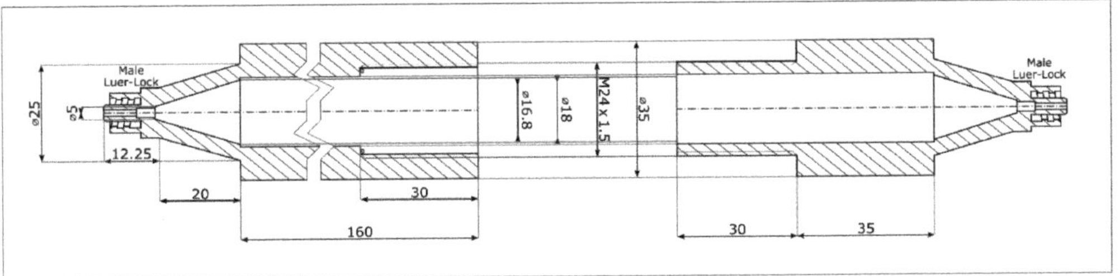

Figure 5. Dimensioning of column B (screwed column inner length 195 mm, inner diameter 18 mm and volume 49.6 mL).

3.10. Setup for Continuous Flow

Column A was filled with overgrown Diaion particles before screwing both parts of it together. Column B was filled with a 2:1 mixture of untreated Diaion particles and overgrown Diaion particles. For evaluation purposes, CV was defined as the volume inside of a packed column not occupied by the media. The column was connected using PEEK-tubing (inner diameter 0.75 mm, outer diameter 1/16″, from GE Healthcare (Uppsala, Sweden) to a LC-10ADVP-HPLC-pump from Shimadzu (Duisburg, Germany) and a F9R fraction collector from GE Healthcare.

Substrate was prepared with 50 mM dopamine, 50 mM pyruvate, 4 g L^{-1} ascorbic acid and 25 µg mL^{-1} gentamicin in 50 mM HEPES buffer (pH 7.4). A 0.9% NaCl solution and a 5 mL syringe were used to prefill columns and determine inner system volumes. Column and substrate feed were placed in an incubator at 37 °C while the further setup was stored at 25 °C.

3.11. Sample Preparation

Samples of 1 mL were taken at predetermined times and added to 200 µL of 1 M H_2SO_4. The samples were centrifuged at 10,000× g for 5 min to remove cells and cell debris. A total of 100 µL of supernatant was evaporated and used for phenylisothiocyanate (PITC) derivatization. This was done following the established procedure of Gonzales et al. [46]. PITC-derivatized samples were then dissolved in 80:20 HEPES buffer:MeOH before measurement. The remaining part of the sample was frozen at −20 °C for 16 h and again centrifuged after thawing. Before chiral norlaudanosoline measurements, samples were diluted 1:10 with 50 mM HEPES buffer (pH 7.4), and 20% MeOH was added before measurement.

3.12. Analytics

For all analytics, the Shimadzu Prominence HPLC 20A (Shimadzu, Duisburg, Germany) equipped with an UV detector was used. For norlaudanosoline analytics, an Astec Chirobiotic T column (25 cm) (Merck, Darmstadt, Germany) was used with 20:20:60 MeOH:ACN:25 mM phosphate buffer (pH 4.2) as mobile phase at a flow rate of 0.2 mL min^{-1} at 20 °C. Retention times were 21 min for (S)-norlaudanosoline and 30 min for (R)-norlaudanosoline (Supporting Figures S4 and S5). The detector was set to 230 nm. For pyruvate analytics, an Aminex HPX-87H column (30 cm) (Bio-Rad, Feldkirchen, Germany) was used with 5 mM sulfuric acid (pH 2.3) as mobile phase at a flow rate of 0.4 mL min^{-1} at 30 °C. Retention time of pyruvate was 14.5 min. The detector was set to 210 nm. For alanine analytics, a Nucleodur C18 Gravity-SB column (15 cm) (Merck, Darmstadt, Germany) was used with 20:20:60 MeOH:ACN:25 mM phosphate buffer (pH 4.2) as mobile phase at a flow rate of 0.3 mL min^{-1} at 35 °C. Retention time of alanine was 6.4 min. The detector was set to 254 nm.

3.13. Biomass Quantitation via Metabolic Activity

For the 3-(4,5-dimethylthiazol-2-yl)-2,5-diphenyltetrazoliumbromide (MTT)-test of outflow of the rescaled packed bed reactor, 3 mL of each sample fraction was centrifuged at 11.000× g for 15 min, and the supernatant was carefully removed, before resuspending the pellet in 500 µL of 0.9% NaCl. Each sample was then split into 4 subsamples, each further diluted by half (undiluted, 1:2, 1:4, 1:8) with 0.9% NaCl. A total of 80 µL of each subsample was transferred into a microtiter plate and 20 µL MTT-solution (5 mg mL^{-1}) was added to each well. After 1 h of incubation, the plates were centrifuged at 11.000× g for 5 min and the supernatant was removed and added to 400 µL solubilization buffer (405 µL isopropanol + 75 µL 20% w/v sodium dodecyl sulfate + 20 µL 1M HCl) and the solution was mixed thoroughly to solubilize the formazan crystals. Then, the absorbance at 575 nm and 660 nm was measured in an Infinite M Nano$^+$ plate reader (Tecan, Männedorf, Switzerland). The difference of the absorbance at 575 nm and at 660 nm was calculated. The resulting signal was compared to the signal of catalyst suspended in 0.9% NaCl. This generated a relative value for the metabolic biomass activity. The samples up to 2.8 CV were processed immediately, while subsequent samples were processed the next day. For quantification of cell mass loss during the experiment, cells treated equally and of the same ratio as in the experiment were measured as reference during the experiment.

4. Conclusions

In the presented study, we have shown that the two-step enzymatic synthesis of (S)-NLS, an important precursor of opioids, can be performed with lyophilized whole cell biocatalysts containing the enzymes TAm and NCS. The decisive factor for a high yield and optical purity of the chiral product was the addition of a high amount of cells containing the

NCS enzyme. This is probably due to the associated low concentration of the intermediate DOPAL, which has a twofold positive effect: it prevents the unselective formation of racemic (S)-NLS by a spontaneous Pictet–Spengler reaction and the occurrence of further undesired side reactions of the highly reactive aldehyde. To allow for a comparison of the reaction in batch and flow systems, the biocatalysts were first immobilized on Diaion HP-2MG particles by adsorption and then filled into 3D-printed columns. No adverse effects of the used plastic material were found, and a fast cycle of design and redesign was possible. Although with 318 mg of (S)-NLS a decent product mass was synthesized, the system was not stable enough for long-time operation due to biocatalyst leaching and inactivation. Further investigation will have to show whether the biocatalyst inactivation and the wash out of cells can be reduced to avoid a serious yield loss over time.

Supplementary Materials: The following supporting information can be downloaded at: https://www.mdpi.com/article/10.3390/catal13101347/s1, Figure S1: Analysis of the TAm expression in *E. coli* cells using SDS-PAGE (15% gel); Figure S2: Analysis of the NCS expression in *E. coli* cells using SDS-PAGE (18% gel); Figure S3: Time course of NCS activity in whole cells incubated at 37 °C before substrate addition; Figure S4: Chromatogram of (S)-norlaudanosoline analytics; Figure S5: Chromatogram of rac-norlaudanosoline analytics.

Author Contributions: K.C.: Conceptualization, methodology, resources, and writing—review and editing; A.H.A.: investigation, experimental, and writing—original draft preparation. All authors have read and agreed to the published version of the manuscript.

Funding: This research received no external funding.

Data Availability Statement: Data is contained within the article or Supplementary materials.

Acknowledgments: We thank Wolfgang Kroutil (University of Graz) for providing us with the plasmids and initial guidance for their use, as well as Christine Schülein and Anette Amtmann for their assistance in the laboratory, and Andreas Perlick for supporting the production of the manuscript. The authors gratefully thank Andreas Hofmann from the Institute of Microwaves and Photonics (FAU Erlangen-Nürnberg) for supporting this work by printing and supplying the columns.

Conflicts of Interest: The authors declare no conflict of interest.

References

1. Hagel, J.M.; Facchini, P.J. Benzylisoquinoline Alkaloid Metabolism: A Century of Discovery and a Brave New World. *Plant Cell Physiol.* **2013**, *54*, 647–672. [CrossRef] [PubMed]
2. WHO. *WHO Model List of Essential Medicines—22nd List*; WHO: Geneva, Switzerland, 2021.
3. Board of International Narcotics Control. *Estimated World Requirements for 2022*; United Nations: New York, NY, USA, 2021.
4. Bernath, J. *Poppy: The Genus Papaver*; CRC Press: Budapest, Hunguary, 1999.
5. Josephine, W.; Reed, T.H. The Quest for a Practical Synthesis of Morphine Alkaloids and Their Derivatives by Chemoenzymatic Methods. *Acc. Chem. Res.* **2015**, *3*, 674–687.
6. Galanie, S.; Thodey, K.; Trenchard, I.; Interrante, M.F.; Smolke, C. Complete biosynthesis of opioids in yeast. *Synth. Biol.* **2015**, *349*, 1095–1100. [CrossRef] [PubMed]
7. Lichman, B.R.; Lamming, E.D.; Pesnot, T.; Smith, J.M.; Hailes, H.C.; Ward, J.M. One-pot triangular chemoenzymatic cascades for the syntheses of chiral alkaloids from dopamine. *Green. Chem. Commun.* **2015**, *17*, 852–855. [CrossRef]
8. Stadler, R.; Kutchan, T.M.; Zenk, M.H. (S)-Norcoclaurine is the central intermediate in benzylisoquinoline alkaloid biosynthesis. *Phytochemistry* **1989**, *28*, 1083–1086. [CrossRef]
9. Battersby, A.; Binks, R.; Francis, R.; McCaldin, D.; Ramuz, H. 1-c as precursors of thebaine, codeine, and morphine. *J. Chem. Soc.* **1964**, *679*, 3600–3610. [CrossRef]
10. Berglund, P.; Guo, F. Transaminase biocatalysis: Optimization and application. *Green Chem.* **2017**, *19*, 333–360.
11. Anderson, D.G.; Mariappan, S.V.S.; Buettner, G.R.; Doorn, J.A. Oxidation of 3,4-Dihydroxyphenylacetaldehyde, a Toxic Dopaminergic Metabolite, to a Semiquinone Radical and an ortho-Quinone. *J. Biol. Chem.* **2011**, *286*, 26978–26986. [CrossRef]
12. Wang, W.; Wu, X.; Yang, C.S.; Zhang, J. An Unrecognized Fundamental Relationship between Neurotransmitters: Glutamate Protects against Catecholamine Oxidation. *Antioxidants* **2021**, *10*, 1564. [CrossRef]
13. Lai, C.-T.; Yu, P.H. Dopamine- and l-β-3,4-dihydroxyphenylalanine hydrochloride (l-Dopa)-induced cytotoxicity towards catecholaminergic neuroblastoma SH-SY5Y Cells: Effects of oxidative stress and antioxidative factors. *Biochem. Pharmacol.* **1997**, *3*, 363–372. [CrossRef]

14. Zucca, F.A.; Segura-Aguilar, J.; Ferrari, E.; Muñoz, P.; Paris, I.; Sulzer, D.; Sarna, T.; Casella, L.; Zecca, L. Interactions of iron, dopamine and neuromelanin pathways in brain aging and Parkinson's disease. *Progress. Neurobiol.* **2017**, *155*, 96–119.

15. Lechner, H.; Soriano, P.; Poschner, R.; Hailes, H.C.; Ward, J.M.; Kroutil, W. Library of Norcoclaurine Synthases and Their Immobilization for Biocatalytic Transformations. *Biotechnol. J.* **2018**, *13*, 1700542. [CrossRef] [PubMed]

16. Pesnot, T.; Gershater, M.C.; Ward, J.M.; Hailes, H.C. Phosphate mediated biomimetic synthesis of tetrahydroisoquinoline alkaloids. *Chem. Commun.* **2011**, *47*, 3242–3244. [CrossRef]

17. Rees, J.N.; Florang, V.R.; Eckert, L.L.; Doorn, J.A. Protein Reactivity of 3,4-Dihydroxyphenylacetaldehyde, a Toxic Dopamine Metabolite, Is Dependent on Both the Aldehyde and the Catechol. *Chem. Res. Toxicol.* **2009**, *22*, 1256–1263. [CrossRef]

18. Jimenez-Gonzalez, C.; Poechlauer, P.; Broxterman, Q.B.; Yang, B.S.; Ende, D.A.; Baird, J.; Bertsch, C.; Hannah, R.E.; Dell'Orco, P.; Noorrnan, H.; et al. Key Green Engineering Research Areas for Sustainable Manufacturing: A Perspective from Pharmaceutical and Fine Chemicals Manufacturers. *Org. Process Res. Dev.* **2011**, *4*, 900–911. [CrossRef]

19. Woodley, J.M. Accelerating the implementation of biocatalysis in industry. *Appl. Microbiol. Biotechnol.* **2019**, *12*, 4733–4739. [CrossRef] [PubMed]

20. Andrade, L.H.; Kroutil, W.; Jamison, T.F. Continuous Flow Synthesis of Chiral Amines in Organic Solvents: Immobilization of E. coli Cells Containing Both ω Transaminase and PLP. *Org. Lett.* **2014**, *16*, 6092–6095. [CrossRef]

21. Molnár, Z.; Farkas, E.; Lakó, Á.; Erdélyi, B.; Kroutil, W.; Vértessy, B.G.; Paizs, C.; Poppe, L. Immobilized Whole-Cell Transaminase Biocatalysts for Continuous-Flow Kinetic Resolution of Amines. *Catalysts* **2019**, *9*, 438. [CrossRef]

22. Santi, M.; Sancineto, L.; Nascimento, V.; Azeredo, J.B.; Orozco, E.; Andrade, L.; Gröger, H.; Santi, C. Flow Biocatalysis: A Challenging Alternative for the Synthesis of APIs and Natural Compounds. *Int. J. Mol. Sci.* **2021**, *3*, 990. [CrossRef]

23. Ngo, T.D.; Kashani, A.; Imbalzano, G.; Nguyen, K.T.Q.; Hui, D. Additive manufacturing (3D printing): A review of materials, methods, applications and challenges. *Compos. Part. B Eng.* **2018**, *143*, 172–196. [CrossRef]

24. Causey, T.B.; Shanmugam, K.T.; Yomano, L.P.; Ingram, L.O. Engineering Escherichia coli for efficient conversion of glucose to pyruvate. *Proc. Natl. Acad. Sci. USA* **2004**, *8*, 2235–2240. [CrossRef]

25. Masato, A.; Plotegher, N.; Boassa, D.; Bubacco, L. Impaired dopamine metabolism in Parkinson's disease pathogenesis. *Mol. Neurodegener.* **2019**, *14*, 35. [CrossRef] [PubMed]

26. Chang, A.; Jeske, L.; Ulbrich, S.; Hofmann, J.; Koblitz, J.; Schomburg, I.; Neumann-Schaal, M.; Jahn, D.; Schomburg, D. BRENDA, the ELIXIR core data resource in 2021: New developments and updates. *Nucleic Acids Res.* **2021**, *D1*, 498–508. [CrossRef] [PubMed]

27. Samanani, N.; Facchini, P. Purification and Characterization of Norcoclaurine Synthase: The First Committed Enzyme in Benzylisoquinoline Alkaloid Biosynthesis In Plants. *J. Biol. Chem.* **2002**, *37*, 33878–33883. [CrossRef]

28. Luk, L.; Bunn, S.; Liscombe, D.; Facchini, P.; Tanner, M. Mechanistic studies on norcoclaurine synthase of benzylisoquinoline alkaloid biosynthesis: An enzymatic Pictet-Spengler reaction. *Biochemistry* **2007**, *4*, 10153–10161. [CrossRef] [PubMed]

29. Bonamore, A.; Rovardi, I.; Gasparrini, F.; Baiocco, P.; Barba, M.; Molinaro, C.; Botta, B.; Boffiab, A.; Macone, A. An enzymatic, stereoselective synthesis of (S)-norcoclaurine. *Green. Chem.* **2010**, *12*, 1623–1627. [CrossRef]

30. Matsumura, E.; Nakagawa, A.; Tomabechi, Y.; Ikushiro, S.; Sakaki, T.; Katayama, T.; Yamamoto, K.; Kumagai, H.; Sato, F.; Minami, H. Microbial production of novel sulphated alkaloids for drug discovery. *Sci. Rep.* **2018**, *8*, 7980. [CrossRef]

31. Nakagawa, A.; Minami, H.; Kim, J.-S.; Koyanagi, T.; Katayama, T.; Sato, F.; Kumagai, H. Bench-top fermentative production of plant benzylisoquinoline alkaloids using a bacterial platform. *Bioengineered* **2012**, *3*, 49–53. [CrossRef] [PubMed]

32. Stöckigt, J.; Chen, Z.; Ruppert, M. Enzymatic and Chemo-Enzymatic Approaches Towards Natural and Non-Natural Alkaloids: Indoles, Isoquinolines, and Others. In *Natural Products via Enzymatic Reactions*; Topics in Current Chemistry; Springer: Berlin, Germany, 2010; pp. 67–103.

33. Maresh, J.J.; Crowe, S.O.; Ralko, A.A.; Aparece, M.D.; Murphy, C.M.; Krzeszowiec, M.; Mullowney, M.W. Facile one-pot synthesis of tetrahydroisoquinolines from amino acids via hypochlorite-mediated decarboxylation and Pictet–Spengler condensation. *Tetrahedron Lett.* **2014**, *36*, 55. [CrossRef]

34. Tufvesson, P.; Lima-Ramos, J.; Nordblad, M.; Woodley, J. Guidelines and Cost Analysis for Catalyst Production in Biocatalytic Processes. *Org. Process Res. Dev.* **2011**, *15*, 266–274. [CrossRef]

35. DiCosimo, R.; McAuliffe, J.; Poulose, A.; Bohlmann, G. Industrial use of immobilized enzymes. *Chem. Soc. Rev.* **2013**, *42*, 6437–6474. [CrossRef]

36. Manohar, S.; Kim, C.K.; Karegoudar, T.B. Enhanced degradation of naphthalene by immobilization of Pseudomonas sp. strain NGK1 in polyurethane foam. *Appl. Microbiol. Biotechnol.* **2001**, *55*, 311–316. [CrossRef] [PubMed]

37. Klein, J.; Ziehr, H. Immobilization of microbial cells by adsorption. *J. Biotechnol.* **1990**, *16*, 1–15. [CrossRef] [PubMed]

38. Li, B.; Logan, B.E. Bacterial adhesion to glass and metal-oxide surfaces. *Colloids Surf. B: Biointerfaces* **2004**, *2*, 81–90. [CrossRef]

39. Shriver-Lake, L.C.; Gammeter, W.B.; Bang, S.S.; Pazirandeh, M. Covalent binding of genetically engineered microorganisms to porous glass beads. *Anal. Chim. Acta* **2002**, *1*, 711–778. [CrossRef]

40. Klerk, A. Voidage Variation in Packed Beds at Small Column to Particle Diameter Ratio. *AIChE J.* **2003**, *49*, 2022–2029. [CrossRef]

41. Lin, B.; Tao, Y. Whole-cell biocatalysts by design. *Microb. Cell Fact.* **2017**, *16*, 106. [CrossRef]

42. Findeisen, A.; Thum, O.; Ansorge-Schumacher, M.B. Biocatalytically active *silCoat* -composites entrapping viable *Escherichia coli*. *Appl. Microbiol. Biotechnol.* **2014**, *98*, 1557–1566. [CrossRef] [PubMed]

43. Riedlberger, P.; Brüning, S.; Weuster-Botz, D. Characterization of stirrers for screening studies of enzymatic biomass hydrolyses on a milliliter scale. *Bioprocess Biosyst. Eng.* **2013**, *36*, 927–935. [CrossRef]
44. *ISO10993-5*; Biological Evaluation of Medical Devices—Part 5: Tests for In Vitro Cytotoxicity. International Organization for Standardization: Rome, Italy, 2009. Available online: https://www.iso.org/standard/36406.html (accessed on 1 October 2023).
45. *ISO10993-10*; Biological Evaluation of Medical Devices—Part 10: Tests for Skin Sensitization. International Organization for Standardization: Rome, Italy, 2021. Available online: https://www.iso.org/standard/75279.html (accessed on 1 October 2023).
46. Gonzales-Castro, M.J.; López-Hernández, J.; Simal-Lozano, J.; Oruna-Concha, M.J. Determination of Amino Acids in Green Beans by Derivatization with Phenylisothiocianate and High-Performance Liquid Chromatography with Ultraviolet Detection. *J. Chromatogr. Sci.* **1997**, *35*, 181–185. [CrossRef]

Review

Agro-Industrial Food Waste as a Low-Cost Substrate for Sustainable Production of Industrial Enzymes: A Critical Review

Vishal Sharma [1,2,†], Mei-Ling Tsai [2,†], Parushi Nargotra [2,†], Chiu-Wen Chen [1,3], Chia-Hung Kuo [2], Pei-Pei Sun [2,*] and Cheng-Di Dong [1,3,*]

1   Department of Marine Environmental Engineering, National Kaohsiung University of Science and Technology, Kaohsiung 81157, Taiwan
2   Department of Seafood Science, National Kaohsiung University of Science and Technology, Kaohsiung 81157, Taiwan
3   Institute of Aquatic Science and Technology, National Kaohsiung University of Science and Technology, Kaohsiung 81157, Taiwan
*   Correspondence: ppsun@nkust.edu.tw (P.-P.S.); cddong@nkust.edu.tw (C.-D.D.)
†   These authors contributed equally to this work.

Abstract: The grave environmental, social, and economic concerns over the unprecedented exploitation of non-renewable energy resources have drawn the attention of policy makers and research organizations towards the sustainable use of agro-industrial food and crop wastes. Enzymes are versatile biocatalysts with immense potential to transform the food industry and lignocellulosic biorefineries. Microbial enzymes offer cleaner and greener solutions to produce fine chemicals and compounds. The production of industrially important enzymes from abundantly present agro-industrial food waste offers economic solutions for the commercial production of value-added chemicals. The recent developments in biocatalytic systems are designed to either increase the catalytic capability of the commercial enzymes or create new enzymes with distinctive properties. The limitations of low catalytic efficiency and enzyme denaturation in ambient conditions can be mitigated by employing diverse and inexpensive immobilization carriers, such as agro-food based materials, biopolymers, and nanomaterials. Moreover, revolutionary protein engineering tools help in designing and constructing tailored enzymes with improved substrate specificity, catalytic activity, stability, and reaction product inhibition. This review discusses the recent developments in the production of essential industrial enzymes from agro-industrial food trash and the application of low-cost immobilization and enzyme engineering approaches for sustainable development.

Keywords: enzymes; agro-industrial food waste; biocatalysis; immobilization; enzyme engineering

Citation: Sharma, V.; Tsai, M.-L.; Nargotra, P.; Chen, C.-W.; Kuo, C.-H.; Sun, P.-P.; Dong, C.-D. Agro-Industrial Food Waste as a Low-Cost Substrate for Sustainable Production of Industrial Enzymes: A Critical Review. *Catalysts* **2022**, *12*, 1373. https://doi.org/10.3390/catal12111373

Academic Editor: Evangelos Topakas

Received: 11 October 2022
Accepted: 3 November 2022
Published: 5 November 2022

Publisher's Note: MDPI stays neutral with regard to jurisdictional claims in published maps and institutional affiliations.

1. Introduction

As global concern over food and agricultural sustainability, environmental resilience, and food safety has grown over the years, the food industry has been exploring more environmentally friendly ways to produce food and nutritional supplements. Enzyme biocatalysis, which operates at the nexus of microbiology, molecular biology, biochemistry, and organic chemistry is an illustration of a sustainable multidisciplinary technology. Biocatalysis has gained enormous industrial potential owing to its influential and eco-friendly prospects with effective kinetics and commercial benefits [1]. Enzymes are macromolecular biocatalysts with extensive applications due to their ability to operate in milder reaction conditions, high catalytic efficiency, superior product selectivity, and their negligible toxicity to the environment and the body [2,3]. They have been the subject of numerous research projects around the world in order to generate novel significant industrial processes. A variety of microbial species (bacteria, fungi) have been traditionally employed in the production of diverse products of commercial importance from different organic substrates converting them into simpler forms through enzymes [4]. Many microbes that are typically utilized as a source of enzymes have had their microbial endogenous and exogenous enzymes thoroughly researched [5]. A significant portion of microbial enzymes are utilized in a variety of industrial processes, including those for food processing, animal feed, biofuels, paper and pulp industries, pharmaceutical industries, textiles, polymer synthesis, and detergent industries [6]. Enzymes such as cellulases, xylanases, amylases, lipases, proteases, and pectinases have been commercially utilized in a wide range of industrial processes, especially in the food industry.

In recent times, the idea of 'circular economy', which refers to the use of organic waste from one industry as a source of raw materials for another, has gained much popularity [7]. It is based on the principle of sustainable development known as the '5Rs' (reduce, recycle, reuse, recovery, and restore) and replaces the traditional linear economic model (make–use–throw) with a more efficient circular one [8]. The food and agro-industrial sectors have been revolutionized owing to modernization and industrialization, which has dramatically increased the production of huge amounts of agro-industrial food waste [9]. The United Nations' Food and Agriculture Organization (FAO) estimates that every year, about 1.3 billion tons of food, which is one-third of global production, is wasted [10]. In addition to the food waste, various agro-industrial residues and crop waste in the form of lignocellulosic biomass (LCB) is generated annually around the globe [11–16]. Most of this plant-based waste is either landfilled or burned alongside other municipal combustible trash in an effort to recover energy [17]. Apparently, this organic refuse, which is a rich source of carbohydrates, proteins, lipids, organic acids, and other necessary minerals, can be channelized towards its value addition [9,18]. It could serve as an inexpensive fermentation source for microbes in the food industries, which digest it via enzymes into key components of circular economies. The industrial applications of enzymes have significantly risen in the last decade, primarily in the food modification, biofuels production, biomedical and pharmaceutical research, and the transformation of agro-industrial waste [18–20].

Even though enzymes offer many more benefits over traditional chemical catalysts to valorize organic waste, a major bottleneck in their commercial viability is their non-reusability, high sensitivity, and poor catalytic activity and stability in extreme environmental conditions of temperature and pH [18]. These challenges therefore need to be critically removed through the development of stable biocatalytic systems. Enzyme immobilization has received significant attention in the past few years as an important engineering approach to customize and enhance a wide range of catalytic features of enzymes, including their activity, specificity, selectivity, and tolerance to inhibitors [12,19]. The development of flexible carriers, including agro-food and crop-based materials, metal organic frameworks, and nanomaterials, allows for the cost-economic immobilization of enzymes with better enzymological properties, enabling catalytic reactions to be carried out under rigorous processing environments [21].

The production of engineered enzymes by promising protein engineering tools such as directed evolution, rational design, and computational methods, aids in improving the enzymological properties with increased purity, catalytic efficiency, specificity, and expression yield, owing to the altered amino acid sequence [22]. The application of tailored enzymes for food processing enables their cost-effective production to achieve sustainable development. Figure 1 shows schematic representation of enhancing the value of enzymes by immobilization and protein engineering approaches through valorization of agro-industrial food/crop waste.

This review encompasses the valorization possibilities of agro-industrial food waste through microbial enzymes. The review also highlights the novel strategies for food enzyme immobilization and their potential applications in the food industry. Moreover, the deeper insights on development of engineered enzymes for sustainable and green processing of the waste biomass into diverse bioproducts are highlighted.

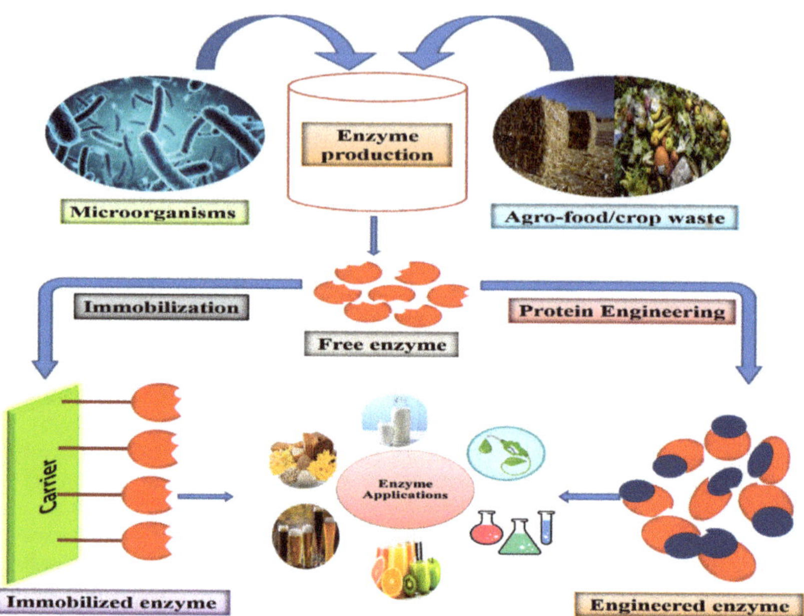

Figure 1. Schematic representation of developing efficient biocatalysts.

2. Types of Agro-Industrial Food Waste

The ever-increasing world population and the rapid urbanization and lifestyle changes have overburdened the food processing industries that generate abundant amounts of plant, animal, and agricultural residue wastes. The massive build-up of such natural food waste is a big challenge for mankind due to the lack of efficient waste management techniques. Commonly, within the food supply chain, food which is abandoned, unused, thrown out, and burnt from the harvest source or slaughter source is referred to as food loss. Most of the food waste is generated during its production, harvesting, transportation, industrial processing, and consumption [7]. The different types of organic wastes are majorly derived as agricultural wastes or those from the food processing industries. The agricultural waste can be majorly categorized as crop residues after harvesting, agro-industrial wastes, and fruit and vegetable waste [4,23], whereas the food processing industries generate enormous amounts of waste from cereals and pulses, fruits and vegetables, dairy, poultry, meat, egg waste, aquatic life waste, and seafood waste during their processing. Moreover, commercial and household kitchen waste also adds to the overall concerns of waste management [24]. Therefore, the apprehensions of deteriorating environment and waste management can be better overcome by transforming the agro-food waste into a substrate for enzyme production, besides a raw material for generation of diverse bioproducts such as biofuels, xylitol, xylooligosaccharides, bioplastics, organic acids, etc. (Figure 2). Emphasis must be laid on implementing modern technologies as well as alternative strategies for the effective and successful use of non-consumable portions from the agro-food wastes leading to proper waste management and sustainable development.

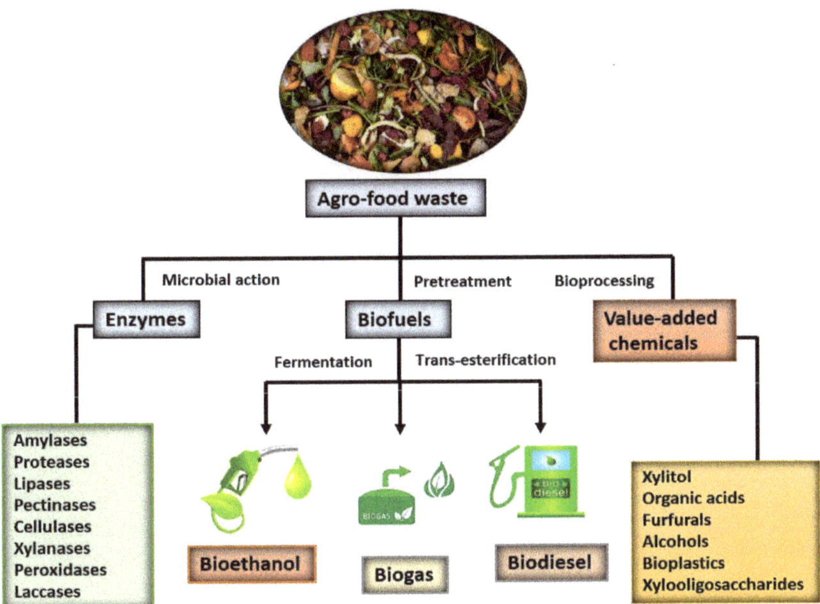

Figure 2. Agro-industrial food waste as a raw material for the production of different bioproducts.

2.1. Food Processing Waste

Different food processing and packaging industries produce tons of food items to meet the constant demands of the food market. However, the side generation of unprocessed food, leftover materials, spoilage, or contamination of raw food materials is an unavoidable practice in the food industries. The food waste is subcategorized majorly as starch-based food waste, dairy waste, meat and fish waste, vegetable trimmings, fruit peels, spent grains, and pulp trash. Traditionally, the leftover food from the food processing industries is dumped, incinerated, landfilled, composted, or anaerobically digested to enrich the soil. These food wastes, however, are extremely nutritious, which makes them substrates for microbial growth and production of diverse enzymes [25]. Moreover, these wastes can be used as inexpensive raw materials for the production of primary and secondary metabolites using microorganisms, with commercial value.

2.1.1. Fruit and Vegetable Waste

Fruit and vegetable waste (FVW) is produced in the food processing industries as a result of cutting, thermal treatment, handling, processing, packaging, transportation, and natural ripening [26]. Additionally, wastes are generated by microbial attack, discoloration, and by a number of biochemical reactions involving enzymes, antioxidants, phenolic chemicals, and oxidation [4]. The landfilling and incineration of food waste is a common practice preferred by industries, despite being unaffordable due to emissions of greenhouse gases and high capital and operating costs [17]. The anaerobic digestion of FVW into biofertilizers is an alternate option that might indirectly minimize environmental pollution and further increase soil nutrition. In many developing nations, FVW consisting of pineapple, papaya, banana leaves, orange peels, spinach, sugarcane tops, cabbage, etc., are converted into animal feed to address the current animal feed scarcity [8,9,27].

2.1.2. Edible Oil Waste

Cooking or edible oil waste is produced by the edible oil industries throughout a number of processing phases, including neutralization, degumming, deodorization, bleaching, and hydrolytic/oxidative rancidity. The hydrolytic rancidity of the waste cooking oil results from lipid oxidation, moisture, age, oxidation, and addition of effluents from industry rich in carbohydrates, fatty acids, and proteins [28]. Waste cooking oil is a left over, dark colored liquid formed by repeated deep-frying processes, which is unfit for human consumption because of its free short-chain fatty acids, aldehydes, ketones, mono/di-glycerides, aromatic compounds, polymers, and many other properties [26]. Edible oil industrial waste is a cheap substrate for microbial lipid production that can be produced by its co-fermentation with food waste [29].

2.1.3. Kitchen Waste

With a rapid increase in population, urbanization, and economic development over the years, an abundant amount of kitchen waste (KW) is generated every day from households, restaurants, public catering rooms, and hotels. KW includes cooked food wastes, leftover fruits and vegetables, meat, shells and pits, egg shells, used oil, and grease [26]. It is typically thought of as organic waste that decomposes quickly and has unpleasant aromas that attract insects and rodents. KW residues are rich in carbohydrates, proteins, lipids, lignin, organic acids, inorganic salts, and other bioactive substances and can therefore be attractive substrates for enzyme production by microorganisms [30]. Humidified bread waste was used as a low-cost substrate for the coproduction of α-amylase and protease through solid state fermentation (SSF) by *Rhizopus oryzae* [31]. The utilization of KW with the goal of producing value-added goods by enzymatic reactions has the potential to improve the food supply chain and, therefore, food security.

2.1.4. Poultry, Slaughterhouse and Egg Processing Waste

A significant amount of waste is generated from the poultry, egg, meat, and related food processing industries, in addition to solid/liquid waste from slaughterhouses. A considerable amount of chicken feather waste is produced by the poultry industry, which could be an excellent source of protein (~90% keratin) for various industrial applications [32]. Bacteria such as *Bacillus* sp. FPF-1, *Brevibacillus* sp., *Chryseobacterium* sp. FPF-8, and Nnolim-K2 were isolated as showing an excellent keratinase-producing ability employing chicken feathers as the substrate [33]. The waste from slaughterhouses includes skin, hairs, feathers, hooves, horns, deboning residues, and other materials that have a high level of organic matter, protein, and animal fat [4,34]. Slaughterhouse waste might pose major health and environmental risks if left untreated, which can instead be easily applied as a raw material in the production of a variety of commercial materials.

2.2. Agricultural Waste

Agricultural waste (AW) is produced in millions of tons annually and its improper management and disposal pose negative effects on the environment, including damaging the ecosystem. Therefore, transformation of AW into valuable products using economical, environmentally friendly, and sustainable methods has been the persistent objective of governments, environmentalists, and other stakeholders. AW, when properly managed and utilized, has the potential to be a significant contributor to ecological sustainability and energy security. Generally, AW in the form of bagasse, bran, husks, peels, leaves, seeds, stems, stalks, etc. (Figure 3) are utilized as soil improvers, fertilizers, animal fodder, and in various other processes [4,23]. Agro-waste as lignocellulosic biomass (LCB) is majorly constituted by cellulose backbone, with hemicellulose and lignin as other vital carbonaceous fractions [13]. These polymers from the LCB have been researched as potential substrates for lignocellulosic biorefineries for the production of second-generation biofuels and other valuable materials [35–39]. The enzymatic saccharification of complex polysaccharides into simple fermentable sugars using cellulolytic enzymes like cellulases and xylanases is an inevitable step in the systematic conversion of LCB into biofuels [37,38]. Solid-state fermentation (SSF) is also an attractive method for the exploitation of agricultural residues for production of a consortium of industrial enzymes.

Figure 3. Classification of different types of agro-industrial food waste.

3. Production of Microbial Enzymes from Agro-Food Wastes

Traditionally, the microbial enzymes have been of much importance in food preparation techniques. The state-of-the-art developments in enzyme technology over the last few years have led to the creation of novel enzymes with a broad range of applications in several industries. These industries are majorly associated with biofuel production, food modification, agro-industrial waste transformation, laundry, and pharmaceutical and biomedical research [18]. Microorganisms including fungi, yeast, and bacteria, as well as their enzymes, are frequently utilized in a variety of food preparations to enhance flavor and texture, and they also provide enormous economic advantages to various enterprises [40]. The majority of the world's enzyme application is categorized into two categories as special enzymes used in research, therapeutics and diagnostics, and industrial enzymes for food and animal feed industries. The market for these enzymes is estimated to reach at about $7 billion USD by the year 2023 and is expected to increase at a 7.1% compound annual growth rate from 2020 to 2027 [41,42]. The valorization of agro-industrial food wastes by the production of low-cost enzymes under solid-state fermentation is a promising and extensively explored method [17]. Different sets of enzymes such as amylases, proteases, lipases, laccases, cellulases, xylanases, and pectinases, among other enzymes, have been produced by SSF using inexpensive food wastes. SSF offers a number of benefits, including lower cost, higher yield, less waste, and simpler equipment and culture media derived from organic, solid agricultural products or waste.

3.1. Amylases

Amylases are one of the most important industrial enzymes divided into two major subclasses, i.e., α-amylase (EC 3.2.1.1) and glucoamylase (EC 3.2.1.3). α-amylase cleaves 1,4-α-D-glucosidic linkages in starch to convert it into maltose, glucose, and maltotriose, whereas glucoamylase specifically converts amylose and amylopectin's non-reducing ends to glucose [43]. A-amylases, however, have been widely explored and applied in food, clinical, brewing, detergent, textiles, and paper industries [17,44]. Additionally, they have also been widely explored for the valorization of agro-industrial residues and organic by-products to improve the generation of bioproducts. Different microbial strains may produce high activity amylases under optimized fermentation conditions using agro-food wastes, such as kitchen waste, potato peels, watermelon rinds, and crop residues such as rice husks [45] and corn cobs, coffee waste, and tomato pomace. Iqbalsyah et al. [45] investigated the production of amylase using rice husk substrate via solid-state fermentation by *Geobacillus* sp. with maximum amylase activity of 1.85 U/g at 48 h. Mojumdar and Deka [43] produced α-amylase by *Bacillus amyloliquefaciens* in SSF using rice bran, wheat bran, and potato peels as common agro-industrial feedstocks. The medium containing wheat bran as a substrate yielded the highest titer of amylase activity (112 U/mL), which was followed by that of potato peels (89 U/mL) and rice bran (77 U/mL). A response surface methodology was used for high α-amylase production (880 U/g) by *Trichoderma virens* under optimized conditions using watermelon rinds waste biomass [44]. These findings suggest that agro-industrial crop and food wastes can be used as inexpensive raw materials for industrial amylase production, replacing the cost-intensive synthetic media.

3.2. Proteases

Proteases are amongst the most significant commercial enzymes with wide applications in food, dairy, detergent, pharmaceutical, and leather industries. Some bacterial strains of the genus *Bacillus* and many fungal strains, including *Rhizopus, Aspergillus, Penicillium,* etc. have been reported to be the most active protease producers [46,47]. The alkaline protease enzyme covers about 65% of the global enzyme market that can proficiently convert proteins to biopeptides [48]. A thermophilic fungi, *Mycothermus thermophilus*, from hydrothermal springs was used for protease production under SSF using wheat bran with an improved enzyme production (1187.03 U/mL) under optimal conditions [49]. Similarly, microbial protease was produced from a newly isolated *Neurospora crassa* under SSF which used soybean okara waste as the substrate under optimal fermentation conditions [50]. A high protease activity of 1959.82 U/g was achieved with optimal activity at pH 9 and 55 °C and preferably hydrolyzed casein protein. Camargo et al. [51] used processed orange and grape wastes for determining their bromatological characteristics and production of proteases from them employing *Aspergillus niger* CBMAI 2084. The mixed grape wastes showed specific protease activity of 174.94 U/mg, whereas a protease activity of 16 U/g·min^{-1} was observed in fermented orange waste.

Additionally, the microbial proteases produced from agro-food wastes can be exploited for various biorefining applications in addition to their potential usage in numerous food applications. Rawoof et al. [52] evaluated the effect of *Lactobacillus manihotivorans* lactic acid production from food waste by simultaneous saccharification and fermentation with high substrate utilization and less processing time. The *Lactobacillus* sp. produced appreciable protease and α-glucosidase enzymes during the hydrolysis of complex molecules in the food waste. The valorization of food waste biomass and in situ enzyme production in such a biorefining strategy could significantly decrease the operating costs of the bioprocess as compared to current industrial practices that use expensive substrates.

3.3. Lipases

Lipases, also called triacylglycerol hydrolases (EC 3.1.1.3), are vital enzymes used in numerous industrial food processes [53,54]. These enzymes are involved in various biochemical reactions that improve product quality, durability, and solubility, and provide superior organoleptic features [55]. Lipases primarily hydrolyze triglycerides to obtain free fatty acids, glycerol, monoacylglycerols, and diacylglycerols [56]. Alternatively, they catalyze the synthesis of new products via aminolysis, alcoholysis, acidolysis, esterification, and transesterification methods [17]. Lipases can be simply produced from lignocellulosic feedstocks and waste from other sources by using various microorganisms. The majority of the previous research studies have concentrated on the production of extracellular lipase with high activity by a range of microbial strains, including fungi, yeast, and bacteria, employing both SSF and submerged fermentation (SmF) [57]. Putri et al. [57] performed lipase production from *Aspergillus niger* under optimized conditions by SSF of rice bran and *Jatropha* seed cake. The results exhibited the yield of a very dry lipase extract (282 U/mL enzyme) using 1% NaCl and 0.5% Tween 80 as the best extractants. Similarly, Pereira et al. [56], using industrially processed mango peel and seed waste, evaluated the lipase production using *Yarrowia lipolytica* by SmF. A lipase production as high as 3500 U/L of extracellular lipase was achieved under optimum conditions of temperature (27.9 °C), pH (5.0), and substrate concentrations.

Lipases can be easily produced using microorganisms under SSF by employing lignocellulosic residues as cheap feedstocks. A recent study investigated the production of lipase through *Penicillium roqueforti* growth via SSF using cocoa bran residues [53]. A maximum lipase activity of 33.33 $\pm$ 3.33 U/g was achieved using palm oil (30%) after 72 h of fermentation with the fungus. Likewise, in a promising experiment, *P. roqueforti* was used to optimize lipase production (48 U/g) employing inexpensive cocoa shell waste biomass applying an artificial neural system combined with a genetic algorithm system [54].

3.4. Lignocellulolytic Enzymes

Lignocellulolytic enzymes such as cellulases, xylanases, lignin peroxidases, laccases, etc. have received much attention in the past decade for use in the lignocellulosic biorefineries for the production of biofuels and other green organic solvents [13,14,36,58]. These enzymes have the capacity to break the complex linkages between polysaccharides (cellulose, hemicellulose) and lignin and convert them into simpler forms [35,39,59]. Due to their ability to assist in the disruption of the lignocellulose structure by degrading cellulose's β-1,4 linkages, these enzymes can improve the nutritional value of feed items with a high fiber content.

3.4.1. Cellulases and Xylanases

Cellulases are the most important class of lignocellulolytic enzymes comprising exoglucanase (E.C. 3.2.1.176), endoglucanase (E.C. 3.2.1.4), and β-glucosidases (E.C. 3.2.1.21) [24,25]. These enzymes are essential for the hydrolysis of plant biomass as they cause complete cellulose hydrolysis by their consecutive actions to form a glucose monomer for bioethanol production [60]. Additionally, cellulases have been widely applied in brewing, bread, detergents, textiles, pulp, and paper industries. Numerous bacterial and fungal species have been reported to produce cellulases using cellulose rich agro-industrial food wastes. Srivastava et al. [61], reported improved cellulase production by a novel fungus, *Cladosporium cladosporioides* NS2, under SSF employing sugarcane bagasse. The fungal isolate exhibited enhanced cellulase production with maximum filter paper cellulase (16.9 IU/gds), endoglucanase (150 IU/gds), and β-glucosidase (200 IU/gds) activity. In another study, Leite et al. [62] carried out the simultaneous production of cellulase and xylanases from different fungal strains using brewer's spent grain as cheap agro-industrial waste. *Aspergillus niger* was the best producer of the β-glucosidase enzyme with 94 $\pm$ 4 U/g activity, while *A. ibericus* achieved maximum cellulase and xylanase activities of 51–62 U cellulase/g and 300 to 313 U xylanase/g, respectively.

In the last few years, the research on xylanases has risen to the forefront due to their wide range of potential applications in numerous industries, including dairy, food, bakery, feed, paper and pulp, and lignocellulosic biorefineries [24,36]. The microbial xylanases are favored among the different sources of xylanolytic enzymes because they may possess desirable processing features and may be produced in large quantities efficiently and economically. Recently, Intasit and co-workers [63] produced a fungal xylanase from lignocellulosic palm wastes by combining SSF and SmF using *Aspergillus tubingensis* TSIP9 in a bioreactor. The combinatorial SSF–SmF endorsed higher xylanase production with high purification (7.4-fold), activity, and stability at different pH (3–8) and temperatures (30–60 °C). Singh et al. [64] conducted a study to explore rice straw, wheat straw, sugarcane bagasse, sawdust, cotton stalk, and rice husk waste as carbon sources for improved xylanase and cellulase production by *Aspergillus flavus* under SSF. Rice straw was determined to be the best waste biomass for use as a carbon source for the production of enzymes with maximum xylanase 180 IU/gds, CMCase 235 IU/gds, FPase (12.5 IU/gds) and β-glucosidase 190 IU/gds, activities under optimum conditions. These studies demonstrated that strain microbial strains can potentially be used for cellulase and xylanase production that provides an economical method to produce high-value enzymes using agro-food wastes by SSF and SmF techniques.

3.4.2. Lignin Degrading Enzymes

Lignin is an abundantly present complex aromatic heteropolymer, which is one of the three major components of lignocellulosic biomass [16]. The lignification process is achieved through polymer–polymer coupling reactions by free radicals formed by oxidases or by the cross-linking of lignin monomers [65]. However, lignin degradation and depolymerization is necessary in many aspects for effective biomass valorization into bioproducts and carbon recycling. The agro-food waste is majorly represented as lignocellulosic biomass that acts as an ideal substrate for the action of microbes and the production of lignolytic enzymes in the food industry. Lignin peroxidases, laccases, and manganese peroxidases have been considered as suitable biological catalysts for lignin depolymerization and degradation into lignin monomers.

In a recent study, two thermotolerant lignin-degrading *Bacillus* sp. LD2 and *Aneurinibacillus* sp. LD3 exhibited an improved lignin degradation rate (61.28%) with high ligninolytic enzyme activities. The lignin degrading enzymes, such as laccase, manganese peroxidase, and lignin peroxidase from these bacteria possessed maximum activities of 1484.5, 1770.75, 3117.25, and U L^{-1}, respectively [66]. Bagewadi et al. [67] used rice straw, corn cobs, sugarcane bagasse, wheat bran, and groundnut shells as potential substrates for laccase production under SSF using *Trichoderma harzianum*. The results revealed wheat bran to be suitable feedstock for maximum laccase production (510 U/g) with an 8.09-fold increase under optimized conditions. It is pertinent to note here that the high-cost commercial enzyme production systems have limited their wide scale applications in the lignocellulosic biorefineries. Therefore, the development of indigenous enzyme production processes using agro-food wastes as the inexpensive raw materials may help in bringing down the overall costs of production of second-generation biofuels and platform biochemicals.

3.5. Pectinolytic Enzymes

The pectinolytic enzymes have been widely explored in the food industry that catalyze the fragmentation of pectin-containing biomaterials forming an integral component of plant cell walls [68]. Pectin disintegration requires a set of pectinolytic enzymes including exo- and endo-polygalacturonases, and pectate- and pectin lyases [17]. Pectinases find applications in wine and fruit juice making to clarify the final product and eliminate turbidity. By using de-esterification and depolymerization reactions, the pectinases break down complex pectins in a sequential and synergistic fashion [69,70]. Most of the microorganisms can produce pectinase enzymes, however, fungi are favored for commercial applications since more than 90% of the enzyme produced is secreted into the culture medium by fungi.

Among different fungal species, *Penicillium*, *Aspergillus niger*, *Trichoderma viride*, and *Rhizopus* offer many advantages as pectinase producers since they are designated as Generally Regarded as Safe (GRAS) strains and produce extracellular bioproducts that may be readily recovered from the fermented medium [68]. Núñez Pérez et al. [69] isolated a wild strain of *Aspergillus* sp. To produce pectinases from pectin-rich dehydrated coffee pulp and husk under SSF with 29.9 IU/g of enzyme activity. Sethi et al. [70] reported the use of natural substrates, such as neem oil cake, mustard oil cake, groundnut oil cake, green gram peels, black gram peels, chickling vetch peels, pearl millet residues, broken rice, wheat bran, finger millet waste, apple pomace, banana peels, and orange peels for pectinase biosynthesis from *Aspergillus terreus* NCFT4269. The results displayed that a maximum pectinase activity of 6500 $\pm$ 1116.21 U/g was achieved under SSF, while the liquid static surface fermentation gave the enzymatic activity of 400 ± 21.45 U/mL under optimized conditions. Future research studies on pectinolytic enzymes should be focused on determining the molecular processes that control enzyme secretion in addition to modes of action of diverse pectinolytic enzymes on agro-food pectic substrates. It would offer a strong platform for controlling microorganisms to produce large amounts of effective and affordable enzyme systems.

4. Low-Cost Enzyme Immobilization Strategies

The bioconversion of agro-industrial food and crop waste into valuable fuels and bioproducts often depends on the cost of the method, equipment, and infrastructure, as well as the market value of the finished products. Enzymes offer an array of advantages in the valorization of abundantly generated agro-food wastes [19,21,41,42,69]. They perform biochemical catalytic reactions with exceptional specificity and decent stereo-selectivity, and, under very mild reaction conditions, enable the development of more environmentally friendly, green, and sustainable biochemical processes [21]. The operating range of enzymes is, however, rather constrained, and enzymes originating from natural sources are particularly effective only under their optimal conditions [18]. Moreover, a lack of maintenance of their stability, catalytic activity, and recovery under variable industrial bioprocessing conditions of pH, temperature, water activity, and solvent properties is a major hurdle [71]. Therefore, the development of steady biocatalytic systems involving enzymes is crucial for expanding their commercial applications. In the last few years, enzyme immobilization has received significant attention as an important bio-engineering approach to customize and enhance a wide range of enzyme catalytic characteristics, including activity, physicochemical stability, specificity, selectivity, and tolerance of inhibitors [12,72,73]. The overall cost of the enzymatic production process can be reduced by immobilizing the industrially significant enzyme support matrixes [74]. To create reusable, long-standing, and stable immobilized biocatalytic systems, the choice of the supporting matrix and the immobilization technique is essential [4]. A number of emerging support materials have been recently used for enzyme immobilization, such as magnetic nanoparticles, graphene oxide, polyurethane foam, or chitosan [75]. The inexpensive immobilization matrixes possess several advantageous properties that help in developing highly efficient biocatalysts (Figure 4).

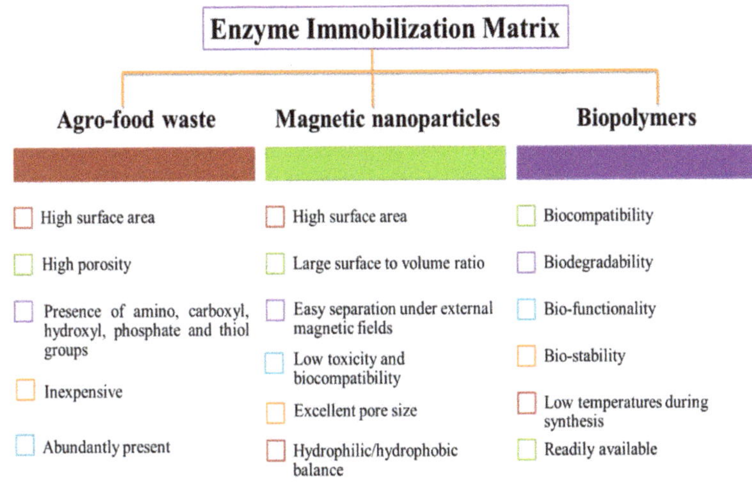

Figure 4. Properties of different enzyme immobilization matrixes.

These support materials can be used to immobilize and entrap the enzymes through different techniques, such as adsorption, crosslinking, covalent binding, and entrapment [21,75–79]. Table 1 shows the list of promising enzyme immobilization supports with major results and applications.

Table 1. Enzyme immobilization by different supporting materials.

Supporting Material	Immobilization Method	Immobilized Enzyme	Major Results	Application	Reference
Rice husk ash	Adsorption	Lipase	Higher adsorption capacity of biocatalyst	Biodiesel production	[80]
Magnetic rice straw (MRS)	Adsorption	Lipase	36% higher esterification yield with Lipase-MRS composite	Biodiesel production	[73]
Rice husk	Covalent	Lipase	Immobilized lipase exhibited high esterification yield (88.0%)	Esterification reaction for biodiesel production	[81]
Rice husk, sugarcane bagasse, babassu mesocarpus, corn cobs, coffee grounds, coconut bark	Adsorption	Lipase	High immobilization efficiencies (>98%)	Hexyl laurate production	[77]
Green coconut fiber	Adsorption and cross-linking	Laccase	>80% initial activity retained up to six cycles	Oxidation of aromaticorganic and inorganic compounds	[82]
Spent coffee grounds	Covalent	β-glucosidase	Enriched aglycone content (67.14 ± 0.60%) by enzymatic treatment	Isoflavone conversion in black soy milk	[83]
Spent grain	Adsorption	Trypsin	High operational and thermal stability of immobilized enzyme	Protein hydrolysis	[84]
Rice straw biochar	Adsorption	Laccase	Increased enzyme stability and reusability	Anthracene biodegradation.	[85]
Rice husk, egg shell membrane	Covalent	Lipase	High immobilization efficiency for rice husk (81%) and eggshell membrane (87%)	-	[86]
Tomato peels	Covalent	Pectinase	High thermal and pH stability, reusability, and storage stability of immobilized biocatalyst	Lycopene production	[87]
Fruits peels and scraps	Covalent	Pectinase and cellulase	80% of residual activity retained for magnetic biocatalyst after ten cycles	Antioxidants production	[88]
Dairy waste	Cross-linkage	Lactase, glucose isomerase	Improved transglycosylation activity of enzyme for improve lactulose yield	Prebiotics production	[89]
Green coconut fiber	Covalent	Laccase	100% activity after 10 cycles	Clarification of apple juice	[90]
Chicken Feather	Covalent	Laccase	94.32% after 3 weeks of storage	Oxidation of Veratryl alcohol	[76]
Carboxymethyl cellulose nanoparticles	Covalent	Lipase	Immobilized enzyme with higher activity and stability	Diverse applications in food industry	[91]
Dialdehyde starch nanoparticles	Adsorption	Lipase	High stability (82.5%) and recycling rate (53.6%)	Food processing	[92]
Graphene oxide-magnetite naoparticles	Covalent	α-amylase	Enhanced stability and half life of immobilized enzyme	Maltose containing syrup production	[93]
Alginate beads	Covalent	Pectinase	Improved thermostability of immobilzed enzyme	Juice processing	[94]
Alginate-gelatin hydrogel matrix	Cross-linkage	Lipase	High thermal stability (96% activity)	Fatty acid production	[79]

4.1. Agro-Food Wastes for Enzyme Immobilization

The exploitation of agricultural residues and food wastes as novel immobilization or supporting matrices for enzymes offers economic solutions for industrial bioprocess, in addition to solving the issues of environment pollution and waste disposal. Undeniably, agro-industrial food and crop wastes contain a variety of characteristics with intriguing potential applications, including high surface area, high porosity, and the presence of several chemical groups, such as amino, carboxyl, hydroxyl, phosphate, and thiol groups [95]. Otari et al. [73] reported a one-step novel and robust method of lipase immobilization on magnetic rice straw using Fe_2O_3 nanoparticles. The results exhibited high lipase immobilization efficiency of 94.3% with 91.3 mg·g^{-1} of enzyme loading, increased enzymatic stability by 8-fold at a high temperature (70 °C), and reusability. Lira et al. [77] used rice husk, sugarcane bagasse, babassu mesocarp, corn cobs, coffee grounds, and coconut bark residual biomasses as supports for lipase immobilization extracted from *Thermomyces lanuginosus*. The results showed an immobilization efficiency of more than 98% with high hydrolytic activity of 4.608 U/g using rice husks as immobilization supports. The activation of agro-industrial food wastes as supports/carriers for enzyme immobilization with low-cost materials postures an upper hand for researchers with its constant availability and environment friendliness. Recently, the stability of *Candida rugosa* lipase was enhanced by its covalent immobilization by glutaraldehyde-activated agricultural wastes. Rice husk support provided the highest stability to the enzyme with the highest retention of initial activity (94.1%), followed by sugarcane bagasse (90.3%) and coconut fiber carrier (89.3%) [81].

Lignocellulolytic enzymes have also been immobilized using agro-food based biomasses as support materials. In a previous study [85], the ligninolytic enzyme-laccase was immobilized on the surface of rice straw biochar for anthracene biodegradation. The laccase exhibited an effective immobilization yield of 66% with high stability up to six cycles and retention of 40% of initial activity. Similarly, Ghosh and Ghosh [82] successfully immobilized purified laccase from *Aspergillus flavus* PUF5 on coconut fiber while retaining 80% of its initial activity after using it for six repeated cycles. These findings support the notion that agro-industrial crop and food waste biomass can produce stable and robust biocatalysts containing immobilized industrially important enzymes with better results than commercial preparations.

4.2. Magnetic Nanoparticles for Enzyme Immobilization

The development of strong enzymatic systems with high stability and catalysis at several extremities is required for commercial applications. Magnetic nanoparticles are regarded as potential support materials when compared to conventional immobilization carriers because of their high surface area, small size, and large surface to volume ratio [74,75]. The nanostructural support materials have huge potential to develop nanobiocatalyst systems involving enzymes displaying high catalytic properties in both aqueous and non-aqueous environments [96]. Suo et al. [91] immobilized lipase on ionic liquid-modified magnetic carboxymethyl cellulose nanoparticles that exhibited strong specific activity, which was 1.43 fold higher than that of the free lipase enzyme. In similar research, Yang et al. [92] achieved high storage stability (82.5%) and recycling rate (53.6%), and better stability and durability of lipase enzymes immobilized by magnetic dialdehyde starch nanoparticles. Moreover, the immobilized lipase displayed better enzymatic properties and improved acid-base tolerance and thermal stability as compared to the free enzyme.

In another research study, Desai et al. [93] prepared graphene oxide–magnetite nanoparticles for amylase immobilization through covalent bonding, which increased the half-life of the immobilized enzyme (20 h) as compared to the free enzyme (13 h). The immobilized amylase also demonstrated high reusability up to eleven subsequent cycles during the production of high maltose containing syrup. The co-immobilization strategy by nanoparticle composites involving two or more enzymes have also been attempted. Nadar and Rathod [88] simultaneously co-immobilized pectinase and cellulase enzymes onto amino functionalized magnetic nanoparticles for antioxidants extraction from waste fruit peel residues. When compared to free form, the magnetic nano-biocatalyst demonstrated a two-fold improved half-life in the temperature range of 50–70 °C and retained up to 80% of residual enzyme activity even after ten repeated cycles. Immobilized laccase on Fe_3O_4 magnetic nanoparticles was employed in a novel detoxifying method to enhance *Rhodotorula glutinis* lipid synthesis from rice straw hydrolysate [96]. The immobilized laccase presented better stability, retaining 56% of its original activity at pH 2 and 76% at 70 °C compared with the free laccase. These findings suggest that enzyme immobilization on magnetic nanoparticles has immense potential to valorize the agro-food waste hydrolysate for improving the production of valuable bioproducts.

4.3. Biopolymers for Enzyme Immobilization

Biopolymers are quite valuable among supports for enzyme immobilization because of their 'green' properties, such as biocompatibility, biodegradability, bio-functionality, and bio-stability [74]. Additionally, biopolymers offer a wide range of chemical and structural properties, need low temperatures during synthesis, and are readily available [97]. Chitin and chitosan are remarkable immobilization carriers of interest among the support materials explored for immobilizing enzymes. A study was conducted on immobilization of pectinase onto chitosan magnetic nanoparticles using dextran polyaldehyde as a macro-molecular cross-linking agent [78]. The immobilized pectinase showed high reusability and permanence retaining 85% and 89% of its original activity after seven cycles and fifteen days of storage, respectively. Recent reports suggest the use of new immobilized food enzymes in food applications. In a novel and sustainable method, a fluidized bed reactor with immobilized propyl endopeptidase protease enzyme from *Aspergillus niger* on food-grade chitosan beads was used to continuously generate high-quality, gluten-free beer from barley malt [98].

The use of hydrogel matrices, such as alginate, for enzyme immobilization is quite popular because they are environmentally benign enzyme carriers with excellent gel porosity and biocompatibility. Recently, Abdel-Mageed et al. [79] synthesized an alginate-gelatin hydrogel matrix as immobilization support for lipase from *Mucor racemosus* (Lip). The immobilized biocatalyst exhibited improved thermal stability while retaining 96% of its activity after four cycles and 90% of its original activity at 4 °C stored for 60 days. Likewise, *Aspergillus aculeatus* pectinase immobilized on alginate beads ensured satisfactory thermostability of the enzyme with low values of activation entropy for juice processing [94]. A recent study focused on the use of co-immobilized pectinase, amylase, and cellulase using different immobilization matrices, such as chitosan, silica gel, and sodium-alginate for fruit juice clarification. The highest activity for amylase (90.9 ± 4.3 U/g carrier), pectinase (85.7 ± 4.0 U/g carrier), and cellulase (23.2 ± 1.1 U/g carrier) was observed when they were co-immobilized onto silica gel with immobilization efficiency of 67.9%, 53.6%, and 72.9%, respectively [99]. These research developments suggest the potential of these biopolymers in enzyme immobilization processes in paving the way for wide scale applications of enzymes in food analysis, food bioprocessing, and food control, among other areas.

5. Enzyme Engineering Approaches

Though enzymes catalyze a wide variety of biochemical reactions, yet they are not suited for many essential catalytic bioprocesses or other industrially relevant substrates that are beyond their natural cellular micro-environments. The desired attributes of diverse industries can be satisfied by using tailored enzymes in novel, cutting-edge enzyme engineering and stabilization techniques, opening up new opportunities for their use in biocatalysis. In the present scenario, the high value of protein engineering has been well recognized in industrial-level biotransformation [100]. Protein engineering, with assistance from molecular approaches or directed evolution, rational design, or computational methods, enables the accelerated designing of biocatalysts that are ideal for any desired bioprocess with commercial applications [101]. Figure 5 explains the different strategies of enzyme engineering for improved biocatalysts.

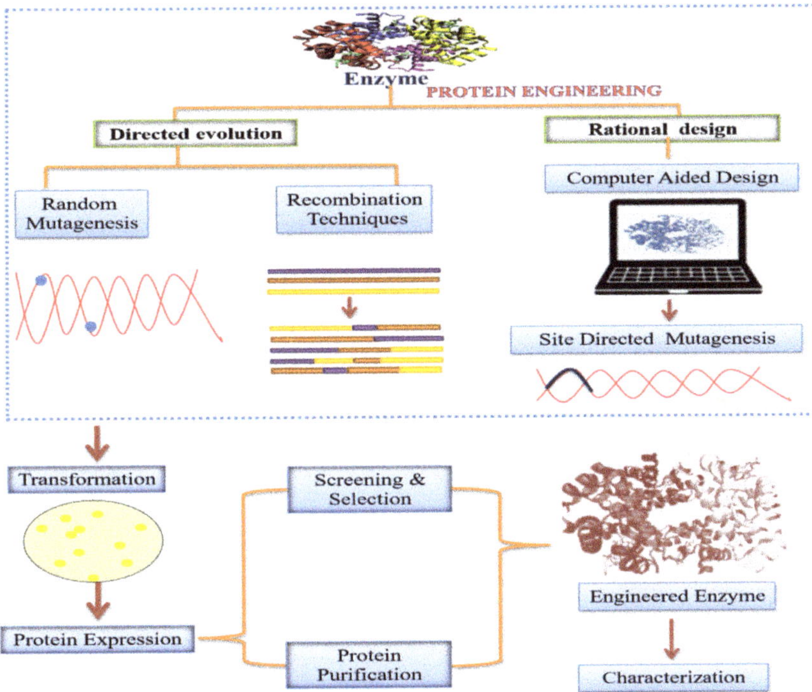

Figure 5. The schematics of different enzyme engineering methods.

The enzymatic diversity created by these methods through the application of semi-rational designs based on artificial intelligence, protein structure, and sequence information yields superior specificity and functionality of industrial enzymes. Different approaches for the enzyme engineering are presented in Table 2.

Table 2. Different enzyme engineering strategies and engineered biocatalysts.

Method	Enzyme	Source	Improved Properties	References
Directed evolution	α-amylase	*Bacillus cereus* GL96	Higher pH and thermostability	[102]
	β-glucosidase A	*Clostridium thermocellum*	Higher thermostability and catalytic activity	[103]
	Lipase	*Proteus mirabilis*	4.3-fold catalytic efficiency and enhanced stability of enzyme	[104]
	Amylase	*Bacillus licheniformis* R-53	Enzyme with better effects in delaying recrystallization, reducing hardness, improving elasticity of bread dough	[105]
	β-glucosidase	*Pichia pastoris*	Better saccharification efficiency	[106]

Table 2. *Cont.*

Method	Enzyme	Source	Improved Properties	References
Rational design	Serine peptidase	*Pseudomonas aeruginosa*	High thermostability and catalytic activity	[107]
	Lipase	*Candida rugosa*	Higher enzyme esterification yield (88.0%) and retained 72.7% of initial activity	[81]
	β-glucosidase	*Talaromyces leycettanus*	Improved substrate affinity and catalytic efficiency	[108]
	α-amylase	*Bacillus licheniformis*	High specific activity and thermostability (70 °C)	[109]
	α-amylase	*Bacillus subtilis*	Improved hydrolytic pattern of engineered enzyme	[110]
Computational designs	Lipase	*Rhizopus chinensis*	High catalytic efficiency (700%), potential for propeptide to shift the substrate specificity	[111]
	Lipase	*Yarrowia lipolytica*	Increased reaction rate and enzyme recyclability	[112]
	Lipase	*Pseudomonas aeruginosa*	Increased enzyme stability and activity by creation of α-helix hotspots after immobilization	[113]
	Glucose oxidase	*Aspergillus niger*	Improved catalytic efficiency with high gluconic acid yield (324 g/L)	[114]

5.1. Directed Evolution

The enzyme biocatalysts must exhibit superior productivity and selectivity with minimal catalyst loadings at high substrate concentrations, which can be accomplished by creating mutant enzyme libraries and using directed evolution. Directed evolution (DE) is a potent method for producing efficient enzyme catalysts that can carry out a variety of biocatalytic tasks without in-depth knowledge of structure–function correlations [115]. DE, also referred to as molecular evolution, does not require information of the sequence or three-dimensional structure of the enzymes [116]. Consistent high-throughput screening of the mutant library and effective library construction are required for successful DE experiments. The major highlights of DE include the creation of random gene libraries, expression of genes in an appropriate host, and screening libraries of mutant enzymes according to the need [117].

Recently, Pouyan et al. [102] used an in silico approach to redesign α-amylase from *Bacillus cereus* GL96 for higher thermostability and characterized the engineered enzyme using directed evolution. The engineered α-amylase exhibited superior properties over wide temperature (70 °C) and pH (4–11) ranges. In another study, lipase from *Proteus mirabilis* was genetically fused to self-crystallizing protein (Cry3Aa) for immobilized lipase crystal production by DE in *Bacillus thuringiensis* [104]. The immobilized lipase mutant exhibited 4.3-fold greater catalytic efficiency and enhanced stability that could efficiently catalyse waste cooking oil into biodiesel for at least 15 cycles with reasonable conversion efficiency.

Similarly, in another study, the site-directed mutagenesis of a lipase from *Pseudomonas fluorescens* provided enhanced thermostability to the enzyme which could be applied to food applications [118]. The transglutaminase enzyme from *Streptomyces mobaraensis*, which catalyses the cross-linking modification of proteins and other biotechnological fields, was subjected to site-directed mutagenesis [119]. The variant enzyme possessed higher thermostability with improved specific activity. These studies provide deeper insights of the structure–function relationship for improving the thermostability of different enzymes through directed evolution. It also offers a theoretical framework and background knowledge for designing enzymes with improved properties to satisfy industrial demands.

5.2. Rational Design

Rational design (RD) is a classical protein engineering strategy for obtaining tailored enzymes with improved catalytic properties, kinetics, thermostability, substrate specificity, and resistance to organic solvents [107,116]. RD brings precise variations in the amino acid sequence via site-directed mutagenesis and is used when the structure, function, and mechanism of action of the target enzyme is already known [120]. The enzyme of interest can be engineered by RD involving targeted mutagenesis, computational techniques, and a de novo design [121]. In a study by Ashraf et al. [107], a serine peptidase from *Pseudomonas aeruginosa* was engineered using rational design for improved thermal stability and catalytic efficiency. The mutant enzyme exhibited higher T_m and increased residual activity at elevated temperature compared to the wild type.

Some researchers have recently used computational design techniques to alter important industrial enzymes used with improved enzymatic activity. Costa et al. [122] employed a computational protein design method to redesign Cel9A-68 cellulase from *Thermobifida fusca* through linker mutations that facilitated higher enzymatic activity for cellulose degradation. In another study by Elatico et al. [113], the computational technique was used to reverse engineer the lipase from *Pseudomonas aeruginosa* PAO1 using proline mutations. The technique helped in creating variants with possible α-helix hotspots for augmented enzyme activity and stability. Given the promising traits the mutant enzyme displays, these protein engineering methods could be taken into consideration for additional research to fulfil the industrial requirements.

6. Current Challenges and Future Prospects

Sustainable development based on the idea of a circular economy could possibly assist in achieving the targets of global waste minimization, valorization, and its recycling. The agro-industrial food waste which is an underutilized resource ideally fulfills the criteria for circular economy for conversion into useful bioproducts. Agro-food waste contains a significant amount of latent nutrients that can be efficiently extracted, recycled, repurposed, and used as substrates for enzyme production. Enzymes have been widely explored in the food industry and lignocellulosic biorefineries for producing numerous value-added biochemicals. However, the scaling up of enzyme production still faces a huge research gap to meet the industrial requirements. The significant challenges and barriers, including high production costs, low stability, and long reaction times, among others, still persist in the commercial applicability of enzymes. Moreover, the market cost of enzymes is quite high owing to the fact that expensive synthetic substrates and processes are used for their production. For enzyme prices to be competitive, they would need to be an average of $0.10 per gallon [123]. The development of biocatalytic enzyme systems from low-cost agro-food wastes represent a distinctive technological approach for environmental and economical sustainability. Different strategies to improve enzyme production costs have been proposed through comprehensive research efforts over the past few decades. Additionally, the shortlisting of agro-food wastes as the carriers for enzyme immobilization with various operational requirements is challenging, but exciting in terms of further mitigating the cost-related issues of enzyme applicability at industrial levels.

There are certain challenges also related to enzyme immobilization practices. These include enzyme distortion during immobilization, steric hindrance of enzymes with substrate, rapid consumption of the substrate, etc. The distortion of the enzyme during immobilization occurs when the enzyme is being used under more severe conditions than normal conditions. However, stabilizing the enzyme during immobilization might allow for higher activity than the soluble enzyme. The steric interference of enzymes with the substrate may depend upon the enzyme loading on the immobilization support. If the surface of the support material does not completely block the active site, a reasonable activity against large substrates can be found with minimal enzyme load, resulting in enzyme molecules with free space around them to bind with the substrate. Apparently, the substrates with different molecular sizes and using different enzyme loadings could help in understanding if the issues are caused by steric hindrances or enzyme distortion. In certain cases, the enzyme is physically adsorbed on the surface of the immobilization matrix and may release from the support, resulting in lower efficiency. This issue may be discovered by measuring the activity in the washing solutions, particularly the initial ones. Therefore, a deeper understanding of the mechanisms of enzyme immobilization on the support matrixes is necessary to mitigate these inadequacies.

Looking forward, the improvements in the science and engineering knowledge for choice of microorganism, enzyme production under SSF and SmF systems, and maintenance of optimum chemical, physical, and biological parameters could develop the sustainable bioprocess. The development of stable biocatalytic systems by novel immobilization technologies using agro-food wastes as carriers could also elevate the industrial applications of enzymes. Over the last several years, the increase in the market demand for enzymes to establish new technological bioprocesses has substantially driven the need for engineered enzymes with unique biocatalytic and economic attributes. Research on engineering of the local enzyme environments and their catalytic regions using exciting computational and machine learning technologies is expected to further increase in coming years and involve multi-step reaction cascades, economizing the overall bioprocess. Moreover, powerful tools like life-cycle assessments and techno-economic analyses could be used for the evaluation of the viable commercial-scale biocatalytic processes. Further understanding of enzymes can be more effectively used in a range of industrial processes, which will come from research studies of both known and yet-to-be discovered enzymes.

7. Conclusions

The management of food and agricultural trash is one of the most pressing issues for modern civilization. The proper repurposing of agro-food wastes utilizing green technology is critical to reduce the negative and destructive consequences of waste disposal that produce compounds with added value, aiding towards implementing circular economy. Microbial enzymes play a key role in the valorization of agro-industrial crop and food wastes compared to conventional chemical catalysts. The utilization of agro-food waste to produce commercially important enzymes by microorganisms offers great promise for efficient waste utilization and sufficient biocatalytic systems with high conversion efficiencies, thereby allowing achievement of the targets of sustainable development. Furthermore, using novel, inexpensive enzyme immobilization supports and engineered enzymes can exhibit improved catalytic performance when applying them to industrial food applications.

Author Contributions: Conceptualization, V.S. and P.N.; Validation, V.S., M.-L.T., C.-W.C., P.N. and C.-D.D.; Formal Analysis, V.S., M.-L.T., P.N. and C.-W.C.; Investigation, V.S., M.-L.T. and C.-D.D.; Writing—Original Draft Preparation, V.S.; Writing—Review & Editing, P.N. and C.-D.D.; Visualization, V.S., M.-L.T. and C.-D.D.; Resources, M.-L.T. and C.-H.K.; Supervision, M.-L.T., P.-P.S. and C.-D.D.; Funding Acquisition, C.-D.D. All authors have read and agreed to the published version of the manuscript.

Funding: This work was funded by Taiwan MOST for funding support (Ref. No. 109-2222-E-992-002).

Data Availability Statement: Not applicable.

Acknowledgments: Authors V.S. and P.N. gratefully acknowledge National Kaohsiung University of Science and Technology, Kaohsiung City, Taiwan, for providing the post-Doctoral fellowship. Authors M.-L.T., C.-W.C., and C.-D.D. acknowledge Ministry of Science and Technology, Taiwan MOST for funding support (Ref. No. 109-2222-E-992-002).

Conflicts of Interest: The authors declare no conflict of interest.

Abbreviations

AW	Agricultural waste
FAO	Food and agriculture organization
FVW	Fruit and vegetable waste
GRAS	Generally Regarded as Safe
KW	Kitchen waste
LCB	Lignocellulosic biomass
SmF	Submerged fermentation
SSF	Solid state fermentation

References

1. Fasim, A.; More, V.S.; More, S.S. Large-Scale Production of Enzymes for Biotechnology Uses. *Curr. Opin. Biotechnol.* **2021**, *69*, 68–76. [CrossRef] [PubMed]
2. Kuo, C.-H.; Huang, C.-Y.; Shieh, C.-J.; Dong, C.-D. Enzymes and Biocatalysis. *Catalysts* **2022**, *12*, 993. [CrossRef]
3. Chapman, J.; Ismail, A.E.; Dinu, C.Z. Industrial Applications of Enzymes: Recent Advances, Techniques, and Outlooks. *Catalysts* **2018**, *8*, 238. [CrossRef]
4. Arya, P.S.; Yagnik, S.M.; Rajput, K.N.; Panchal, R.R.; Raval, V.H. Valorization of Agro-Food Wastes: Ease of Concomitant-Enzymes Production with Application in Food and Biofuel Industries. *Bioresour. Technol.* **2022**, *361*, 127738. [CrossRef] [PubMed]
5. Golgeri M., D.B.; Mulla, S.I.; Bagewadi, Z.K.; Tyagi, S.; Hu, A.; Sharma, S.; Bilal, M.; Bharagava, R.N.; Ferreira, L.F.R.; Gurumurthy, D.M.; et al. A Systematic Review on Potential Microbial Carbohydrases: Current and Future Perspectives. *Crit. Rev. Food Sci. Nutr.* **2022**, *62*, 1–18. [CrossRef] [PubMed]
6. Thapa, S.; Li, H.; OHair, J.; Bhatti, S.; Chen, F.-C.; Nasr, K.A.; Johnson, T.; Zhou, S. Biochemical Characteristics of Microbial Enzymes and Their Significance from Industrial Perspectives. *Mol. Biotechnol.* **2019**, *61*, 579–601. [CrossRef] [PubMed]
7. Singh, A.; Singhania, R.R.; Soam, S.; Chen, C.-W.; Haldar, D.; Varjani, S.; Chang, J.-S.; Dong, C.-D.; Patel, A.K. Production of Bioethanol from Food Waste: Status and Perspectives. *Bioresour. Technol.* **2022**, *360*, 127651. [CrossRef]
8. Rojas, L.F.; Zapata, P.; Ruiz-Tirado, L. Agro-Industrial Waste Enzymes: Perspectives in Circular Economy. *Curr. Opin. Green Sustain. Chem.* **2022**, *34*, 100585. [CrossRef]
9. Kumar, V.; Sharma, N.; Umesh, M.; Selvaraj, M.; Al-Shehri, B.M.; Chakraborty, P.; Duhan, L.; Sharma, S.; Pasrija, R.; Awasthi, M.K.; et al. Emerging Challenges for the Agro-Industrial Food Waste Utilization: A Review on Food Waste Biorefinery. *Bioresour. Technol.* **2022**, *362*, 127790. [CrossRef] [PubMed]
10. | Food Loss Reduction CoP | Food and Agriculture Organization of the United Nations. Available online: https://www.fao.org/platform-food-loss-waste/resources/detail/en/c/1378978/ (accessed on 7 September 2022).
11. Sharma, V.; Nargotra, P.; Sharma, S.; Bajaj, B.K. Efficacy and Functional Mechanisms of a Novel Combinatorial Pretreatment Approach Based on Deep Eutectic Solvent and Ultrasonic Waves for Bioconversion of Sugarcane Bagasse. *Renew. Energy* **2021**, *163*, 1910–1922. [CrossRef]
12. Sharma, S.; Nargotra, P.; Sharma, V.; Bangotra, R.; Kaur, M.; Kapoor, N.; Paul, S.; Bajaj, B.K. Nanobiocatalysts for Efficacious Bioconversion of Ionic Liquid Pretreated Sugarcane Tops Biomass to Biofuel. *Bioresour. Technol.* **2021**, *333*, 125191. [CrossRef] [PubMed]
13. Nargotra, P.; Sharma, V.; Gupta, M.; Kour, S.; Bajaj, B.K. Application of Ionic Liquid and Alkali Pretreatment for Enhancing Saccharification of Sunflower Stalk Biomass for Potential Biofuel-Ethanol Production. *Bioresour. Technol.* **2018**, *267*, 560–568. [CrossRef] [PubMed]
14. Sharma, V.; Nargotra, P.; Bajaj, B.K. Ultrasound and Surfactant Assisted Ionic Liquid Pretreatment of Sugarcane Bagasse for Enhancing Saccharification Using Enzymes from an Ionic Liquid Tolerant *Aspergillus assiutensis* VS34. *Bioresour. Technol.* **2019**, *285*, 121319. [CrossRef] [PubMed]
15. Raina, D.; Kumar, V.; Saran, S. A Critical Review on Exploitation of Agro-Industrial Biomass as Substrates for the Therapeutic Microbial Enzymes Production and Implemented Protein Purification Techniques. *Chemosphere* **2022**, *294*, 133712. [CrossRef] [PubMed]
16. Sharma, V.; Tsai, M.-L.; Chen, C.-W.; Sun, P.-P.; Patel, A.K.; Singhania, R.R.; Nargotra, P.; Dong, C.-D. Deep Eutectic Solvents as Promising Pretreatment Agents for Sustainable Lignocellulosic Biorefineries: A Review. *Bioresour. Technol.* **2022**, *360*, 127631. [CrossRef] [PubMed]
17. Uçkun Kiran, E.; Trzcinski, A.P.; Ng, W.J.; Liu, Y. Enzyme Production from Food Wastes Using a Biorefinery Concept. *Waste Biomass Valor.* **2014**, *5*, 903–917. [CrossRef]
18. Bilal, M.; Iqbal, H.M.N. Sustainable Bioconversion of Food Waste into High-Value Products by Immobilized Enzymes to Meet Bio-Economy Challenges and Opportunities—A Review. *Food Res. Int.* **2019**, *123*, 226–240. [CrossRef]
19. Basso, A.; Serban, S. Industrial Applications of Immobilized Enzymes—A Review. *Mol. Catal.* **2019**, *479*, 110607. [CrossRef]
20. Nargotra, P.; Sharma, V.; Sharma, S. Purification of an Ionic Liquid Stable Cellulase from Aspergillus Aculeatus PN14 with Potential for Biomass Refining. *Environ. Sustain.* **2022**, *5*, 313–323. [CrossRef]

21. Xie, J.; Zhang, Y.; Simpson, B. Food Enzymes Immobilization: Novel Carriers, Techniques and Applications. *Curr. Opin. Food Sci.* **2022**, *43*, 27–35. [CrossRef]

22. Zhang, Y.; Geary, T.; Simpson, B.K. Genetically Modified Food Enzymes: A Review. *Curr. Opin. Food Sci.* **2019**, *25*, 14–18. [CrossRef]

23. Yaashikaa, P.R.; Senthil Kumar, P.; Varjani, S. Valorization of Agro-Industrial Wastes for Biorefinery Process and Circular Bioeconomy: A Critical Review. *Bioresour. Technol.* **2022**, *343*, 126126. [CrossRef] [PubMed]

24. Haldar, D.; Shabbirahmed, A.M.; Singhania, R.R.; Chen, C.-W.; Dong, C.-D.; Ponnusamy, V.K.; Patel, A.K. Understanding the Management of Household Food Waste and Its Engineering for Sustainable Valorization-A State-of-the-Art Review. *Bioresour. Technol.* **2022**, *358*, 127390. [CrossRef]

25. Ravindran, R.; Hassan, S.S.; Williams, G.A.; Jaiswal, A.K. A Review on Bioconversion of Agro-Industrial Wastes to Industrially Important Enzymes. *Bioengineering* **2018**, *5*, 93. [CrossRef] [PubMed]

26. Sharma, P.; Gaur, V.K.; Kim, S.-H.; Pandey, A. Microbial Strategies for Bio-Transforming Food Waste into Resources. *Bioresour. Technol.* **2020**, *299*, 122580. [CrossRef]

27. Sahoo, A.; Sarkar, S.; Lal, B.; Kumawat, P.; Sharma, S.; De, K. Utilization of Fruit and Vegetable Waste as an Alternative Feed Resource for Sustainable and Eco-Friendly Sheep Farming. *Waste Manag.* **2021**, *128*, 232–242. [CrossRef] [PubMed]

28. Okino-Delgado, C.H.; Prado, D.Z.; Facanali, R.; Marques, M.M.O.; Nascimento, A.S.; Fernandes, C.J.D.C.; Zambuzzi, W.F.; Fleuri, L.F. Bioremediation of Cooking Oil Waste Using Lipases from Wastes. *PLoS ONE* **2017**, *12*, e0186246. [CrossRef] [PubMed]

29. Gao, Z.; Ma, Y.; Liu, Y.; Wang, Q. Waste Cooking Oil Used as Carbon Source for Microbial Lipid Production: Promoter or Inhibitor. *Environ. Res.* **2022**, *203*, 111881. [CrossRef]

30. Sharma, A.; Kuthiala, T.; Thakur, K.; Thatai, K.S.; Singh, G.; Kumar, P.; Arya, S.K. Kitchen Waste: Sustainable Bioconversion to Value-Added Product and Economic Challenges. *Biomass Conv. Bioref.* **2022**, *12*, 1–22. [CrossRef]

31. Benabda, O.; M'hir, S.; Kasmi, M.; Mnif, W.; Hamdi, M. Optimization of Protease and Amylase Production by *Rhizopus oryzae* Cultivated on Bread Waste Using Solid-State Fermentation. *J. Chem.* **2019**, *2019*, 3738181. [CrossRef]

32. Minh, P.N.N.; Le, T.T.; Camp, J.V.; Raes, K. Valorization of Waste and By-Products from the Agrofood Industry Using Fermentation Processes and Enzyme Treatments. In *Utilisation of Bioactive Compounds from Agricultural and Food Waste*; CRC Press: Boca Raton, FL, USA, 2017; ISBN 978-1-315-15154-0.

33. Nnolim, N.E.; Okoh, A.I.; Nwodo, U.U. Bacillus Sp. FPF-1 Produced Keratinase with High Potential for Chicken Feather Degradation. *Molecules* **2020**, *25*, 1505. [CrossRef] [PubMed]

34. Singh, S.; Gupta, P.; Sharma, V.; Koul, S.; Kour, K.; Bajaj, B.K. Multifarious Potential Applications of Keratinase of Bacillus Subtilis K-5. *Biocatal. Biotransform.* **2014**, *32*, 333–342. [CrossRef]

35. Nargotra, P.; Sharma, V.; Sharma, S.; Kapoor, N.; Bajaj, B.K. Development of Consolidated Bioprocess for Biofuel-Ethanol Production from Ultrasound-Assisted Deep Eutectic Solvent Pretreated Parthenium Hysterophorus Biomass. *Biomass Conv. Bioref.* **2020**, *10*, 1–16. [CrossRef]

36. Sharma, S.; Sharma, V.; Nargotra, P.; Bajaj, B.K. Bioprocess Development for Production of a Process-Apt Xylanase with Multifaceted Application Potential for a Range of Industrial Processes. *SN Appl. Sci.* **2020**, *2*, 739. [CrossRef]

37. Sharma, V.; Nargotra, P.; Sharma, S.; Bajaj, B.K. Efficient Bioconversion of Sugarcane Tops Biomass into Biofuel-Ethanol Using an Optimized Alkali-Ionic Liquid Pretreatment Approach. *Biomass Conv. Bioref.* **2020**, *10*, 1–14. [CrossRef]

38. Nargotra, P.; Sharma, V.; Bajaj, B.K. Consolidated Bioprocessing of Surfactant-Assisted Ionic Liquid-Pretreated *Parthenium hysterophorus* L. Biomass for Bioethanol Production. *Bioresour. Technol.* **2019**, *289*, 121611. [CrossRef]

39. Sharma, V.; Bhat, B.; Gupta, M.; Vaid, S.; Sharma, S.; Nargotra, P.; Singh, S.; Bajaj, B.K. Role of Systematic Biology in Biorefining of Lignocellulosic Residues for Biofuels and Chemicals Production. In *Sustainable Biotechnology-Enzymatic Resources of Renewable Energy*; Singh, O.V., Chandel, A.K., Eds.; Springer International Publishing: Cham, Switzerland, 2018; pp. 5–55. ISBN 978-3-319-95480-6.

40. Raveendran, S.; Parameswaran, B.; Ummalyma, S.B.; Abraham, A.; Mathew, A.K.; Madhavan, A.; Rebello, S.; Pandey, A. Applications of Microbial Enzymes in Food Industry. *Food Technol. Biotechnol.* **2018**, *56*, 16–30. [CrossRef]

41. Menezes, D.B.; de Andrade, L.R.M.; Vilar, D.; Vega-Baudrit, J.R.; Torres, N.H.; Bilal, M.; Silva, D.P.; López, J.A.; Hernández-Macedo, M.L.; Bharagava, R.N.; et al. Synthesis of Industrial Enzymes from Lignocellulosic Fractions. In *Enzymes for Pollutant Degradation*; Mulla, S.I., Bharagava, R.N., Eds.; Microorganisms for Sustainability; Springer Nature: Singapore, 2022; pp. 19–48. ISBN 9789811645747.

42. Leite, P.; Sousa, D.; Fernandes, H.; Ferreira, M.; Costa, A.R.; Filipe, D.; Gonçalves, M.; Peres, H.; Belo, I.; Salgado, J.M. Recent Advances in Production of Lignocellulolytic Enzymes by Solid-State Fermentation of Agro-Industrial Wastes. *Curr. Opin. Green Sustain. Chem.* **2021**, *27*, 100407. [CrossRef]

43. Mojumdar, A.; Deka, J. Recycling Agro-Industrial Waste to Produce Amylase and Characterizing Amylase–Gold Nanoparticle Composite. *Int. J. Recycl. Org. Waste Agricult.* **2019**, *8*, 263–269. [CrossRef]

44. Abdel-Mageed, H.M.; Barakat, A.Z.; Bassuiny, R.I.; Elsayed, A.M.; Salah, H.A.; Abdel-Aty, A.M.; Mohamed, S.A. Biotechnology Approach Using Watermelon Rind for Optimization of α-Amylase Enzyme Production from *Trichoderma virens* Using Response Surface Methodology under Solid-State Fermentation. *Folia Microbiol.* **2022**, *67*, 253–264. [CrossRef]

45. Iqbalsyah, T.M.; Amna, U.; Utami, R.S.; Oesman, F. Febriani—Concomitant Cellulase and Amylase Production by a Thermophilic Bacterial Isolate in a Solid-State Fermentation Using Rice Husks. *Agric. Nat. Resour.* **2019**, *53*, 327–333.

46. Chimbekujwo, K.I.; Ja'afaru, M.I.; Adeyemo, O.M. Purification, Characterization and Optimization Conditions of Protease Produced by *Aspergillus brasiliensis* Strain BCW2. *Sci. Afr.* **2020**, *8*, e00398. [CrossRef]

47. Escaramboni, B.; Garnica, B.C.; Abe, M.M.; Palmieri, D.A.; Fernández Núñez, E.G.; de Oliva Neto, P. Food Waste as a Feedstock for Fungal Biosynthesis of Amylases and Proteases. *Waste Biomass Valor.* **2022**, *13*, 213–226. [CrossRef]

48. Balachandran, C.; Vishali, A.; Nagendran, N.A.; Baskar, K.; Hashem, A.; Abd_Allah, E.F. Optimization of Protease Production from *Bacillus halodurans* under Solid State Fermentation Using Agrowastes. *Saudi J. Biol. Sci.* **2021**, *28*, 4263–4269. [CrossRef]

49. Talhi, I.; Dehimat, L.; Jaouani, A.; Cherfia, R.; Berkani, M.; Almomani, F.; Vasseghian, Y.; Chaouche, N.K. Optimization of Thermostable Proteases Production under Agro-Wastes Solid-State Fermentation by a New Thermophilic *Mycothermus thermophilus* Isolated from a Hydrothermal Spring Hammam Debagh, Algeria. *Chemosphere* **2022**, *286*, 131479. [CrossRef]

50. Zheng, L.; Yu, X.; Wei, C.; Qiu, L.; Yu, C.; Xing, Q.; Fan, Y.; Deng, Z. Production and Characterization of a Novel Alkaline Protease from a Newly Isolated Neurospora Crassa through Solid-State Fermentation. *LWT* **2020**, *122*, 108990. [CrossRef]

51. Camargo, D.A.; Pereira, M.S.; dos Santos, A.G.; Fleuri, L.F. Isolated and Fermented Orange and Grape Wastes: Bromatological Characterization and Phytase, Lipase and Protease Source. *Innov. Food Sci. Emerg. Technol.* **2022**, *77*, 102978. [CrossRef]

52. Rawoof, S.A.A.; Kumar, P.S.; Devaraj, K.; Devaraj, T.; Subramanian, S. Enhancement of Lactic Acid Production from Food Waste through Simultaneous Saccharification and Fermentation Using Selective Microbial Strains. *Biomass Conv. Bioref.* **2020**, *10*, 1–12. [CrossRef]

53. Araujo, S.C.; Ramos, M.R.M.F.; do Espírito Santo, E.L.; de Menezes, L.H.S.; de Carvalho, M.S.; de Carvalho Tavares, I.M.; Franco, M.; de Oliveira, J.R. Optimization of Lipase Production by *Penicillium roqueforti* ATCC 10110 through Solid-State Fermentation Using Agro-Industrial Residue Based on a Univariate Analysis. *Prep. Biochem. Biotechnol.* **2022**, *52*, 325–330. [CrossRef]

54. Sales de Menezes, L.H.; Carneiro, L.L.; Maria de Carvalho Tavares, I.; Santos, P.H.; Pereira das Chagas, T.; Mendes, A.A.; Paranhos da Silva, E.G.; Franco, M.; Rangel de Oliveira, J. Artificial Neural Network Hybridized with a Genetic Algorithm for Optimization of Lipase Production from *Penicillium roqueforti* ATCC 10110 in Solid-State Fermentation. *Biocatal. Agric. Biotechnol.* **2021**, *31*, 101885. [CrossRef]

55. Reyes-Reyes, A.L.; Valero Barranco, F.; Sandoval, G. Recent Advances in Lipases and Their Applications in the Food and Nutraceutical Industry. *Catalysts* **2022**, *12*, 960. [CrossRef]

56. Pereira, A.S.; Fontes-Sant'Ana, G.C.; Amaral, P.F.F. Mango Agro-Industrial Wastes for Lipase Production from *Yarrowia lipolytica* and the Potential of the Fermented Solid as a Biocatalyst. *Food Bioprod. Process.* **2019**, *115*, 68–77. [CrossRef]

57. Putri, D.N.; Khootama, A.; Perdani, M.S.; Utami, T.S.; Hermansyah, H. Optimization of *Aspergillus niger* Lipase Production by Solid State Fermentation of Agro-Industrial Waste. *Energy Rep.* **2020**, *6*, 331–335. [CrossRef]

58. Singhania, R.R.; Ruiz, H.A.; Awasthi, M.K.; Dong, C.-D.; Chen, C.-W.; Patel, A.K. Challenges in Cellulase Bioprocess for Biofuel Applications. *Renew. Sust. Energy Rev.* **2021**, *151*, 111622. [CrossRef]

59. Singhania, R.R.; Patel, A.K.; Singh, A.; Haldar, D.; Soam, S.; Chen, C.-W.; Tsai, M.-L.; Dong, C.-D. Consolidated Bioprocessing of Lignocellulosic Biomass: Technological Advances and Challenges. *Bioresour. Technol.* **2022**, *354*, 127153. [CrossRef]

60. Al-Mardeai, S.; Elnajjar, E.; Hashaikeh, R.; Kruczek, B.; Van der Bruggen, B.; Al-Zuhair, S. Membrane Bioreactors: A Promising Approach to Enhanced Enzymatic Hydrolysis of Cellulose. *Catalysts* **2022**, *12*, 1121. [CrossRef]

61. Srivastava, N.; Elgorban, A.M.; Mishra, P.K.; Marraiki, N.; Alharbi, A.M.; Ahmad, I.; Gupta, V.K. Enhance Production of Fungal Cellulase Cocktail Using Cellulosic Waste. *Environ. Technol. Innov.* **2020**, *19*, 100949. [CrossRef]

62. Leite, P.; Silva, C.; Salgado, J.M.; Belo, I. Simultaneous Production of Lignocellulolytic Enzymes and Extraction of Antioxidant Compounds by Solid-State Fermentation of Agro-Industrial Wastes. *Ind. Crops Prod.* **2019**, *137*, 315–322. [CrossRef]

63. Intasit, R.; Cheirsilp, B.; Suyotha, W.; Boonsawang, P. Purification and Characterization of a Highly-Stable Fungal Xylanase from *Aspergillus tubingensis* Cultivated on Palm Wastes through Combined Solid-State and Submerged Fermentation. *Prep. Biochem. Biotechnol.* **2022**, *52*, 311–317. [CrossRef]

64. Singh, A.; Bajar, S.; Devi, A.; Bishnoi, N.R. Adding Value to Agro-Industrial Waste for Cellulase and Xylanase Production via Solid-State Bioconversion. *Biomass Conv. Bioref.* **2021**, *11*, 1–10. [CrossRef]

65. Pollegioni, L.; Tonin, F.; Rosini, E. Lignin-Degrading Enzymes. *FEBS J.* **2015**, *282*, 1190–1213. [CrossRef] [PubMed]

66. Wu, X.; Amanze, C.; Wang, J.; Yu, Z.; Shen, L.; Wu, X.; Li, J.; Yu, R.; Liu, Y.; Zeng, W. Isolation and Characterization of a Novel Thermotolerant Alkali Lignin-Degrading Bacterium *Aneurinibacillus* Sp. LD3 and Its Application in Food Waste Composting. *Chemosphere* **2022**, *307*, 135859. [CrossRef] [PubMed]

67. Bagewadi, Z.K.; Mulla, S.I.; Ninnekar, H.Z. Optimization of Laccase Production and Its Application in Delignification of Biomass. *Int. J. Recycl. Org. Waste Agricult.* **2017**, *6*, 351–365. [CrossRef]

68. Almowallad, S.A.; Aljobair, M.O.; Alkuraieef, A.N.; Aljahani, A.H.; Alsuhaibani, A.M.; Alsayadi, M.M. Utilization of Agro-Industrial Orange Peel and Sugar Beet Pulp Wastes for Fungal Endo- Polygalacturonase Production. *Saudi J. Biol. Sci.* **2022**, *29*, 963–969. [CrossRef] [PubMed]

69. Núñez Pérez, J.; Chávez Arias, B.S.; de la Vega Quintero, J.C.; Zárate Baca, S.; Pais-Chanfrau, J.M. Multi-Objective Statistical Optimization of Pectinolytic Enzymes Production by an *Aspergillus* Sp. on Dehydrated Coffee Residues in Solid-State Fermentation. *Fermentation* **2022**, *8*, 170. [CrossRef]

70. Sethi, B.K.; Nanda, P.K.; Sahoo, S. Enhanced Production of Pectinase by Aspergillus Terreus NCFT 4269.10 Using Banana Peels as Substrate. *3 Biotech* **2016**, *6*, 36. [CrossRef]

71. Sharma, S.; Vaid, S.; Bhat, B.; Singh, S.; Bajaj, B.K. Chapter 17—Thermostable Enzymes for Industrial Biotechnology. In *Advances in Enzyme Technology*; Singh, R.S., Singhania, R.R., Pandey, A., Larroche, C., Eds.; Biomass, Biofuels, Biochemicals; Elsevier: Amsterdam, The Netherlands, 2019; pp. 469–495. ISBN 978-0-444-64114-4.

72. Ng, H.S.; Kee, P.E.; Yim, H.S.; Chen, P.-T.; Wei, Y.-H.; Chi-Wei Lan, J. Recent Advances on the Sustainable Approaches for Conversion and Reutilization of Food Wastes to Valuable Bioproducts. *Bioresour. Technol.* **2020**, *302*, 122889. [CrossRef]

73. Otari, S.V.; Patel, S.K.S.; Kalia, V.C.; Lee, J.-K. One-Step Hydrothermal Synthesis of Magnetic Rice Straw for Effective Lipase Immobilization and Its Application in Esterification Reaction. *Bioresour. Technol.* **2020**, *302*, 122887. [CrossRef]

74. Zdarta, J.; Meyer, A.S.; Jesionowski, T.; Pinelo, M. A General Overview of Support Materials for Enzyme Immobilization: Characteristics, Properties, Practical Utility. *Catalysts* **2018**, *8*, 92. [CrossRef]

75. Bilal, M.; Zhao, Y.; Rasheed, T.; Iqbal, H.M.N. Magnetic Nanoparticles as Versatile Carriers for Enzymes Immobilization: A Review. *Int. J. Biol. Macromol.* **2018**, *120*, 2530–2544. [CrossRef]

76. Suman, S.K.; Patnam, P.L.; Ghosh, S.; Jain, S.L. Chicken Feather Derived Novel Support Material for Immobilization of Laccase and Its Application in Oxidation of Veratryl Alcohol. *ACS Sustain. Chem. Eng.* **2019**, *7*, 3464–3474. [CrossRef]

77. de S. Lira, R.K.; Zardini, R.T.; de Carvalho, M.C.C.; Wojcieszak, R.; Leite, S.G.F.; Itabaiana, I. Agroindustrial Wastes as a Support for the Immobilization of Lipase from *Thermomyces lanuginosus*: Synthesis of Hexyl Laurate. *Biomolecules* **2021**, *11*, 445. [CrossRef] [PubMed]

78. Sojitra, U.V.; Nadar, S.S.; Rathod, V.K. Immobilization of Pectinase onto Chitosan Magnetic Nanoparticles by Macromolecular Cross-Linker. *Carbohydr. Polym.* **2017**, *157*, 677–685. [CrossRef] [PubMed]

79. Abdel-Mageed, H.M.; Nada, D.; Radwan, R.A.; Mohamed, S.A.; Gohary, N.A.E.L. Optimization of Catalytic Properties of *Mucor racemosus* Lipase through Immobilization in a Biocompatible Alginate Gelatin Hydrogel Matrix for Free Fatty Acid Production: A Sustainable Robust Biocatalyst for Ultrasound-Assisted Olive Oil Hydrolysis. *3 Biotech* **2022**, *12*, 285. [CrossRef] [PubMed]

80. Bonet-Ragel, K.; López-Pou, L.; Tutusaus, G.; Benaiges, M.D.; Valero, F. Rice Husk Ash as a Potential Carrier for the Immobilization of Lipases Applied in the Enzymatic Production of Biodiesel. *Biocatal. Biotransform.* **2018**, *36*, 151–158. [CrossRef]

81. Costa-Silva, T.A.; Carvalho, A.K.F.; Souza, C.R.F.; Freitas, L.; De Castro, H.F.; Oliveira, W.P. Highly Effective Candida Rugosa Lipase Immobilization on Renewable Carriers: Integrated Drying and Immobilization Process to Improve Enzyme Performance. *Chem. Eng. Res. Des.* **2022**, *183*, 41–55. [CrossRef]

82. Ghosh, P.; Ghosh, U. Immobilization of Purified Fungal Laccase on Cost Effective Green Coconut Fiber and Study of Its Physical and Kinetic Characteristics in Both Free and Immobilized Form. *Curr. Biotechnol.* **2019**, *8*, 3–14. [CrossRef]

83. Chen, K.-I.; Lo, Y.-C.; Liu, C.-W.; Yu, R.-C.; Chou, C.-C.; Cheng, K.-C. Enrichment of Two Isoflavone Aglycones in Black Soymilk by Using Spent Coffee Grounds as an Immobiliser for β-Glucosidase. *Food Chem.* **2013**, *139*, 79–85. [CrossRef]

84. Rocha, C.; Gonçalves, M.P.; Teixeira, J.A. Immobilization of Trypsin on Spent Grains for Whey Protein Hydrolysis. *Process Biochem.* **2011**, *46*, 505–511. [CrossRef]

85. Imam, A.; Suman, S.K.; Singh, R.; Vempatapu, B.P.; Ray, A.; Kanaujia, P.K. Application of Laccase Immobilized Rice Straw Biochar for Anthracene Degradation. *Environ. Pollut.* **2021**, *268*, 115827. [CrossRef]

86. Abdulla, R.; Sanny, S.A.; Derman, E. Stability Studies of Immobilized Lipase on Rice Husk and Eggshell Membrane. *IOP Conf. Ser. Mater. Sci. Eng.* **2017**, *206*, 012032. [CrossRef]

87. Ladole, M.R.; Nair, R.R.; Bhutada, Y.D.; Amritkar, V.D.; Pandit, A.B. Synergistic Effect of Ultrasonication and Co-Immobilized Enzymes on Tomato Peels for Lycopene Extraction. *Ultrason. Sonochem.* **2018**, *48*, 453–462. [CrossRef] [PubMed]

88. Nadar, S.S.; Rathod, V.K. A Co-Immobilization of Pectinase and Cellulase onto Magnetic Nanoparticles for Antioxidant Extraction from Waste Fruit Peels. *Biocatal. Agric. Biotechnol.* **2019**, *17*, 470–479. [CrossRef]

89. Hua, X.; Yang, R.; Zhang, W.; Fei, Y.; Jin, Z.; Jiang, B. Dual-Enzymatic Synthesis of Lactulose in Organic-Aqueous Two-Phase Media. *Food Res. Int.* **2010**, *43*, 716–722. [CrossRef]

90. de Souza Bezerra, T.M.; Bassan, J.C.; de Oliveira Santos, V.T.; Ferraz, A.; Monti, R. Covalent Immobilization of Laccase in Green Coconut Fiber and Use in Clarification of Apple Juice. *Process Biochem.* **2015**, *50*, 417–423. [CrossRef]

91. Suo, H.; Xu, L.; Xue, Y.; Qiu, X.; Huang, H.; Hu, Y. Ionic Liquids-Modified Cellulose Coated Magnetic Nanoparticles for Enzyme Immobilization: Improvement of Catalytic Performance. *Carbohydr. Polym.* **2020**, *234*, 115914. [CrossRef] [PubMed]

92. Yang, X.; Chen, Y.; Yao, S.; Qian, J.; Guo, H.; Cai, X. Preparation of Immobilized Lipase on Magnetic Nanoparticles Dialdehyde Starch. *Int. J. Biol. Macromol.* **2019**, *218*, 324–332. [CrossRef]

93. Desai, R.P.; Dave, D.; Suthar, S.A.; Shah, S.; Ruparelia, N.; Kikani, B.A. Immobilization of α-Amylase on GO-Magnetite Nanoparticles for the Production of High Maltose Containing Syrup. *Int. J. Biol. Macromol.* **2021**, *169*, 228–238. [CrossRef]

94. de Oliveira, R.L.; da Silva, O.S.; Converti, A.; Porto, T.S. Thermodynamic and Kinetic Studies on Pectinase Extracted from *Aspergillus aculeatus*: Free and Immobilized Enzyme Entrapped in Alginate Beads. *Int. J. Biol. Macromol.* **2018**, *115*, 1088–1093. [CrossRef]

95. Girelli, A.M.; Astolfi, M.L.; Scuto, F.R. Agro-Industrial Wastes as Potential Carriers for Enzyme Immobilization: A Review. *Chemosphere* **2020**, *244*, 125368. [CrossRef]

96. Yin, L.; Chen, J.; Wu, W.; Du, Z.; Guan, Y. Immobilization of Laccase on Magnetic Nanoparticles and Application in the Detoxification of Rice Straw Hydrolysate for the Lipid Production of *Rhodotorula glutinis*. *Appl. Biochem. Biotechnol.* **2021**, *193*, 998–1010. [CrossRef] [PubMed]

97. Yushkova, E.D.; Nazarova, E.A.; Matyuhina, A.V.; Noskova, A.O.; Shavronskaya, D.O.; Vinogradov, V.V.; Skvortsova, N.N.; Krivoshapkina, E.F. Application of Immobilized Enzymes in Food Industry. *J. Agric. Food Chem.* **2019**, *67*, 11553–11567. [CrossRef] [PubMed]

98. Benucci, I.; Caso, M.C.; Bavaro, T.; Masci, S.; Keršienė, M.; Esti, M. Prolyl Endopeptidase from *Aspergillus niger* Immobilized on a Food-Grade Carrier for the Production of Gluten-Reduced Beer. *Food Control* **2020**, *110*, 106987. [CrossRef]

99. Ozyilmaz, G.; Gunay, E. Clarification of Apple, Grape and Pear Juices by Co-Immobilized Amylase, Pectinase and Cellulase. *Food Chem.* **2023**, *398*, 133900. [CrossRef] [PubMed]

100. Bilal, M.; Iqbal, H.M.N. Tailoring Multipurpose Biocatalysts via Protein Engineering Approaches: A Review. *Catal. Lett.* **2019**, *149*, 2204–2217. [CrossRef]

101. Bilal, M.; Cui, J.; Iqbal, H.M.N. Tailoring Enzyme Microenvironment: State-of-the-Art Strategy to Fulfill the Quest for Efficient Bio-Catalysis. *Int. J. Biol. Macromol.* **2019**, *130*, 186–196. [CrossRef]

102. Pouyan, S.; Lagzian, M.; Sangtarash, M.H. Enhancing Thermostabilization of a Newly Discovered α-Amylase from *Bacillus cereus* GL96 by Combining Computer-Aided Directed Evolution and Site-Directed Mutagenesis. *Int. J. Biol. Macromol.* **2022**, *197*, 12–22. [CrossRef]

103. Yoav, S.; Stern, J.; Salama-Alber, O.; Frolow, F.; Anbar, M.; Karpol, A.; Hadar, Y.; Morag, E.; Bayer, E.A. Directed Evolution of *Clostridium thermocellum* β-Glucosidase A Towards Enhanced Thermostability. *Int. J. Mol. Sci.* **2019**, *20*, 4701. [CrossRef]

104. Heater, B.S.; Chan, W.S.; Lee, M.M.; Chan, M.K. Directed Evolution of a Genetically Encoded Immobilized Lipase for the Efficient Production of Biodiesel from Waste Cooking Oil. *Biotechnol. Biofuels* **2019**, *12*, 165. [CrossRef]

105. Ruan, Y.; Zhang, R.; Xu, Y. Directed Evolution of Maltogenic Amylase from *Bacillus licheniformis* R-53: Enhancing Activity and Thermostability Improves Bread Quality and Extends Shelf Life. *Food Chem.* **2022**, *381*, 132222. [CrossRef]

106. Kao, M.-R.; Yu, S.-M.; Ho, T.-H.u.D. Improvements of the Productivity and Saccharification Efficiency of the Cellulolytic β-Glucosidase D2-BGL in *Pichia pastoris* via Directed Evolution. *Biotechnol. Biofuels* **2021**, *14*, 126. [CrossRef] [PubMed]

107. Ashraf, N.M.; Krishnagopal, A.; Hussain, A.; Kastner, D.; Sayed, A.M.M.; Mok, Y.-K.; Swaminathan, K.; Zeeshan, N. Engineering of Serine Protease for Improved Thermostability and Catalytic Activity Using Rational Design. *Int. J. Biol. Macromol.* **2019**, *126*, 229–237. [CrossRef] [PubMed]

108. Xia, W.; Bai, Y.; Shi, P. Improving the Substrate Affinity and Catalytic Efficiency of β-Glucosidase Bgl3A from *Talaromyces leycettanus* JCM12802 by Rational Design. *Biomolecules* **2021**, *11*, 1882. [CrossRef] [PubMed]

109. Cui, X.; Yuan, X.; Li, S.; Hu, X.; Zhao, J.; Zhang, G. Simultaneously Improving the Specific Activity and Thermostability of α-Amylase BLA by Rational Design. *Bioprocess Biosyst. Eng.* **2022**, *45*, 1839–1848. [CrossRef]

110. Wang, C.-H.; Lu, L.-H.; Huang, C.; He, B.-F.; Huang, R.-B. Simultaneously Improved Thermostability and Hydrolytic Pattern of Alpha-Amylase by Engineering Central Beta Strands of TIM Barrel. *Appl. Biochem. Biotechnol.* **2020**, *192*, 57–70. [CrossRef] [PubMed]

111. Wang, S.; Xu, Y.; Yu, X.-W. Propeptide in Rhizopus Chinensis Lipase: New Insights into Its Mechanism of Activity and Substrate Selectivity by Computational Design. *J. Agric. Food Chem.* **2021**, *69*, 4263–4275. [CrossRef] [PubMed]

112. Li, L.; Wu, W.; Deng, Z.; Zhang, S.; Guan, W. Improved Thermostability of Lipase Lip2 from *Yarrowia lipolytica* through Disulfide Bond Design for Preparation of Medium-Long-Medium Structured Lipids. *LWT* **2022**, *166*, 113786. [CrossRef]

113. Elatico, A.J.J.; Nellas, R.B. Computational Reverse Engineering of the Lipase from *Pseudomonas aeruginosa* PAO1: α-Helices. *J. Mol. Graph. Model.* **2020**, *100*, 107657. [CrossRef]

114. Mu, Q.; Cui, Y.; Tian, Y.; Hu, M.; Tao, Y.; Wu, B. Thermostability Improvement of the Glucose Oxidase from *Aspergillus niger* for Efficient Gluconic Acid Production via Computational Design. *Int. J. Biol. Macromol.* **2019**, *136*, 1060–1068. [CrossRef]

115. Contreras, F.; Pramanik, S.; Rozhkova, A.M.; Zorov, I.N.; Korotkova, O.; Sinitsyn, A.P.; Schwaneberg, U.; Davari, M.D. Engineering Robust Cellulases for Tailored Lignocellulosic Degradation Cocktails. *Int. J. Mol. Sci.* **2020**, *21*, 1589. [CrossRef]

116. Sharma, V.K.; Sharma, M.; Usmani, Z.; Pandey, A.; Singh, B.N.; Tabatabaei, M.; Gupta, V.K. Tailored Enzymes as Next-Generation Food-Packaging Tools. *Trends Biotechnol.* **2022**, *40*, 1004–1017. [CrossRef] [PubMed]

117. Kumar, A.; Singh, S. Directed Evolution: Tailoring Biocatalysts for Industrial Applications. *Crit. Rev. Biotechnol.* **2013**, *33*, 365–378. [CrossRef] [PubMed]

118. Guan, L.; Gao, Y.; Li, J.; Wang, K.; Zhang, Z.; Yan, S.; Ji, N.; Zhou, Y.; Lu, S. Directed Evolution of *Pseudomonas fluorescens* Lipase Variants With Improved Thermostability Using Error-Prone PCR. *Front. Bioeng. Biotechnol.* **2020**, *8*, 1034. [CrossRef] [PubMed]

119. Liu, Y.; Huang, L.; Shan, M.; Sang, J.; Li, Y.; Jia, L.; Wang, N.; Wang, S.; Shao, S.; Liu, F.; et al. Enhancing the Activity and Thermostability of *Streptomyces mobaraensis* Transglutaminase by Directed Evolution and Molecular Dynamics Simulation. *Biochem. Eng. J.* **2019**, *151*, 107333. [CrossRef]

120. Sharma, A.; Gupta, G.; Ahmad, T.; Mansoor, S.; Kaur, B. Enzyme Engineering: Current Trends and Future Perspectives. *Food Rev. Int.* **2021**, *37*, 121–154. [CrossRef]

121. Madhavan, A.; Arun, K.B.; Binod, P.; Sirohi, R.; Tarafdar, A.; Reshmy, R.; Kumar Awasthi, M.; Sindhu, R. Design of Novel Enzyme Biocatalysts for Industrial Bioprocess: Harnessing the Power of Protein Engineering, High Throughput Screening and Synthetic Biology. *Bioresour. Technol.* **2021**, *325*, 124617. [CrossRef]

122. Costa, M.G.S.; Silva, Y.F.; Batista, P.R. Computational Engineering of Cellulase Cel9A-68 Functional Motions through Mutations in Its Linker Region. *Phys. Chem. Chem. Phys.* **2018**, *20*, 7643–7652. [CrossRef]
123. Ramos, M.D.N.; Milessi, T.S.; Candido, R.G.; Mendes, A.A.; Aguiar, A. Enzymatic Catalysis as a Tool in Biofuels Production in Brazil: Current Status and Perspectives. *Energy Sustain. Dev.* **2022**, *68*, 103–119. [CrossRef]

 *catalysts*

Review

Natural Sun-Screening Compounds and DNA-Repair Enzymes: Photoprotection and Photoaging

Amit Gupta [1], Ashish P. Singh [1], Varsha K. Singh [1], Prashant R. Singh [1], Jyoti Jaiswal [1], Neha Kumari [1], Vijay Upadhye [2], Suresh C. Singh [3] and Rajeshwar P. Sinha [1,4,*]

[1] Laboratory of Photobiology and Molecular Microbiology, Centre of Advanced Study in Botany, Institute of Science, Banaras Hindu University, Varanasi 221005, Uttar Pradesh, India
[2] Center of Research for Development, Parul University, Vadodara 391760, Gujarat, India
[3] Taurmed Technologies Private Limited, Manesar 122051, Haryana, India
[4] University Center for Research & Development (UCRD), Chandigarh University, Chandigarh 140413, Punjab, India
[*] Correspondence: rpsinhabhu@gmail.com; Tel.: +91-542-2307147; Fax: +91-542-2366402

Abstract: Ultraviolet radiation (UVR) has been scientifically proven to cause skin disorders such as sunburn, skin cancer and the symptoms of chronic exposure. Natural sun screening compounds have recently gained tremendous attention from the cosmetic and cosmeceutical sectors for treating skin disorders such as hyperpigmentation and aging. A wide range of natural UV-absorbing compounds have been used to replace or reduce the number of synthetic sunscreen molecules. One of the primary causes of photoaging is DNA damage, mainly caused by UVR. Photoprotection provided by traditional sunscreens is purely preventative and has no efficacy after DNA damage has been initiated. As a result, the quest for DNA-repair mechanisms that block, reverse, or postpone pathologic processes in UV-exposed skin has stimulated anti-photoaging research and methods to increase the effectiveness of traditional sunscreens. This review summarizes many natural compounds from microalgae, lichens, and plants that have demonstrated potential photoprotection effects against UV radiation-induced skin damage. Furthermore, it offers an overview of current breakthroughs in DNA-repair enzymes utilized in sunscreens and their influence on photoaging.

Keywords: natural sunscreen; cosmetic; photoprotection; photoaging; DNA-repair enzymes

Citation: Gupta, A.; Singh, A.P.; Singh, V.K.; Singh, P.R.; Jaiswal, J.; Kumari, N.; Upadhye, V.; Singh, S.C.; Sinha, R.P. Natural Sun-Screening Compounds and DNA-Repair Enzymes: Photoprotection and Photoaging. *Catalysts* **2023**, *13*, 745. https://doi.org/10.3390/catal13040745

Academic Editors: Chia-Hung Kuo, Chwen-Jen Shieh and Hui-Min David Wang

Received: 28 February 2023
Revised: 8 April 2023
Accepted: 11 April 2023
Published: 13 April 2023

1. Introduction

Prolonged sun exposure has long been recognized to induce photoaging [1], a process in which the skin suffers changes in epidermal thickness, significantly increasing pigment heterogeneity and dermal collagen deterioration, dermal elastosis and keratinocyte and melanocyte mutagenesis. [2]. Regular UV exposure has a variety of biological impacts on mammals, including the start of early aging, hyperpigmentation (dark spots), erythema, immune system suppression and DNA damage. Recent qualitative research further distinguished between hypertrophic and atrophic skin aging, with the latest characterized by increasing skin thickness and sallowness and the latter by erythema and an increased risk of skin cancer [3]. The multibillion-dollar anti-aging goods market [4] is a reflection of the priority today's society places on looking young. Ultraviolet (UV) radiation has been shown to be responsible for 80% of skin aging on the face [5]. Based on this, the strongest defence against cutaneous age-related abnormalities is prevention with strict photoprotection [4], despite the market's emphasis on skin-aging reversal. A wide-brimmed hat, photoprotective apparel, sunglasses, and the use of sun protection factor (SPF) 30 broad-spectrum hued sunscreen to exposed areas are all necessary for adequate photoprotection.

In order to defend themselves, microalgae produce useful compounds that attract the attention of the beauty industry and make them prospective candidates for cosmetic or cosmeceutical products, particularly those that protect skin from UV damage. Solar energy is converted by microalgal cells into chemical energy, some of which is then stored as bioactive molecules [6]. These secondary metabolites are referred to as "bioactive molecules" since they exhibit biological activity [6]. Researchers concluded that microalgal extracts or bioactive compounds derived from microalgae have enormous potential for innovative new bio-based products such as cosmetics, medications, and nutraceuticals after several studies revealed that microalgae have high antioxidant capacities [6]. Several skin care products for humans include microalgae. For instance, the company Soliance employs an entire *Arthrospira* species, and personal skin care products include the *Chlorella* species-derived peptide sequence LVMH. Alguronic acid is another ingredient in the anti-aging skin care products produced by the company Solazyme [7]. The production of more reliable and secure formulations at a cheaper cost is made possible by the use of *Spirulina* extracts in dermo-cosmetic formulations. Long-term research is necessary to confirm its anti-aging efficacy. *Dunaliella salina* is an esteemed microalga that is abundant in β-carotene, which has anti-aging potencial [8]. Similar to other organisms, exposure to UV light in humans can have negative effects, including short-term erythema or "sunburn" and long-term premature aging of the skin [9]. In this review, we discussed the value of cosmetics made from promising microalgae and their prospective applications in sunscreen [10–15], anti-aging [16], and skin-whitening products.

Furthermore, DNA damage is one of the primary causes of photoaging and is mostly caused by ultraviolet radiation (UV-R) [17]. In an attempt to identify the biological targets of UV-R and the subsequent cascade of impaired cell functioning and tissue deterioration, recent research examined the harmful effects of UV-R at the cellular and molecular levels [18]. Although endogenous DNA repair mechanisms, persistent DNA damage from prolonged UV exposure can build up and result in photoaging and the development of skin cancer [19,20]. Conventional sunscreen protection is preventative and useless once DNA damage has already occurred. While the majority of the literature to date has focused on how DNA-repair enzymes affect the development of cancers linked to UV exposure [20], the goal of the current review is to examine recent research on the biological process of photoaging and the function of DNA-repair enzymes. The top sunscreens with DNA-repair enzymes that are currently on the market are mentioned below [21].

2. UV-Induced Skin Damage

The prevalence of several diseases and conditions linked to solar UV radiation has dramatically risen and is still rising. Mammalian skin exposed to UV radiation on a regular basis has a variety of biological effects, such as the onset of early aging, hyperpigmentation (dark patches), erythema, DNA damage, and immune system suppression [2,18]. These changes can directly or indirectly contribute to the growth of skin cancer [22,23]. The least prevalent but most active component of solar radiation is UV-B (280-315 nm). As the "burning ray", UV-B radiation comprises up to 4–5% of total solar radiation, and sunburn brought on by UV-B is 1000 times more intensive than UV-A. Moreover, UV-B seems to be more genotoxic than UV-A. Ionizing radiation may directly damage atomic structures in live cells, resulting in alterations in chemical and biological processes. It can also work in an indirect manner by radiolyzing water, which leads to endogenous bursts of reactive oxygen species (ROS) in the intercellular matrix as well as in and around the radiation track that could harm lipids, proteins, and nucleic acids [24]. Proteins and genes involved in oxidative metabolism are susceptible to direct or indirect effects from bursts of ROS. It mostly affects the basal cell layer of the skin's epidermis. The formation of the isomerization of trans- to cis-urocanic acid, pyrimidine photoproducts, the upregulation of ornithine decarboxylase activity, the production of free radicals in the skin, photoaging, cell cycle growth arrest, and photo carcinogenesis are just a few of the harmful biological effects it causes. UV-B exposure most likely initiates the production of free radicals, which depletes the skin's antioxidant reserves, making the skin less able to defend itself against free radicals produced by exposure to daylight. DNA damage is thought to be the cause of squamous and basal cell carcinoma in the skin. Moreover, it may also weaken the skin's immunological defences [25,26]. Even very brief exposures to UV-C are extremely harmful to all forms of life. The skin suffers severe damage as a result. Thankfully, ozone in the Earth's atmosphere entirely absorbs UV-C from the sun, and no solar radiation with wavelengths below 290 nm reaches the planet's surface [25,26]. In this review, we will briefly touch on how UV radiation causes the development of ROS [27] and how these ROS can cause skin ailments [28], including hyperpigmentation (dark patches), skin aging, and photoaging [29].

2.1. Hyperpigmentation

Hyperpigmentation is a process in which patches of the skin become darker in color than the surrounding skin. The pigment melanin was overproduced and accumulated, changing the hue of the skin. Lentigines, liver spots, and age spots are other names for these hyperpigmented lesions. They appear on sun-exposed areas of the body, especially the arms and face, and are thus most likely caused by long-term UV exposure [30]. The dermal melanophages that have been seen histologically to lay beneath the lentigines may also contribute, at least in part, to their appearance of darkness. These findings probably point to a difference in the genetic expression of the keratinocytes (KC) and melanocyte (MC) within the spot as compared to the MC in the non-spot skin around it. The epidermal architecture is frequently drastically altered, which may be partially due to the damage caused by persistent UV exposure that is linked to spot formation. Tyrosinase, a glycoprotein [30] found in the melanosome membrane that catalyzes the conversion of *L*-tyrosine to melanin, is one enzyme that plays a major role in regulating melanogenesis [31].

Melanogenesis is controlled at the level of tyrosinase maturation and translocation. The presence of certain carbohydrate moieties regulates the translocation of tyrosinases [32]. Eumelanin and pheomelanin, two different kinds of melanin, are produced by melanosomes. Tyrosinase and associated proteins (TRP-1 and TRP-2) are created by MITF's phosphorylation, which is triggered by a number of signaling pathways, including the ERK, cAMP, and Wnt pathways. Upstream of the receptor molecules such as KIT (ligand SCF) and MC1R (ligands α-MSH, ASP, and ACTH) upregulate these signaling pathways, and the cAMP pathway is activated by the KIT receptor, as well as the MC1R which stimulates the ERK and cAMP pathways simultaneously. This further phosphorylates the MITF (Figure 1). Finally, tyrosinase-related enzymes are expressed as a result, which further facilitates the formation of melanin [33–35]. Likewise, when exposed to UV, the skin produces ROS, which activate the MC1R and α-MSH receptors and increase the synthesis of tyrosinase, resulting in the excess production of melanin [36] (Figure 1).

However, actinic lentigos exhibit a rise in the mRNA levels of melanogenesis-related genes, such as tyrosinase, TYRP1 (tyrosinase-related protein 1), POMC (proopiomelanocortin), DCT (dopachrome tautomerase), and MITF (microphthalmia-associated transcription factor). Moreover, the epidermal endothelin cascade (which includes endothelin-1, endothelin converting enzyme 1-a, and endothelin-B receptor) is accentuated, and a role for stem cell factors in hyperpigmentation (particularly solar lentigos) has been discovered [37]. Changes in the factor XIII1 melanophages and epidermal–melanin axis have also been noted. As many of these alterations continue even after avoiding more UV exposure, they appear to be irreversible [38]. The specifics of these apparent alterations in genomic expression have not yet been established.

2.2. Skin Aging

Higher mammals go through the intricate process of skin aging, which is brought on by two things. One is intrinsic, in which genetics, epigenetics, and hormonal status have a role in aging [39]. The latter is extrinsic, where skin aging results from exposure to UV, visible, and IR radiations, climate, air pollution, chemicals, smoking, drugs, lifestyle, diet, emotional factors and stress, etc. [40]. Premature aging or photoaging are terms used to describe this process [41]. All skin regions are impacted by intrinsic aging. Aged skin is thin, translucent, dry, prone to fine lines and wavy hair growth, unable to produce enough sweat, and loses subcutaneous fat tissue, which results in hollowed cheeks and eye sockets, and inadequate sweating [42]. Each bodily site's symptoms could be different. According to Davis et al., the degree of pigmentation and maybe other, as of yet unidentified, contributing variables cause the intrinsic aging process, which is found to vary by ethnic group [43]. Indeed, there is no such thing as skin that ages only due to intrinsic reasons. Those who have lived solely indoors their entire lives may develop skin conditions that are quite similar to that. Individuals often have skin that displays distinct extrinsic aging phases overlaid on the degree of intrinsic aging.

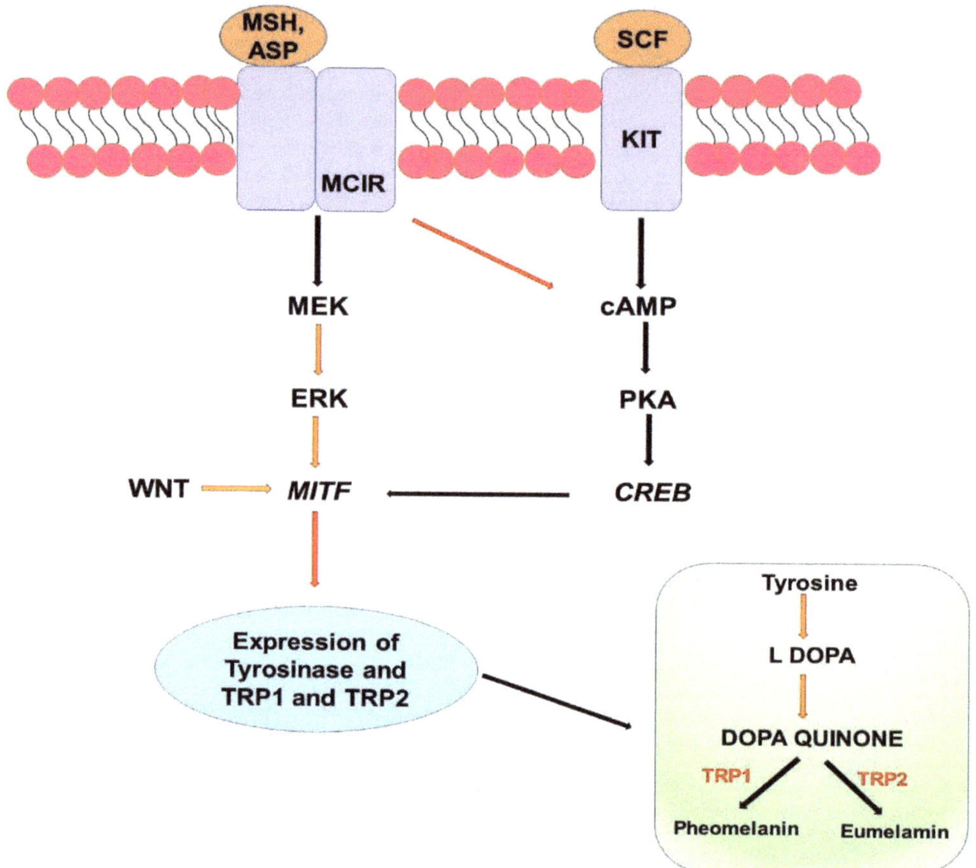

Figure 1. Signaling channels that are involved in melanin production. Adenosine 3′,5′-cyclic AMP (cAMP), cAMP response element-binding (CREB), tyrosinase-related protein (TRP), microphthalmia-associated transcription factor (MITF), melanocortin 1 receptor (MC1R), wingless-related integration site (Wnt), -melanocyte-stimulating hormone (MSH), agonist stimulating protein (ASP), and stem cell factor (SC).

The aging of the skin is significantly influenced by ROS. Around 1.5–5% of the oxygen that is ingested by the skin is converted into ROS by radiolyzing water [44]. Inherent aging is thought to be mostly caused by ROS, which is continually created as a byproduct of aerobic metabolism in the mitochondria's electron transport chain [45]. In the skin, keratinocytes (KC) and fibroblasts (FB) are the primary sources of "mitochondrial" ROS. The reactive superoxide anion radical ($^{\bullet}O_2{}^-$) is a ROS that is primarily created in mitochondria from oxygen by adding an electron to each oxygen (O_2) molecule. Superoxide anions, which are ROS particles, can damage cellular structures when produced in large quantities [44].

The numerous growth factor and cytokine receptors are activated by ROS, and this further stimulates the P13/AKT pathway and mitogen-activated protein kinase (MAPK) signal transmission. However, the FoxO is rendered inactive by the AKT pathway, which prevents the cell's production of antioxidant enzymes. Activator protein-1 (AP-1) and NF-κB are both nuclear proteins that MAPK regulates, and the expression of MMP is caused by AP-1 stimulation [46]. MMPs are a group of extracellular proteinases that include zinc that breaks down extracellular substances, including collagen and elastic fibers, which results in the development of wrinkles [46,47]. Moreover, the hyaluronidase enzyme, which breaks down hyaluronic acid, is activated by ROS. The extracellular matrix contains hyaluronic acid, which helps to maintain the skin smooth, wet, and lubricated by absorbing and holding onto water molecules [48,49]. Yet, recent experimental data demonstrate that low amounts of ROS might serve as a beneficial signaling agent, particularly when superoxide anions are transformed into hydrogen peroxide. The principal mitochondrial neutralizer of constantly generated superoxide anions is manganese superoxide dismutase (MnSOD). Outside of the mitochondria, there are other types of SOD, but only MnSOD has been proven to be crucial for the survival of aerobic life [50]. Although, the extracellular matrix degrades more quickly, and cells have a decreased capacity for cell replication, which are two additional key phenomena linked to intrinsic skin aging. All dividing cells lose some of their capacity to replicate over time. This particularly impacts MC, FB, and KC in the skin. Cellular senescence is the name of this process. In older skin, senescent, non-dividing cells are more prevalent. [51].

Extrinsic aging is brought on by oxidative environmental variables such as sunlight [52,53], cigarette smoke, or other pollutants [54]. According to epidemiological research, prolonged exposure to cigarette smoke and UV-A radiation both hasten the aging process of the skin [55]. The main cause of extrinsic skin aging, also known as photoaging, is exposure to UV light [53]. The rate of skin aging caused by UV radiation relies on the frequency, length, and intensity of sun exposure as well as the natural defense provided by skin pigmentation [56]. Deep wrinkles, loss of elasticity, dryness, laxity, an appearance of having a rough texture, telangiectasias, and pigmentation disorders are characteristics of photo-aged skin, which is considerably different from the skin that is mostly intrinsically old [57]. According to the Fitzpatrick skin phototype, photoaging is more pronounced in people with fair skin (skin types I and II) and less pronounced in those with skin types III or above [58]. As a result, the degree of photoaging is mostly determined by the total UV radiation received as well as by the skin's level of pigmentation. Damage to the structural elements of the connective tissue of the dermis is the main factor responsible for the aged look of photodamaged skin [59]. The three main groups of biomolecules that make up connective tissue structural proteins (collagen and elastin), glycosaminoglycans (GAGs), proteoglycans, and special macromolecules are created by FB (laminin fibrillin, hyaluronan, and fibronectin) [59]. In recent years, a great deal of work has been carried out to understand the molecular alterations in photoaged skin [60]. The direct and indirect effects of UV radiation on epidermal and dermal structures in connection to aging and the function of oxidative processes herein are discussed in the next portion of this review.

Principles of Sun-Induced Photoaging

The word "photoaging" refers to certain functional, clinical, and histological characteristics of skin that has had repeated sun exposure [61]. It originated from a number of words, including rapid skin aging and heliodermatosis. Nonetheless, several characteristics of photoaged skin are unique, making it a distinct process with its own pathophysiology. Particularly among the Western population, longer lifespans, more free time, and excessive exposure to UV radiation from natural sunshine or tanning beds have led to an increase in the need for skin protection against the harmful effects of UV exposure. Photoaging will thus be a growing problem in the future. The clinical and histological signs of photoaged skin have been recognized for some years, but the underlying molecular processes causing the precise macro- and microscopic abnormalities have only recently been identified [61].

"Telomere shortening and damage are recognized causes of cellular senescence and aging" [62,63]. A cell's age may be determined by the length of its telomeres. Many age-related diseases, including benign and malignant neoplasms and photoaged skin, have been linked to genomic instability brought on by telomere shortening [64–67]. Accelerated telomere shortening has been associated with DNA-damaging substances such as ROS in vitro investigations [68]. This contributes to UV-induced damage to important regulatory genes, which results in the classic signs of photoaging [69]. Reactive oxygen species (ROS), which are produced by UV irradiation, activate a variety of cell surface receptors, including those for keratinocyte growth factor, insulin, tumor necrosis factor (TNF-α), and interleukin-1. Protein-tyrosine phosphatase-j, an enzyme that keeps receptors such as the EGF receptor inactive (hypophosphorylated), is inhibited by ROS, which mediates receptor activation in part. By stimulation of the stress-related mitogen-activated protein (MAP) kinases p38 and c-Jun amino-terminal kinase, receptor activation causes intracellular signaling (JNK). The nuclear transcription complex AP-1, which is made up of the proteins c-Jun and c-Fos, is transcriptionally induced by kinase activation [70].

In addition to activating receptors, UV also produces ROS that damage membrane lipids, causing ceramide to be released and AP-1 to be activated (Figure 2) [71]. Cyclooxygenase enzymes convert arachidonic acid into prostaglandins, which attract inflammatory cells to the location. Since it inhibits the effects of transforming growth factor-β (TGF-β), a cytokine that stimulates collagen gene transcription [70,72] and negatively controls keratinocyte proliferation, increased AP-1 transcription and activity prevent the production of the main dermal collagens I and III (Figure 2) [70,73]. Activated receptors drive the transcription of the matrix metalloproteinase (MMP) gene, which in turn encourages signaling cascades that lead to the production of transcription factor AP-1. Furthermore, procollagen gene expression in fibroblasts is suppressed by AP-1. Matrix metalloproteinases, which are produced by KC and FB, degrade collagen and other proteins that make up the dermal extracellular matrix. The structural and functional integrity of the extracellular matrix is threatened by dermal damage that is not effectively healed. Repeated sun exposure causes a build-up of dermal damage, which eventually results in the recognizable wrinkles of photodamaged skin (Figure 2).

The intracellular signaling proteins SMAD2, SMAD3, SMAD3, and SMAD7 are activated by TGF-β to mediate its actions, and SMAD7 inhibits these effects [74]. In fact, UV causes the human skin's SMAD7 protein to be produced, which disrupts the TGF-β SMAD2-3 signaling [75,76]. This causes keratinocyte proliferation, epidermal hyperplasia, and a reduction in type I procollagen synthesis, which results in collagen loss. Further reducing collagen transcription, AP-1 lowers the number of TGF-β receptors in the body. This also counteracts the intrinsic retinoic acid stimulatory action on collagen synthesis. Collagen production is therefore decreased in the skin that has been photodamaged by UV radiation on a regular basis [75]. Moreover, UV-induced PTEN phosphatase inhibition and Akt kinase activation, which both work by stimulating the phosphoinositide 3-kinase signaling pathway, boost AP-1 activity [77]. The cysteine-rich 61 protein (CYR61), a new regulator of collagen production that is triggered by UV irradiation in FB, has recently been demonstrated also to stimulate AP-1. Matrix metalloproteinase (MMP)-1 and other enzymes that break down extracellular matrix components are produced as a result of CYR61 (collagenase). Moreover, CYR61 lowers the amount of TGF-β receptors and reduces the formation of type I procollagen [78]. The levels and activity of MMPs, notably MMP-1, MMP-3 (stromelysin-1), and MMP-9, are increased by CYR61's induction of AP-1 (92-kDa gelatinase) [70,79]. Furthermore, UV irradiation enhances the development of MMPs and activates the nuclear factor (NF)-κB transcription factor, which in turn causes the expression of proinflammatory cytokines such as vascular endothelial growth factor (VEGF), IL-1, IL-6, and TNF-β [80,81]. After neutrophil infiltration into UV-irradiated skin, MMP-8, a collagenase of neutrophil origin, further exacerbates matrix deterioration. In fact, one study contends that infiltrating cutaneous neutrophils is the primary cause of UV-induced MMP and elastase release. Tissue inhibitors of metalloproteinases (TIMPs), which slow down matrix breakdown, cannot completely counteract the negative effects of MMPs. Superoxide and peroxynitrite, a byproduct of the nitric oxide process generated by UV-irradiated KC, can activate NF-κB [82]. The dermis accumulates partially damaged collagen fragments as a result of UV-induced collagen degradation, which is usually incomplete and is thought to lessen the structural integrity of the skin [70]. Large collagen degradation products also prevent the synthesis of new collagen [83], which means that collagen degradation itself adversely controls the production of new collagen.

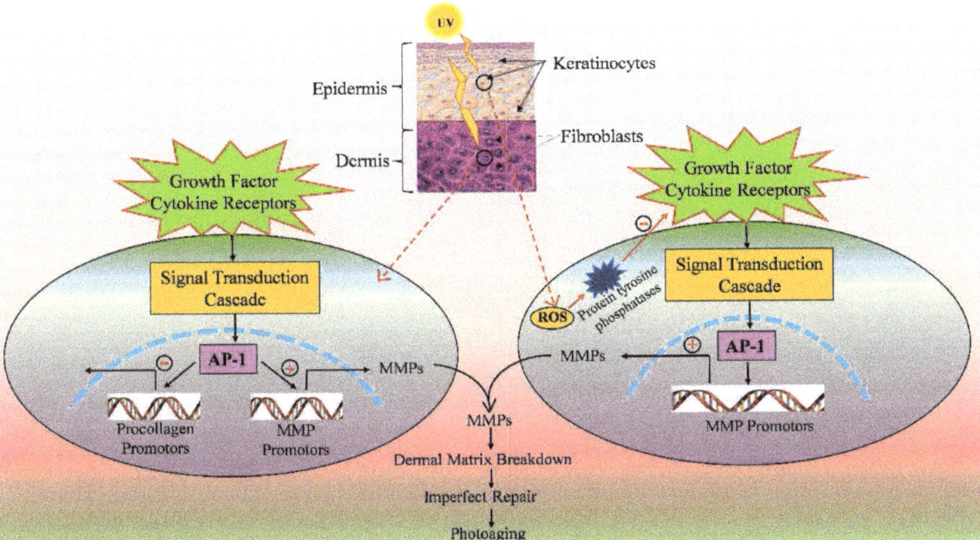

Figure 2. Model showing the effects of solar UV rays on connective skin tissue. The keratinocytes (KC) and fibroblasts (FB) growth factor and cytokine receptors are activated by ultraviolet rays (jagged arrow). Matrix metalloproteinase (MMP) gene transcription is stimulated by activated receptors, which in turn promote signal transduction cascades that result in transcription factor AP-1. AP-1 also suppresses procollagen gene expression in FB. Collagen and other proteins that make up the dermal extracellular matrix are broken down by matrix metalloproteinases, which are released by KC and FB dermal injury that is not properly repaired; this compromises the extracellular matrix's structural and functional integrity. Frequent sun exposure leads to dermal damage build-up, which finally produces the distinctive wrinkles of photodamaged skin (Modified from 70). Dashed lines represent the nucleus.

3. Application of Microalgae in Cosmetics

Microalgae produce a number of secondary metabolites with anti-inflammatory, anti-blemish and antimicrobial activities [5]. Certain microalgal extracts such as *Arthrospira platensis*, *Chlorella vulgaris*, and *Dunaliella salina* can be used for repairing skin aging, healing and preventing wrinkle formation [84–87]. The microalgae or its components' activity is the basis for the creation of several commercially available cosmetics and cosmeceuticals. Anti-aging creams, refreshing/regenerating care items, emollients and anti-irritants in peelers, sunscreen cream, and hair care items are a few examples of commercialized microalgae sources in the skin care industry. These cosmetics and cosmeceuticals contain bioactive microalgae components or algal extract (Table 1 and Figure 3). The products could offer promising and innovative alternatives to existing cosmetics and drive the development of new functions for cosmetic products.

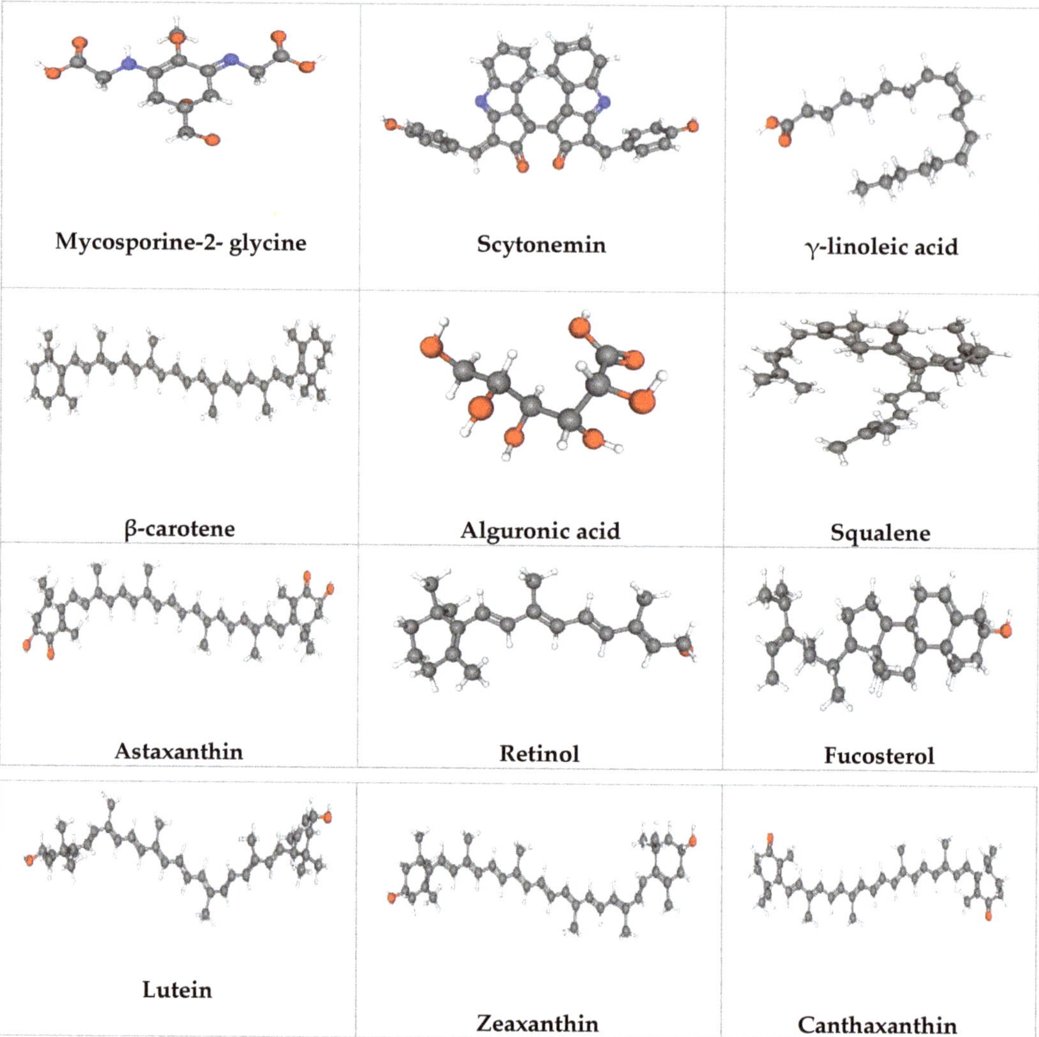

Figure 3. Potential microalgal compounds used in cosmetics such as sunscreen and anti-aging.

Cosmetic and cosmeceutical products must safeguard the natural dermal qualities and improve its appearance of health. They are typically applied when skin becomes dry due to a change in the filaggrin gene, which produces the skin's natural moisturizing component. Additionally, by protecting against both external and internal influences, they could be used to lessen the signs and symptoms of aging [6]. In addition to offering these advantages, they should have anti-bacterial and anti-fungal activities against organisms such as Mucor ramaniannus and Candida albicans to protect the equilibrium of the skin flora. Moreover, extracellular matrix stability, acne management, cell regeneration, skin whitening, inflammation control, stimulation of angiogenesis, and oxidative stress management are all areas where microalgal antimicrobial peptides play distinctive roles. Cosmetics may contain chemicals that have unintended adverse effects, such as triggering hypersensitivity reactions, anaphylactic shock, or fatal poisoning [88]. To select compounds as safe cosmetics, some crucial studies such as genetic toxicity, phototoxicity, photo genotoxicity, toxicokinetics, and carcinogenicity should be conducted. Due to the growing popularity of cosmetics, there is a greater demand for natural and sustainable resources for the manufacture of cosmetics [87]. Cosmetics made from microalgae can replace current goods and are safe for the environment; the FDA has authorized Arthrospira extract as a "safe" food ingredient.

Table 1. Potential application of microalgae as natural cosmetics.

Species	Ingredient	Activity/Uses	References
Anabaena virabilis, Nostoc punctiforme, Gloeocapsa sp.	Mycosporine-like amino acids (MAAs)	Photoprotection	[89,90]
Arthrospira platensis, Dunaliella salina, Chlorella vulgaris, and *Nannochloropsis oculata*	Sporopollenin, scytonemin, mycosporine-like amino acids (MAAs)	Photoprotection	[91]
Odontella aurita	Polyunsaturated fatty acids (PUFAs)	Prevents oxidative stress on skin	[92]
Porphyra sp., *Porphyridium* sp.	Phycoerythrin	Cosmetics (face powder, eye shadow)	[93]
Arthrospira sp.	Phycocyanin	Cosmetics (eye shadow)	[94]
Dunaliella, Nannochloropsis	Carotenoids	Wrinkle reduction, collagen formation and tissue regeneration	[95]
Arthrospira sp.	Phycobiliprotein	Improves moisture balance of the skin, skin complexion, strengthens skin's immunity and reduces wrinkle	[96]
Dunaliella salina	β-carotene	Stimulates cell proliferation	[95]
Arthrospiraplatensis	γ-linoleic acid	Prevents early skin aging and wrinkle formation	[96]
Asterocapsa nidulans	Alguronic acid	Strengthens skin immunity	[97]
Porphyridium sp.	Sulfated polysaccharide	Strengthens skin immunity	[97]
Botryococcus braunii	Squalene	Improves skin elasticity and moisture retention, prevents age spots and hyperpigmentation	[98]
Haematococcus lacustris	Astaxanthin	Photoprotection	[99,100]
Nostoc commune, Anabaena variabilis, Aphanothece halophytica	Mycosporine-2- glycine	Inhibition of advanced glycation end products	[97]
Olisthodiscus luteus, Microchloropsis salina	Fucosterol	Decrease matrix metalloproteinases expression and increase collagen production	[91]
Dunaliella bardawil	β-carotene	Improves bioavailability and antioxidant properties on skin	[101]

Researchers have discovered that compounds derived from microalgae can be utilized as the primary active ingredient in cosmetics, but some of these compounds also have characteristics that allow for them to be employed as excipients, such as stabilizers, dyes, or thickening agents [87,102]. Typically, personal care items, including face lotion, cream, shampoo, body soap, and colorants for cosmetics, are formulated using their extracts or bioactive compounds [6,103,104]. Their sterols can also be utilized in moisturizing lotions [105]. Moreover, their pigments, like the keto carotenoid astaxanthin and β-carotene, are utilized in skin care, anti-aging, and anti-irritant treatments [105]. In addition, β-carotene with non-modified β-ionone rings such as β-carotenes are the precursor molecules for vitamin A [104]. Fucoxanthin protects against UV-B-induced skin damage by reducing intracellular ROS, like astaxanthin, an orange-colored pigment with antioxidant and provitamin A properties. Fucoxanthin has anti-pigmentary action in UV-B-induced melanogenesis in addition to its sunscreen properties [105,106]. Fucoxanthin could be an active component of cosmeceuticals and nutraceuticals utilized in the defense of the skin from photoaging [16,105,107]. Microalgae are also a source of phenolic compounds which are valuable antioxidants and photoprotective compounds [108,109]. The green microalga *Lobosphaera incisa* accumulates triacylglycerols (TAGs) with exceptionally high levels of long-chain polyunsaturated fatty acids (LC-PUFA) and arachidonic acids (ARA) under nitrogen (N)-deprived conditions [110]. The cosmetic industry uses extracts from microalgae which are high in pigments, PUFAs, phycobiliproteins and carbohydrates which can be used to make lotions, creams and ointments. Mycosporine-like amino acids (MAAs) are small <400 Da, water-soluble, colorless UV-absorbing compounds synthesized by marine microbes as an adaptation for their high sunlight environments. They have a unique absorption spectrum between 309 and 362 nm. Structurally, MAAs are divided into two groups: (i) the mycosporines, which have a single modified amino acid residue connected to a cyclohexenone core, and (ii) MAAs, which have two such amino acid substituents [9,14]. MAAs have good antioxidant properties. MAAs produced from *Dunaliella*, *Arthrospira*, and *Chlorella* have the potential to act as sunscreens to reduce the damage induced by ultraviolet rays. Microalgae *Odontella aurita* also showed potential free radical scavenging activity to maintain youthful skin.

Adequate research is currently lacking in order to apply the findings of in vitro experiments on model organisms and the application of efficient compounds to human skin. Additionally, patients and researchers must understand that compliance is crucial when using natural cosmetics since they work more slowly than traditional cosmetics made of synthetic substances. Important research on genetic toxicity, photo genotoxicity, phototoxicity, toxicokinetics, and carcinogenicity should be carried out in order to choose substances as safe cosmetics. To specify compounds as a harmless cosmetic, certain crucial studies such genetic toxicity, photogenotoxicity, phototoxicity, toxicokinetics, and carcinogenicity should be performed [111]. Using genetic toxicity studies, such as the Ames Salmonella test, one may determine the carcinogenic potential of bio compounds produced from microalgae [112]. Moreover, the 3T3 neutral red uptake (3T3 NRU) experiment, the 3-dimensional (3D) epidermis model, and the erythrocyte photohemolysis test can be used to determine the phototoxicity of microalgal bioactive compounds. Nevertheless, as natural products have long been used to promote health and wellbeing, they may have promised nutritional and therapeutic benefits with minimal-to-no negative effects.

3.1. Photoprotection Prospects of Microalgal Products

Organic carbon molecules and oxygen, which are necessary for life on Earth, are produced by cyanobacteria, algae, and plants, which also turn solar energy into chemical energy. UV rays hit the Earth's surface more intensely as a result of ozone layer depletion and the rising emission of atmospheric contaminants. The ozone layer absorbs UV-C radiation, which ranges in wavelength from 100 to 280 nm. The ozone layer blocks most of the ultraviolet-B (UV-B) radiation (280–315 nm) that reaches the Earth's surface, but some do get through. As the ozone layer thins, UV-B radiation intensity rises. However, the ozone layer has little effect on ultraviolet-A (UV-A) light, which ranges in wavelength from 315 to 400 nm (320 to 400 nm). The primary kind of solar radiation that enters our atmosphere is UV-B, which harms living things exposed to the sun by causing DNA strand breaks, membrane rupture, enzyme deactivation, the production of cytotoxic DNA lesions, and other extremely hazardous effects. Because native DNA molecules cannot absorb UV-A radiation, ROS are created, which cause indirect DNA damage [113]. As a result of these consequences, UV-A and UV-B have mutagenic and carcinogenic effects on humans, speed up the skin's aging process, and cause photodermatitis. In order to defend themselves from UV radiation, microalgae have developed a number of defenses, including (i) the expression of DNA-repair enzymes; (ii) the creation of antioxidant enzymes; (iii) the avoidance of the UV; and (iv) the production and accumulation of UV filter metabolites [114]. These mechanisms, along with the manufacture of UV filter metabolites, commonly known as "microbial sunscreens", make microalgae potential candidates for the cosmetic sector to be employed in sunscreens produced from natural sources [113,115].

To defend themselves from sun radiation, microalgae produce a variety of UV filter substances, including sporopollenin, scytonemin, mycosporine-like amino acids, β-carotene, and other substances such as biopterin glucoside, lycopene, and ectoine for UV protection and photoaging [91]. These bioactive substances shield the body from skin cancer, sunburn, and other diseases by inhibiting the manufacture of melanin, among other things.

The skin is shielded from UV damage by lutein, which is generated by Chlorella prototothecoides, Scenedesmus almeriensis, Muriellopsis sp., Neospongiococcus gelatinosum, Chlorococcum citriforme, Chlorella zofingiensis, D. salina, and Galdieria sulphuraria [15,87,102]. The chemicals in microalgal extracts or extracts from microalgae help shield the skin from UV damage [87]. The most significant and extensively researched compounds that are utilized in sunscreens made by microalgae are scytonemin and mycosporine-like amino acids. Cyanobacteria produce scytonemin—a lipophilic, extracellular yellow-brown pigment—in their sheath when exposed to intense sun radiation in order to shield themselves from UV-A radiation with an absorption of up to 90% [10,15,114]. Scytonemin has a maximal absorption range of 252–386 nm [114,116,117]. Coelastrin A and Coelastrin B, two novel MAAs from Coelastrella rubescens, exhibit photoprotective properties [118]. Klebsormidin A and klebsormidin B, the newly identified MAAs from Klebsormidium, showed that their biosynthesis and intracellular enrichment is strongly induced by UVR exposure, supporting the function of these compounds as natural UV-sunscreens [119].

Scytonemin is produced when the gene responsible for its production is activated by UV-A, and it then builds up in the body. Scytonemin and derivatives of scytonemin can be produced by a number of cyanobacterial species such as *Anabaena, Calothrix, Chlorogloeopsis, Diplocolon, Gloeocapsa, Hapalosiphon, Lyngbya, Nostoc, Phormidium, Pleurocapsa, Rivularia, Schizothrix, Scytonema, and Tolypothrix* [113].

The hydrophilic and colorless mycosporine-like amino acids (MAAs) are produced by marine organisms such as cyanobacteria [11,12], microalgae, macroalgae, fungi, etc., that act as an antioxidant by preventing ROS-induced DNA damage as well as a photoprotectant by shielding cells from UV-B and UV-A radiation by absorbing radiation and dissipating excess heat energy into the cell and surroundings [120–122]. Only a small percentage of the physical and chemical filters on the market, referred to as "broad-spectrum sunscreen", can effectively block both UV-A and UV-B rays [123]. Therefore, it is crucial to include MAAs as a UV filter agent in sunscreens since they have a high capacity to absorb UV between 309 and 362 nm, making them a broad-spectrum sunscreen [115]. They can be a very stable cosmetic product because they are also very photostable and very resistant to heat, pH fluctuations, and different solvents [124]. The first sunscreen product, named Helioguard 365, was created by the Swiss company Mibelle AG Biotechnology using a natural UV screening substance called an MAA containing a certain proportion of Porphyra-334 and shinorine obtained from red algae *Porphyra umbilicalis* [125,126].

3.2. Microalgal Compounds as Anti-Aging Therapies

The formation of AGE (advanced glycation end products), the impact of ROS, and matrix metalloproteinases (MMPs) are the most important processes among theories about the aging process. Pharmaceutical companies have recently become interested in substances that are preventing AGE. Recently described as an inhibitor of AGE formation, mycosporine-

2-glycine (M2G), a very uncommon mycosporine-like amino acid, has been proposed as a key component in anti-aging therapies [127]. *Nostoc commune, Anabaena variabilis,* and *Aphanothece halophytica* have all been found to be capable of producing mycosporine-2-glycine. The key strategy to delay the aging of the skin is moisturization. It can support skin elasticity and beauty maintenance and increase environmental harm prevention. Hydroxy acid (HA) benefits the skin and has been utilized in cosmetic goods to moisturize skin. Plants are capable of producing hydroxy acids, but because plant output is restricted, interest in algal polysaccharides is growing. According to studies, *Pediastrum duplex* extract has a significant number of polysaccharides and can be used to preserve and moisturize skin [87]. Salicylic acid, α-HA, and β-HA are different types of HA. Due to the hydroxyl group connected to the carbon atom adjacent to the carboxyl group, α-HA is also known as 2-hydroxy acid. Lactic and glycolic acids are cosmetics' two most widely used 2-hydroxy acids. As a result of the hydroxyl group linked to the carbon atom that comes in second place when counting, starting from the carboxyl group, β-HA is also known as a 3-hydroxy acid. Citric acid is the most well-known 3-hydroxy acid utilized in the cosmetic formulation [128]. *A. variabilis, Anacystis nidulans, Chlorella pyrenoidosa, Chlamydomonas reinhardtii, Cyanidium caldarium, Phormidium foveolarum,* and *Oscillatoria* species have also been shown to create 2-hydroxy acids and 3-hydroxy acids, and their extracts can act as antioxidants [129,130].

3.3. Microalgal Product's Potential as Skin Whitening Agents

The pigment melanin is what gives hair, skin, and eyes their pigmentation, and it is created to protect the skin from UV damage. Nevertheless, melanin overproduction gives the skin a distinct color [87,131]. Tyrosinase is a crucial enzyme that starts the manufacture of melanin, and tyrosinase inhibitors can stop pigmentation in the skin [132]. Tyrosinase inhibitors contain phenolic structures or metal chelating groups, which result in two mechanisms: inhibiting interactions between the substrate and the enzyme and chelating copper within the active site of the enzyme. Finding novel, naturally derived alternatives to these synthetic tyrosinase inhibitors is important due to their high toxicity, low stability, insufficient action, and poor skin penetration, and microalgae can serve as a promising possibility [133]. *S. plantensis* extracts can be employed as tyrosinase inhibitors [133]. According to Oh et al. (2015), *Pavlova lutheri* inhibits melanogenesis [134]. Tyrosinase inhibitory activity was also demonstrated by Oscillapeptin G from *Oscillatoria agardhii*, asthaxanthin from *Haematococcus pluvialis*, and zeaxanthin from *N. oculata* [135]. In addition to inhibiting the tyrosinase enzyme, vitamins C and E can inhibit the skin's synthesis of melanosomes. A well-known NADH (Nicotinamide Adenine Dinucleotide Hydrate)-based process that protects mammalian skin from UV radiation damage is the self-acting, synergistic combination of vitamins E and C. Pediastrum cruentum's high concentration of vitamins E and C make it a potential candidate as a cosmetic to prevent melanoma [136].

4. Excipients in Sunscreen

Despite the increasing use of cosmeceuticals and sunscreen additives, assessing their safety and efficacy is critical. While the specific mechanism of UV-R and VL-induced photoaging is unknown, the downstream consequences of increased ROS, MMPs, and DNA damage have already been widely described [137]. Sunscreen additives have been used or suggested in sunscreens to improve photoprotection and assist in preventing photoaging by battling the negative effects of sunshine on the skin. Antioxidants are crucial in avoiding, alleviating, and damping free radicals and oxidative stress. Despite the fact that our cells manufacture natural antioxidants, UV-R and other stressors frequently outnumber our indigenous supply [138]. To restore depleted antioxidant reserves and reduce oxidative stress on the skin, topical antioxidants have been included in sunscreens. Wang et al. [139] examined the radical skin protection factor (RSF) and antioxidant power (AP) of 12 sunscreen lotions containing vitamin C, vitamin E, or other antioxidant compounds against simulated UVA- and UVB-induced ROS in an ex vivo investigation conducted in 2011.

Recent reviews and research have shown that adding antioxidants to sunscreen formulas has a favorable effect. In one study, sunscreens containing SPF 25 plus a combination of caffeine, vitamin E, vitamin C, Gorgonian extract, *Echinacea pallida* extract, and chamomile essential oil revealed a lower MMP-1 expression than those with simply SPF 25. [140]. The effectiveness of antioxidants in sunscreens may vary depending on the formulation of the sunscreen. It has been argued that in order for antioxidants to be effective, they must have strong antioxidative capabilities, be present in high concentrations, be viable in the final formulation, and be able to permeate the stratum corneum while being active in the epidermis and dermis [138]. With regard to antioxidants studied in topical formulations, vitamin C (l-ascorbic acid) is the most abundant antioxidant in the skin and, due to its water solubility, plays a key role in the skin's aqueous compartments [138]. It also aids in the replenishment of vitamin E, functions as a cofactor in collagen formation, and lowers elastin build-up. Thankfully, it is possible to create a stable formulation by combining it with other antioxidants, such as vitamin E (α-tocopherol) and ferulic acid [138,141]. Murray et al. [142] revealed that once compared to untreated skin, skin pre-treated with sunlight-simulated UV-R after application of a topical formulation of 15% l-ascorbic acid, 0.5% ferulic acid (CE-Fer), and 1% α-tocopherol for 4 days significantly reduced UV-induced erythema, thymine dimers, sunburn cells, and p53 induction. Additionally, vitamin E has been proven in several animal and human trials to be helpful in reducing lipid peroxidation, photoaging, and photocarcinogenesis [138]. This shows that topical CEFer may play a function in photoaging and skin cancer prevention [141,142]. Vitamin A and its analogues, namely, retinoids and β-carotene, have indeed been extensively researched in the field of anti-aging and have demonstrated efficacy in the prevention and restoration of photoaging [143]. To decrease protein-1 and MMP-1 production, they bind to cytoplasmic receptors, including cellular retinoic acid-binding protein types I and II, in addition to nuclear receptors such as nuclear retinoic acid receptors and retinoid X receptors [138]. This causes enhanced epidermal proliferation, epidermal thickness, stratum corneum compression, glycosaminoglycan biosynthesis and deposition, and increased collagen formation [138]. Additionally, there is scientific proof that topical retinoids might play a role in non-melanoma skin cancer prevention by beginning tumor cell growth arrest and normal cellular differentiation [144]. Nevertheless, because retinol and retinoids are somewhat unstable when exposed to UV and visible light, their usage as a sunscreen component is primarily for anti-aging benefits rather than improved photoprotection. Several botanicals contain polyphenols, such as tea leaves, almond seeds, grape seeds (Vitis vinifera), and pomegranate extract [145]. Through one study, sunscreen incorporating polyphenols such as epigallocatechin-3-gallate from tea extracts outperformed sunscreen alone in terms of protecting human skin versus solar-simulated UV-R [140]. Moreover, green tea extract combined with resveratrol, another polyphenol, gave SPF protection irrespective of chemical and physical UV filters.

Melatonin functions as an antioxidant in three distinct but complementary ways. It can serve as a free radical scavenger, reduce free radical production, and increase the activity of antioxidant enzymes [146]. It has shown potential in the treatment of both UV-B and UVA-induced oxidative damage. Melatonin treatment reduced p53 expression, enhanced DNA repair, and lowered CPD production in human MC and KC [147]. Melatonin also inhibited UV-induced erythema and stimulated endogenous enzymes to combat oxidative stress [9]. According to this study, melatonin may be an additive in protecting KC, FB, and MC against UV-induced photoaging. Genistein, a soybean isoflavone, has been shown to block tyrosine kinase—the enzyme that begins epithelial receptor-mediated signaling [148]. It has been shown that applying genistein to human skin prior to delivering an erythemogenic UV irradiation dosage inhibits JNK and MMP-1 overexpression without causing cutaneous erythema, suggesting that genistein does not serve as a sunscreen but rather impacts UV-mediated signaling [149].

Photolyases, in addition to antioxidants, are useful sunscreen ingredients. Photolyases, especially CPDs, are enzymes with the unique capacity to repair DNA damage. These are flavoproteins that require flavonoids as cofactors to absorb UV. In both in vivo and in vitro experiments, UV-absorbed energy is subsequently transferred to damaged DNA to break CPD links [150]. When combined with SPF 50 sunscreen and antioxidants, it dramatically decreases photoaging indicators compared to sunscreen exclusively or sunscreen with antioxidants [151].

5. DNA-Repair Enzymes with Sunscreens: Recent Concepts

Sunscreens are essential for combating photoaging because they block the transmission of UV-R [152]. Numerous studies have shown that using traditional sunscreens on a regular basis can help prevent the development of skin cancer [153,154]. They do not heal skin cells that have already been harmed by sun exposure [155]. Sunscreens that incorporate DNA-repair enzymes, as well as antioxidants in combination with the sun protection factor (SPF), provide "active photoprotection" [155]. Using a dual mechanism of prevention and repair, these chemicals may overcome the existing insufficiency of sun radiation damage management [155]. The two most important examples, photolyase and T4 endonuclease V, are explored further below.

5.1. Photolyase

The genome of 81 strains of the genera *Synechococcus*, *Cyanobium*, and *Prochlorococcus*, obtained from various marine and brackish habitats, were compared. It was observed that these strains together had eight distinct photolyase/cryptochrome protein members [156]. Photolyase, a flavoenzyme containing the flavin adenine dinucleotide molecule, functions as a catalytic cofactor in restoring UV-induced DNA damage in CPD and 6-4PPs [157]. Photolyase identifies damaged thymine dimers and repairs them by directly absorbing blue light, even by flavin adenine dinucleotide molecules or by transferring energy from an activated antenna chromophore to second chromophore [157]. This eventually divides into separate pyrimidines, returning the electron towards the enzyme's redox cofactor [158]. Photolyase use is also related to a decrease in MMP-1 in the skin's dermal and epidermal compartments. MMP-1 overexpression in human skin cells causes collagen breakdown, which is important in photoaging [159]. Because CPD has a far higher mutagenesis potential than 6-4PPs or other lesions and is responsible for most UV-induced mutations, current research has focused exclusively on CPD photolyase as a repair enzyme [160]. CPD photolyase has been shown to be useful in reducing photodamage in both in vitro and in vivo experiments [161]. Numerous clinical trials regarding the use of a topical medication comprising liposome-encapsulated CPD photolyase have now been reported. It has been utilized either in people with no skin lesions or as an adjuvant treatment in patients with actinic keratoses (AK), which are in situ squamous cell carcinomas caused by persistent sun exposure [162]. Sunscreens with chemical UV filters mixed with liposome-encapsulated CPD photolyase were used in all human investigations [161,163]. The encapsulation of liposomes transports enzymes through the human stratum corneum and delivers biologically active proteins into the living epidermis. [164]. This method may open up a new avenue for photoprotection against some types of UV-induced skin damage [163].

5.2. T4-bacteriophage Endonuclease V (T4 Endonuclease V)

T4 endonuclease V is an enzyme discovered in *Escherichia coli* infected with the T4 bacteriophage. It has been demonstrated to repair UV-induced cyclobutane pyrimidine dimers in DNA, which, if left unrepaired, lead to mutations that cause actinic keratoses and non-melanoma skin malignancies (NMSC) [165]. When UV-damaged DNA is identified, it is cleaved by two coupled activities: apurinic-apyrimidinic endonuclease and pyrimidine dimer-DNA glycosylase. This enzyme improves UV-damaged DNA repair and has additional favorable effects on UV-damaged cells. The effect of T4 endonuclease V increases the efficacy and speed of naturally occurring DNA repair by around four-fold. [166]. Moreover, the enzyme promotes skin regeneration and repair while preventing the breakdown of extracellular matrix components, which aids in the prevention of photoaging [167]. T4 endonuclease V encapsulation into liposomes as delivery vehicles, dubbed "T4N5", is essential for sufficient penetration into the stratum corneum. Mouse research found that applying T4N5 to the skin might be a beneficial adjuvant to sunscreens for preventing and decreasing UV-R-related local effects such as sunburn cell development [168].

5.3. Comparison of Sunscreens with and without DNA-Repair Enzymes

Improvements in the study of skin biology have resulted in the creation of a wide range of therapies to prevent aging and enable skin regeneration. The following characteristics should be included in a perfect sunscreen: 1) UV-B and UV-A radiation protection; 2) ROS scavenging capabilities; preferably, 3) filter stability and safety; 4) and the incorporation of enzymes that aid in cellular DNA repair [34]. Recent irradiation experiments have shown that adding DNA-repair enzymes to traditional sunscreens may prevent UV-R-induced molecular damage to exposed skin more than traditional sunscreens alone [169]. For example, in clinical research conducted by Carducci et al. [169], 28 AK patients were randomly assigned to either topically apply sunscreens containing DNA-repair enzymes (n = 14) or sunscreens alone (n = 14) for 6 months. The primary end measures were hyperkeratosis, field cancerization, and changes in CPD levels in skin biopsies. CPD levels were found to be 61% lower in patients who used sunscreens containing DNA-repair enzymes versus 35% lower in patients who used traditional sunscreens ($p < 0.001$), showing their effectiveness in lowering CPD production [169]. The combination of CPD photolyase and topical antioxidants greatly reduced CPD and free radical-induced protein degradation. The scientists found that sunscreens combining antioxidants and photolyase outperform traditional sunscreens to prevent skin aging, most likely due to a synergistic effect [151]. So far, this has been the only clinical study examining the effect of sunscreens containing DNA-repair enzymes on photoaging.

The current study focuses on creating novel sunscreens and their increased protective impact based on previous evidence from comparable investigations. A sunscreen containing CPD photolyase (Eryfotona® AK-NMSC, Isdin SA) has recently been examined as an example. Clinical and histological investigations revealed good effects on field cancerization in AK patients, such as an increase in the number of AK lesions and a reduction in the degree of cancerization in the field [170]. Because DNA photodamage and ROS are both early events in the formation of AK, the impact might also be applied to photoaging. Another new product (Ateia® Kwizda Pharma, Vienna, Austria) combines a traditional sunscreen with a proprietary component combination, Nopasome®. Nopasome® is a liposome-encapsulated CPD photolyase (Photosome®), T4 endonuclease V (Ultrasome®), and nopal cactus extract. Ladival® med (STADA Arzneimittel, Bad Vilbel, Germany) is another branded photolyase-containing sunscreen with SPF 15 or 20 [171]. The following is a list of presently available sunscreens that contain DNA-repair enzymes and have adequate scientific backing (Table 2).

Table 2. A description of the existing range of sunscreens with DNA-repair enzymes. [a] American SPF (Currently, SPF is exclusively calculated for US goods based on UV-B protection).

Name	Composition	SPF	DNA-Repair/Activity	Company	Reference
Ateia®	2% Nopasome, Nopal cactus extract, jojoba oil and vitamin E	50⁺ 50 30 25	Liposome-encapsulated photolyase, endonuclease	Kwizda Pharma GmbH, Vienna, Austria	[163]
Neova Active® (SPF43) Neova Everyday® (SPF44) Neova Silc Sheer®2.0 (SPF 40)	Octinoxate 6.5%, Octisalate 2.5%, Zinc Oxide 8.5%. **Photolysomes. Sodium Hyaluronate.** and Vitamin E	43 [a] 44 [a] 40 [a]	Liposome-encapsulated photolyase, endonuclease	Pharma Cosmetics, Oradell, New Jersey, United States	[172]

Table 2. *Cont.*

Name	Composition	SPF	DNA-Repair/Activity	Company	Reference
Neova Smart Moisture®	Copper Peptide Complex, **Photolysomes and Amino acid blend**	30 [a]	Liposome-encapsulated photolyase	Pharma Cosmetics, Oradell, New Jersey, United States	[172]
Eryfotona® AK-NMSC	Repairsomes, bisabolol, acetato detocoferol, pantenol, etc.	100+	Liposome-encapsulated photolyase	Isdin, SA, Barcelona, Spain	[172]
Heliocare 360° AK Fluid	Fernblock®+ Genorepair®Complex, Sulforaphane, Biomimetic melanin, Arginine, etc.	100+	Liposome-encapsulated photolyase, endonuclease, 8-oxoguanine glycosylase	Cantabria Labs, Madrid, Spain	[161,164]
Priori Tetra®	Zinc Oxide + Titanium Dioxide, mustard seed extracts, Soliberine + Melanin Carnosine, etc.	50 [a]	Liposome-encapsulated photolyase, endonuclease, 8-oxoguanine glycosylase	PRIORI Skincare, San Diego, California, United States	[161,164]
Ladival® med	Alkyl benzoate, pentylene glycol, xanthan gum titanium dioxide, vitis vinifera seed extract isostearic acid, etc.	20 15	Liposome-encapsulated photolyase	STADA Arzneimittel, Bad Vilbel, Germany	[171]
Sesderma Repaskin®	Phytospingosine liposomes, Alkyl benzoate, pentylene glycol, titanium dioxide, tocopheryl acetate and xanthan gum, etc.	50 30	Liposome-encapsulated photolyase	Sesderma, Madrid, Spain	[161]

In conclusion, recent research indicates that including DNA-repair enzymes in traditional sunscreens gives a more effective alternative for avoiding UV-R-generated damage that causes carcinogenesis and photoaging. Combining them with topically applied antioxidants is a good way to enhance this impact further. Unfortunately, there is minimal evidence for these effects in people, notably in the prevention of skin aging. So far, only one clinical trial has shown the effect of DNA-repair enzymes in sunscreens on photoaging [151].

5.4. Limitation of DNA-Repair Enzymes in Sunscreens

This research's limitations include topical liposomal DNA-repair enzymes that were reported to be protective against UV-induced skin cancer in humans, which does not necessarily justify their preventative impact on photoaging. Most of the included studies failed to discriminate between the effects of DNA-repair enzymes on carcinogenesis and photoaging. As a result, it can now not clearly translate improved carcinogenesis outcomes to photoaging treatment or prevention. The tiny size of study cohorts further restricts the significance of their findings. Clinical investigations with a significant number of individuals, concentrating exclusively on anti-photoaging effects, are required.

6. Conclusions and Future Perspectives

The overexposure of human skin to external stressors such as UV and pollution causes an increase in ROS generation, which causes various skin-related issues such as hyperpigmentation, early aging, etc. The cosmeceutical and cosmetic industries have taken notice of the numerous distinctive and significant bioactive metabolites that microalgae synthesize through photosynthesis and other mechanisms. Microalgae are natural and sustainable sources of bioactive compounds and are thus thought to be great possibilities for cosmetics. It is necessary to assess all potential microalgal metabolites for use in cosmetics/cosmeceuticals and optimize production technology for each. Although microalgal bioactive metabolites can be produced using biotechnologically advantageous and environmentally friendly processes, they are also considered "safe". This study explores a number of microalgae species and bioactive compounds obtained from microalgae for use in sunscreen, skin-whitening cosmetics, and anti-aging products. Microalgae are regarded to be attractive prospects for cosmetic products since they are natural and sustainable sources of bioactive compounds. Furthermore, this study presents an outline of contemporary advancements in DNA-repair enzymes employed in sunscreens and their impact on photoaging. In Tables 1 and 2, we have included some of these bioactive compounds together with their producer and the current range of sunscreens that contain DNA-repair enzymes. Since UV-R-induced DNA damage plays a significant role in both processes, it is possible to hypothesize that DNA-repair enzyme's positive effects can also be applied to photoaging. Controlled trials that support this effect and show that DNA- repair enzyme-infused sunscreens are better than regular sunscreens are still missing. Antioxidants, photolyases, and other sunscreen compounds have made it possible to reverse skin aging in addition to providing better photoprotection.

Nevertheless, before clinical guidelines are released, larger-scale, repeatable investigations must be conducted. Although the majority of microalgae are investigated for their cosmetic properties, several species remain unexplored. As a result, there is a need to standardize low-cost, effective, and more productive extraction techniques. In addition to effectiveness, the molecular basis of these compound's actions and safety issues are particularly important for forthcoming difficulties in the cosmetics sector.

Author Contributions: Study concept and design, A.G., V.U. and R.P.S.; manuscript writing, A.G., A.P.S., V.K.S. and P.R.S.; manuscript editing, A.G., A.P.S., J.J., N.K. and S.C.S. All authors have read and agreed to the published version of the manuscript.

Funding: This research did not receive any specific grant from funding agencies in the public, commercial, or not-for-profit sectors.

Data Availability Statement: Not applicable.

Acknowledgments: Amit Gupta (09/013 (0912)/2019-EMR-I), Prashant R. Singh (09/013 (0795)/2018-EMR-I), Neha Kumari (09/013 (0819)/2018-EMR-I) and Varsha K. Singh (09/0013 (12862)/2021-EMR-1) are thankful to the Council of Scientific & Industrial Research (CSIR), New Delhi, India, for the Senior Research Fellowship (SRF) and Junior Research Fellowship (JRF). Ashish P. Singh (NTA Ref. No. 191620014505) and Jyoti Jaiswal (926/CSIR-UGC-JRF DEC, 2018) are thankful to the University Grants Commission (UGC), New Delhi, India, for the financial assistance in the form of Junior Research Fellowship (JRF) and Senior Research Fellowship (SRF). The incentive grant received from IoE (Scheme no. 6031), Banaras Hindu University, Varanasi, India, to Rajeshwar P. Sinha is highly acknowledged.

Conflicts of Interest: The authors declare no conflict of interest.

References

1. Passeron, T.; Krutmann, J.; Andersen, M.L.; Katta, R.; Zouboulis, C.C. Clinical and biological impact of the exposome on the skin. *J. Eur. Acad. Dermatol. Venereol.* **2020**, *34*, 4–25. [CrossRef] [PubMed]
2. Yaar, M.; Gilchrest, B.A. Photoageing: Mechanism, prevention and therapy. *Br. J. Dermatol.* **2007**, *157*, 874–887. [CrossRef] [PubMed]
3. Sachs, D.L.; Varani, J.; Chubb, H.; Fligiel, S.E.G.; Cui, Y.; Calderone, K.; Helfrich, Y.; Fisher, G.J.; Voorhees, J. Atrophic and hypertrophic photoaging: Clinical, histologic, and molecular features of 2 distinct phenotypes of photoaged skin. *J. Am. Acad. Dermatol.* **2019**, *81*, 480–488. [CrossRef] [PubMed]
4. Poon, F.; Kang, S.; Chien, A.L. Mechanisms and treatments of photoaging. *Photodermatol. Photoimmunol. Photomed.* **2015**, *31*, 65–74. [CrossRef] [PubMed]
5. Flament, F.; Bazin, R.; Laquieze, S.; Rubert, V.; Simonpietri, E.; Piot, B. Effect of the sun on visible clinical signs of aging in Caucasian skin. *Clin. Cosmet. Investig. Dermatol.* **2013**, *6*, 221–232. [CrossRef]
6. Ariede, M.B.; Candido, T.M.; Jacome, A.L.M.; Velasco, M.V.R.; de Carvalho, J.C.M.; Baby, A.R. Cosmetic attributes of algae—A review. *Algal Res.* **2017**, *25*, 483–487. [CrossRef]

7. Lucakova, S.; Branyikova, I.; Hayes, M. Microalgal proteins and bioactives for food, feed, and other applications. *Appl. Sci.* **2022**, *12*, 4402. [CrossRef]

8. Gomes, C.; Silva, A.C.; Marques, A.C.; Sousa Lobo, J.; Amaral, M.H. Biotechnology applied to cosmetics and aesthetic medicines. *Cosmetics* **2020**, *7*, 33. [CrossRef]

9. Geraldes, V.; Pinto, E. Mycosporine-like amino acids (MAAs): Biology, chemistry and identification features. *Pharmaceuticals* **2021**, *14*, 63. [CrossRef]

10. Santiesteban-Romero, B.; Martínez-Ruiz, M.; Sosa-Hernández, J.E.; Parra-Saldívar, R.; Iqbal, H.M. Microalgae Photo-Protectants and Related Bio-Carriers Loaded with Bioactive Entities for Skin Applications—An Insight of Microalgae Biotechnology. *Mar. Drugs* **2022**, *20*, 487. [CrossRef]

11. Vega, J.; Schneider, G.; Moreira, B.R.; Herrera, C.; Bonomi-Barufi, J.; Figueroa, F.L. Mycosporine-like amino acids from red macroalgae: UV-photoprotectors with potential cosmeceutical applications. *Appl. Sci.* **2021**, *11*, 5112. [CrossRef]

12. Rosic, N.N. Mycosporine-like amino acids: Making the foundation for organic personalized sunscreens. *Mar. Drugs* **2019**, *17*, 638. [CrossRef] [PubMed]

13. Sen, S.; Mallick, N. Mycosporine-like amino acids: Algal metabolites shaping the safety and sustainability profiles of commercial sunscreens. *Algal Res.* **2021**, *58*, 102425. [CrossRef]

14. Singh, A.; Čížková, M.; Bišová, K.; Vítová, M. Exploring Mycosporine-Like Amino Acids (MAAs) as Safe and Natural Protective Agents against UV-Induced Skin Damage. *Antioxidants* **2021**, *10*, 683. [CrossRef]

15. Milito, A.; Castellano, I.; Damiani, E. From sea to skin: Is there a future for natural photoprotectants? *Mar. Drugs* **2021**, *19*, 379. [CrossRef]

16. Pajot, A.; Hao Huynh, G.; Picot, L.; Marchal, L.; Nicolau, E. Fucoxanthin from algae to human, an extraordinary bioresource: Insights and advances in up and downstream processes. *Mar. Drugs* **2022**, *20*, 222. [CrossRef]

17. Emanuele, E. Reduced ultraviolet-induced DNA damage and apoptosis in human skin with topical application of a photolyase-containing DNA repair enzyme cream: Clues to skin cancer prevention. *Mol. Med. Rep.* **2012**, *5*, 570–574. [CrossRef]

18. Rai, S.; Rai, G.; Kumar, A. Eco-evolutionary impact of ultraviolet radiation (UVR) exposure on microorganisms, with a special focus on our skin microbiome. *Microbiol. Res.* **2022**, *260*, 127044. [CrossRef]

19. Gilchrest, B.A. Photoaging. *J. Investg. Dermatol.* **2013**, *133*, E2–E6. [CrossRef]

20. Cadet, J.; Douki, T.; Ravanat, J.L. Oxidatively generated damage to cellular DNA by UVB and UVA radiation. *Photochem. Photobiol.* **2015**, *91*, 140–155. [CrossRef]

21. Luze, H.; Nischwitz, S.P.; Zalaudek, I.; Müllegger, R.; Kamolz, L.P. DNA repair enzymes in sunscreens and their impact on photoageing-A systematic review. *Photodermatol. Photoimmunol. Photomed.* **2020**, *36*, 424–432. [CrossRef] [PubMed]

22. Tebbe, B. Relevance of oral supplementation with antioxidants for prevention and treatment of skin disorders. *Skin Pharmacol. Physiol.* **2001**, *14*, 296–302. [CrossRef] [PubMed]

23. Afaq, F.; Mukhtar, H. Effects of solar radiation on cutaneous detoxification pathways. *J. Photochem. Photobiol. B Biol.* **2001**, *63*, 61–69. [CrossRef]

24. Azzam, E.I.; Jay-Gerin, J.P.; Pain, D. Ionizing radiation-induced metabolic oxidative stress and prolonged cell injury. *Cancer Lett.* **2012**, *327*, 48–60. [CrossRef] [PubMed]

25. De Gruijl, F.R. Photocarcinogenesis: UVA vs. UVB radiation. *Skin Pharmacol. Physiol.* **2002**, *15*, 316–320. [CrossRef] [PubMed]

26. Clydesdale, G.J.; Dandie, G.W.; Muller, H.K. Ultraviolet light induced injury: Immunological and inflammatory effects. *Immunol. Cell Biol.* **2001**, *79*, 547–568. [CrossRef]

27. Berthon, J.Y.; Nachat-Kappes, R.; Bey, M.; Cadoret, J.P.; Renimel, I.; Filaire, E. Marine algae as attractive source to skin care. *Free Radic. Res.* **2017**, *51*, 555–567. [CrossRef]

28. Schalka, S.; Silva, M.S.; Lopes, L.F.; de Freitas, L.M.; Baptista, M.S. The skin redoxome. *J. Eur. Acad. Dermatol. Venereol.* **2022**, *36*, 181–195. [CrossRef]

29. Roy, A.; Sahu, R.; Matlam, M.; Deshmukh, V.; Dwivedi, J.; Jha, A. In vitro techniques to assess the proficiency of skin care cosmetic formulations. *Pharmacogn. Rev.* **2013**, *7*, 97.

30. Parvez, S.; Kang, M.; Chung, H.S.; Cho, C.; Hong, M.C.; Shin, M.K.; Bae, H. Survey and mechanism of skin depigmenting and lightening agents. *Phytother. Res.* **2006**, *20*, 921–934. [CrossRef]

31. Chang, T.S. Natural melanogenesis inhibitors acting through the down-regulation of tyrosinase activity. *Materials* **2012**, *5*, 1661–1685. [CrossRef]

32. Imokawa, G.; Mishima, Y. Importance of glycoproteins in the initiation of melanogenesis: An electron microscopic study of B-16 melanoma cells after release from inhibition of glycosylation. *J. Investig. Dermatol.* **1986**, *87*, 319–325. [CrossRef] [PubMed]

33. Azam, M.S.; Choi, J.; Lee, M.S.; Kim, H.R. Hypopigmenting effects of brown algae-derived phytochemicals: A review on molecular mechanisms. *Mar. Drugs* **2017**, *15*, 297. [CrossRef] [PubMed]

34. D'Mello, S.A.; Finlay, G.J.; Baguley, B.C.; Askarian-Amiri, M.E. Signaling pathways in melanogenesis. *Int. J. Mol. Sci.* **2016**, *17*, 1144. [CrossRef]

35. Shin, H.; Hong, S.D.; Roh, E.; Jung, S.H.; Cho, W.J.; Hong Park, S.; Yoon, D.Y.; Ko, S.M.; Hwang, B.Y.; Hong, J.T.; et al. cAMP-dependent activation of protein kinase A as a therapeutic target of skin hyperpigmentation by diphenylmethylene hydrazinecarbothioamide. *Br. J. Pharmacol.* **2015**, *172*, 3434–3445. [CrossRef] [PubMed]

36. Chou, T.H.; Ding, H.Y.; Hung, W.J.; Liang, C.H. Antioxidative characteristics and inhibition of α-melanocyte-stimulating hormone-stimulated melanogenesis of vanillin and vanillic acid from Origanum vulgare. *Exp. Dermatol.* **2010**, *19*, 742–750. [CrossRef] [PubMed]
37. Ünver, N.; Freyschmidt-Paul, P.; Hörster, S.; Wenck, H.; Stäb, F.; Blatt, T.; Elsässer, H.P. Alterations in the epidermal–dermal melanin axis and factor XIIIa melanophages in senile lentigo and ageing skin. *Br. J. Dermatol.* **2006**, *155*, 119–128. [CrossRef]
38. Motokawa, T.; Kato, T.; Katagiri, T.; Matsunaga, J.; Takeuchi, I.; Tomita, Y.; Suzuki, I. Messenger RNA levels of melanogenesis-associated genes in lentigo senilis lesions. *Br. J. Dermatol.* **2005**, *37*, 120–123. [CrossRef]
39. Makrantonaki, E.; Adjaye, J.; Herwig, R.; Brink, T.C.; Groth, D.; Hultschig, C.; Lehrach, H.; Zouboulis, C.C. Age-specific hormonal decline is accompanied by transcriptional changes in human sebocytes in vitro. *Aging cell.* **2006**, *5*, 331–344. [CrossRef]
40. Fussell, J.C.; Kelly, F.J. Oxidative contribution of air pollution to extrinsic skin ageing. *Free Radic Biol Med.* **2020**, *151*, 111–122. [CrossRef]
41. Pientaweeratch, S.; Panapisal, V.; Tansirikongkol, A. Antioxidant, anti-collagenase and anti-elastase activities of Phyllanthus emblica, Manilkara zapota and silymarin: An in vitro comparative study for antiaging applications. *Pharm. Biol.* **2016**, *54*, 1865–1872. [CrossRef] [PubMed]
42. Sjerobabski-Masnec, I.; Šitum, M. Skin aging. *Acta Clin. Croat.* **2010**, *49*, 515–518. [PubMed]
43. Davis, E.C.; Callender, V.D. Aesthetic dermatology for aging ethnic skin. *Dermatol. Surg.* **2011**, *37*, 901–917. [CrossRef] [PubMed]
44. Poljšak, B.; Dahmane, R.G.; Godić, A. Intrinsic skin aging: The role of oxidative stress. *Acta Dermatovenerol. Alp. Panon. Adriat.* **2012**, *21*, 33–36.
45. Farage, M.A.; Miller, K.W.; Elsner, P.; Maibach, H.I. Intrinsic and extrinsic factors in skin ageing: A review. *Int. J. Cosmet. Sci.* **2008**, *30*, 87–95. [CrossRef]
46. Leem, K.H. Effects of Olibanum extracts on the collagenase activity and procollagen synthesis in Hs68 human fibroblasts and tyrosinase activity. *Adv. Sci. Technol. Lett.* **2015**, *88*, 172–175.
47. Ndlovu, G.; Fouche, G.; Tselanyane, M.; Cordier, W.; Steenkamp, V. In vitro determination of the antiaging potential of four southern African medicinal plants. *BMC Complement. Altern. Med.* **2013**, *13*, 304–311. [CrossRef]
48. Papakonstantinou, E.; Roth, M.; Karakiulakis, G. Hyaluronic acid: A key molecule in skin aging. *Dermatoendocrinol.* **2012**, *4*, 253–258. [CrossRef]
49. Girish, K.S.; Kemparaju, K. The magic glue hyaluronan and its eraser hyaluronidase: A biological overview. *Life Sci.* **2007**, *80*, 1921–1943. [CrossRef]
50. St Clair, D.; Kasarskis, E. Genetic polymorphism of the human manganese superoxide dismutase: What difference does it make? *Pharmacogenet. Genom.* **2003**, *13*, 129–130. [CrossRef]
51. Dimri, G.P.; Lee, X.; Basile, G.; Acosta, M.; Scott, G.; Roskelley, C.; Medrano, E.E.; Linskens, M.; Rubelj, I.; Pereira-Smith, O. A biomarker that identifies senescent human cells in culture and in aging skin in vivo. *Proc. Natl. Acad. Sci. USA* **1995**, *92*, 9363–9367. [CrossRef] [PubMed]
52. Gilchrest, B.A. Skin aging and photoaging: An overview. *J. Am. Acad. Dermatol.* **1989**, *21*, 610–613. [CrossRef] [PubMed]
53. Wlaschek, M.; Tantcheva-Poór, I.; Naderi, L.; Ma, W.; Schneider, L.A.; Razi-Wolf, Z.; Schüller, J.; Scharffetter-Kochanek, K. Solar UV irradiation and dermal photoaging. *J. Photochem. Photobiol. B Biol.* **2001**, *63*, 41–51. [CrossRef] [PubMed]
54. Bernhard, D.; Moser, C.; Backovic, A.; Wck, G. Cigarette smoke–an aging accelerator? *Exp. Gerontol.* **2007**, *42*, 160–165. [CrossRef]
55. Yin, L.; Morita, A.; Tsuji, T. Skin aging induced by ultraviolet exposure and tobacco smoking: Evidence from epidemiological and molecular studies. *Photodermatol. Photoimmunol. Photomed.* **2001**, *17*, 178–183. [CrossRef]
56. Ortonne, J.P. Photoprotective properties of skin melanin. *Br. J. Dermatol.* **2002**, *146*, 7–10. [CrossRef]
57. Kligman, L.M. The nature of photoaging: Its prevention and repair. *Photodermatology* **1986**, *3*, 215–227.
58. Šitum, M.; Buljan, M.; Čavka, V.; Bulat, V.; Krolo, I.; Lugović Mihić, L. Skin changes in the elderly people–how strong is the influence of the UV radiation on skin aging? *Collegium antropologicum.* **2010**, *34*, 9–13.
59. Naylor, E.C.; Watson, R.E.; Sherratt, M.J. Molecular aspects of skin ageing. *Maturitas* **2011**, *69*, 249–256. [CrossRef]
60. Lim, H.W. *Clinical Photomedicine*; CRC Press: Boca Raton, FL, USA, 1993; Volume 6.
61. Pandel, R.; Poljšak, B.; Godic, A.; Dahmane, R. Skin photoaging and the role of antioxidants in its prevention. *ISRN Dermatol.* **2013**, *2013*, 930164. [CrossRef]
62. Rossiello, F.; Jurk, D.; Passos, J.F.; d'Adda di Fagagna, F. Telomere dysfunction in ageing and age-related diseases. *Nat. Cell Biol.* **2022**, *24*, 35–147. [CrossRef] [PubMed]
63. Shay, J.W.; Wright, W.E. Senescence and immortalization: Role of telomeres and telomerase. *Carcinogenesis* **2005**, *26*, 867–874. [CrossRef] [PubMed]
64. Counter, C.M.; Press, W.; Compton, C.C. Telomere shortening in cultured autografts of patients with burns. *Lancet* **2003**, *361*, 1345–1346. [CrossRef]
65. Nakamura, K.I.; Izumiyama-Shimomura, N.; Takubo, K.; Sawabe, M.; Arai, T.; Aoyagi, Y.; Fujiwara, M.; Tsuchiya, E.; Kobayashi, Y.; Kato, M.; et al. Comparative analysis of telomere lengths and erosion with age in human epidermis and lingual epithelium. *J. Investig. Dermatol.* **2002**, *119*, 1014–1019. [CrossRef] [PubMed]
66. Friedrich, U.; Griese, E.U.; Schwab, M.; Fritz, P.; Thon, K.P.; Klotz, U. Telomere length in different tissues of elderly patients. *Mech. Ageing Dev.* **2000**, *119*, 89–99. [CrossRef]

67. Yin, B.; Jiang, X. Telomere shortening in cultured human dermal fibroblasts is associated with acute photodamage induced by UVA irradiation. *Postepy. Dermatol. Alergol.* **2013**, *30*, 13–18. [CrossRef]
68. Adelfalk, C.; Lorenz, M.; Serra, V.; von Zglinicki, T.; Hirsch-Kauffmann, M.; Schweiger, M. Accelerated telomere shortening in Fanconi anemia fibroblasts–a longitudinal study. *FEBS lett.* **2001**, *506*, 22–26. [CrossRef]
69. Li, G.Z.; Eller, M.S.; Firoozabadi, R.; Gilchrest, B.A. Evidence that exposure of the telomere 3' overhang sequence induces senescence. *Proc. Natl. Acad. Sci. USA* **2003**, *100*, 527–531. [CrossRef]
70. Fisher, G.J.; Kang, S.; Varani, J.; Bata-Csorgo, Z.; Wan, Y.; Datta, S.; Voorhees, J.J. Mechanisms of photoaging and chronological skin aging. *Arch. Dermatol.* **2002**, *138*, 1462–1470. [CrossRef]
71. Garmyn, M.; Yarosh, D.B. The molecular and genetic effects of ultraviolet radiation exposure on skin cells. In *Photodermatology*; CRC Press: Boca Raton, FL, USA, 2007; pp. 41–54.
72. Fisher, G.J.; Datta, S.C.; Talwar, H.S.; Wang, Z.Q.; Varani, J.; Kang, S.; Voorhees, J.J. Molecular basis of sun-induced premature skin ageing and retinoid antagonism. *Nature* **1996**, *379*, 335–339. [CrossRef]
73. Piipponen, M.; Li, D.; Landén, N.X. The immune functions of keratinocytes in skin wound healing. *Int. J. Mol. Sci.* **2020**, *21*, 8790. [CrossRef] [PubMed]
74. Hayashi, H.; Abdollah, S.; Qiu, Y.; Cai, J.; Xu, Y.Y.; Grinnell, B.W.; Richardson, M.A.; Topper, J.N.; Gimbrone, M.A.; Wrana, J.L.; et al. The MAD-related protein Smad7 associates with the TGFβ receptor and functions as an antagonist of TGFβ signaling. *Cell* **1997**, *89*, 1165–1173. [CrossRef] [PubMed]
75. Quan, T.; He, T.; Voorhees, J.J.; Fisher, G.J. Ultraviolet irradiation blocks cellular responses to transforming growth factor-β by down-regulating its type-II receptor and inducing Smad7. *J. Biol. Chem.* **2001**, *276*, 26349–26356. [CrossRef] [PubMed]
76. Quan, T.; He, T.; Voorhees, J.J.; Fisher, G.J. Ultraviolet irradiation induces Smad7 via induction of transcription factor AP-1 in human skin fibroblasts. *J. Biol. Chem.* **2005**, *280*, 8079–8085. [CrossRef] [PubMed]
77. Oh, J.H.; Kim, A.; Park, J.M.; Kim, S.H.; Chung, A.S. Ultraviolet B-induced matrix metalloproteinase-1 and-3 secretions are mediated via PTEN/Akt pathway in human dermal fibroblasts. *J. Cell. Physiol.* **2006**, *209*, 775–785. [CrossRef] [PubMed]
78. Quan, T.; He, T.; Shao, Y.; Lin, L.; Kang, S.; Voorhees, J.J.; Fisher, G.J. Elevated cysteine-rich 61 mediates aberrant collagen homeostasis in chronologically aged and photoaged human skin. *Am. J. Pathol.* **2006**, *169*, 482–490. [CrossRef]
79. Angel, P.; Szabowski, A.; Schorpp-Kistner, M. Function and regulation of AP-1 subunits in skin physiology and pathology. *Oncogene* **2001**, *20*, 2413–2423. [CrossRef]
80. Ruland, J.; Mak, T.W. Transducing signals from antigen receptors to nuclear factor κB. *Immunol. Rev.* **2003**, *193*, 93–100. [CrossRef]
81. Kang, S.; Fisher, G.J.; Voorhees, J.J. Photoaging: Pathogenesis, prevention, and treatment. *Clin. Geriatr. Med.* **2001**, *17*, 643–659. [CrossRef]
82. Cooke, C.L.M.; Davidge, S.T. Peroxynitrite increases iNOS through NF-κB and decreases prostacyclin synthase in endothelial cells. *Am. J. Physiol. Cell Physiol.* **2002**, *282*, C395–C402. [CrossRef]
83. Varani, J.; Spearman, D.; Perone, P.; Fligiel, S.E.; Datta, S.C.; Wang, Z.Q.; Shao, Y.; Kang, S.; Fisher, G.J.; Voorhees, J.J. Inhibition of type I procollagen synthesis by damaged collagen in photoaged skin and by collagenase-degraded collagen in vitro. *Am. J. Clin. Pathol.* **2001**, *158*, 931–942. [CrossRef]
84. Kumar, M.; Enamala, S.; Chavali, M.; Donepudi, J. Production of biofuels from microalgae—A review on cultivation, harvesting, lipid extraction, and numerous applications of microalgae. *Renew. Sustain. Energy Rev.* **2018**, *94*, 49–68.
85. Mishra, S.; Gupta, A.; Upadhye, V.; Singh, S.C.; Sinha, R.P.; Häder, D.P. Therapeutic Strategies against Biofilm Infections. *Life* **2023**, *13*, 172. [CrossRef] [PubMed]
86. Mobin, S.M.A.; Chowdhury, H.; Alam, F. Commercially important bioproducts from microalgae and their current applications-A review. *Energy Procedia* **2019**, *160*, 752–760. [CrossRef]
87. Wang, H.D.; Chen, C.; Huynh, P.; Chang, J. Exploring the Potential of using algae in cosmetics. *Bioresour. Technol.* **2015**, *184*, 355–362. [CrossRef]
88. Alencar-Silva, T.; Braga, M.C.; Santana, G.O.S.; Saldanha-Araujo, F.; Pogue, R.; Dias, S.C.; Franco, O.L.; Carvalho, J.L. Breaking the frontiers of cosmetology with antimicrobial peptides. *Biotechnol. Adv.* **2018**, *36*, 2019–2031. [CrossRef] [PubMed]
89. Balskus, E.P.; Walsh, C.T. The genetic and molecular basis for sunscreen biosynthesis in cyanobacteria. *Science* **2010**, *329*, 1653–1656. [CrossRef]
90. Garciapichel, F.; Wingard, C.E.; Castenholz, R.W. evidence regarding the UV sunscreen role of a mycosporine-like compound in the cyanobacterium *Gloeocapsa* sp. *Appl. Environ. Microbiol.* **1993**, *59*, 170–176. [CrossRef]
91. Priyadarshani, I.; Rath, B. Commercial and industrial applications of micro algae—A review. *J. Algal Biomass Util.* **2012**, *3*, 89–100.
92. Pulz, O.; Gross, W. Valuable products from biotechnology of microalgae. *Appl. Microbiol. Biotechnol.* **2004**, *65*, 635–648. [CrossRef]
93. Arad, S.M.; Yaron, A. Natural pigments from red microalgae for use in foods and cosmetics. *Trends Food Sci. Tech.* **1992**, *3*, 92–97. [CrossRef]
94. Hirata, T.; Tanaka, M.; Ooike, M.; Tsunomura, T.; Sakaguchi, M. Antioxidant activities of phycocyanobilin prepared from *Spirulina platensis*. *J. Appl. Phycol.* **2000**, *12*, 435–439. [CrossRef]
95. Stolz, P.; Obermayer, B. Manufacturing microalgae for skin care. *Cosmet. Toilet.* **2005**, *120*, 99–106.
96. Chakdar, H.; Jadhav, S.D.; Dhar, D.W.; Pabbi, S. potential applications of blue green algae. *J. Sci. Ind. Res.* **2012**, *71*, 13–20.
97. Yarkent, Ç.; Gürlek, C.; Oncel, S.S. Potential of microalgal compounds in trending natural cosmetics: A review. *Sustain. Chem. Pharm.* **2020**, *17*, 100304. [CrossRef]

98. Achitouv, E.; Metzger, P.; Rager, M.N.; Largeau, C. C-31-C-34 methylated squalenes from a bolivian strain of *Botryococcus braunii*. *Phytochemistry* **2004**, *65*, 3159–3165. [CrossRef] [PubMed]

99. Guerin, M.; Huntley, M.E.; Olaizola, M. *Haematococcus astaxanthin*: Applications for human health and nutrition. *Trends Biotechnol.* **2003**, *21*, 210–216. [CrossRef]

100. Tominaga, K.; Hongo, N.; Karato, M.; Yamashita, E. Cosmetic benefits of astaxanthin on humans subjects. *Acta Biochim. Pol.* **2012**, *59*, 43–47. [CrossRef]

101. Walker, T.L.; Purton, S.; Becker, D.K.; Collet, C. Microalgae as bioreactors. *Plant Cell Rep.* **2005**, *24*, 629–664. [CrossRef]

102. Levasseur, W.; Perr'e, P.; Pozzobon, V. A review of high value-added molecules production by microalgae in light of the classification. *Biotechnol. Adv.* **2020**, *41*, 107545. [CrossRef] [PubMed]

103. Rizwan, M.; Mujtaba, G.; Ahmed, S.; Lee, K.; Rashid, N. Exploring the Potential of microalgae for new biotechnology applications and beyond: A review. *Renew. Sustain. Energy Rev.* **2018**, *92*, 394–404. [CrossRef]

104. Tiwari, O.N.; Gayen, K.; Kumar, T. Food and bioproducts processing downstream processing of microalgae for pigments, protein and carbohydrate in industrial application: A review. *Food Bioprod. Process.* **2018**, *110*, 60–84.

105. Balić, A.; Mokos, M. Do we utilize our knowledge of the skin protective effects of carotenoids enough? *Antioxidants* **2019**, *8*, 259. [CrossRef]

106. Chekanov, K. Diversity and Distribution of Carotenogenic Algae in Europe: A Review. *Mar. Drugs* **2023**, *21*, 108. [CrossRef]

107. Chekanov, K.; Fedorenko, T.; Kublanovskaya, A.; Litvinov, D.; Lobakova, E. Diversity of carotenogenic microalgae in the White Sea polar region. *FEMS Microbiol. Ecol.* **2020**, *96*, 183. [CrossRef] [PubMed]

108. Besednova, N.N.; Andryukov, B.G.; Zaporozhets, S.; Kryzhanovsky, S.P.; Kuznetsova, T.A.; Fedyanina, L.N.; Makarenkova, I.D.; Zvyagintseva, T.N. Algae Polyphenolic Compounds and Modern Antibacterial Strategies: Current Achievements and Immediate Prospects. *Biomedicines* **2020**, *8*, 342. [CrossRef]

109. Jimenez-Lopez, C.; Pereira, A.G.; Lourenço-Lopes, C.; Garcia-Oliveira, P.; Cassani, L.; Fraga-Corral, M.; Prieto, M.A.; Simal-Gandara, J. Main bioactive phenolic compounds in marine algae and their mechanisms of action supporting potential health benefits. *Food Chem.* **2021**, *341*, 128262. [CrossRef]

110. Kokabi, K.; Gorelova, O.; Zorin, B.; Didi-Cohen, S.; Itkin, M.; Malitsky, S.; Solovchenko, A.; Boussiba, S.; Khozin-Goldberg, I. Lipidome Remodeling and Autophagic Respose in the Arachidonic-Acid-Rich Microalga *Lobosphaera incisa* Under Nitrogen and Phosphorous Deprivation. *Front. Plant Sci.* **2020**, *11*, 614846. [CrossRef]

111. El Gamal, A.A. Biological importance of marine algae. *Saudi Pharm. J.* **2010**, *18*, 1–25. [CrossRef] [PubMed]

112. Jain, A.K.; Singh, D.; Dubey, K.; Maurya, R.; Mittal, S.; Pandey, A.K. Models and methods for in vitro toxicity. In *In Vitro Toxicology*; Academic Press: Cambridge, MA, USA, 2018; pp. 45–65.

113. Rastogi, R.P.; Sonani, R.R.; Madamwar, D. Cyanobacterial sunscreen scytonemin: Role in photoprotection and biomedical research. *Appl. Biochem. Biotechnol.* **2015**, *176*, 1551–1563. [CrossRef]

114. Balskus, E.P.; Case, R.J.; Walsh, C.T. The biosynthesis of cyanobacterial sunscreen scytonemin in intertidal microbial mat communities. *FEMS Microbiol. Ecol.* **2011**, *77*, 322–332. [CrossRef] [PubMed]

115. Chrapusta, E.; Kaminski, A.; Duchnik, K.; Bober, B.; Adamski, M.; Bialczyk, J. Mycosporine-like amino acids: Potential health and beauty ingredients. *Mar. Drugs* **2017**, *15*, 326. [CrossRef]

116. Pathak, J.; Ahmed, H.; Rajneesh; Singh, S.P.; Hader, D.P.; Sinha, R.P. Genetic regulation of scytonemin and mycosporine-like amino acids (MAAs) biosynthesis in cyanobacteria. *Plant Gene* **2019**, *17*, 100172. [CrossRef]

117. Rajneesh; Pathak, J.; Richa; Hader, D.P.; Sinha, R.P. Impacts of ultraviolet radiation on certain physiological and biochemical processes in cyanobacteria inhabiting diverse habitats. *Environ. Exp. Bot.* **2019**, *161*, 375–387. [CrossRef]

118. Zaytseva, A.; Chekanov, K.; Zaytsev, P.; Bakhareva, D.; Gorelova, O.; Kochkin, D.; Lobakova, E. Sunscreen effect exerted by secondary carotenoids and mycosporine-like amino acids in the aeroterrestrial chlorophyte *Coelastrella rubescens* under high light and UV-A irradiation. *Plants* **2021**, *10*, 2601. [CrossRef]

119. Hartmann, A.; Glaser, K.; Holzinger, A.; Ganzera, M.; Karsten, U. Klebsormidin A and B, two new UV-sunscreen compounds in green microalgal *Interfilum* and *Klebsormidium* species (Streptophyta) from terrestrial habitats. *Front. Microbiol.* **2020**, *11*, 499. [CrossRef]

120. Fernanda, P.M. Algae and aquatic macrophytes responses to cope to ultraviolet radiation—A Review. *Emir. J. Food Agric.* **2012**, *24*, 527–545. [CrossRef]

121. Singh, A.; Tyagi, M.B.; Kumar, A. Cyanobacteria growing on tree barks possess high amount of sunscreen compound mycosporine-like amino acids (MAAs). *Plant Physiol. Biochem.* **2017**, *119*, 110–120. [CrossRef]

122. Tarasuntisuk, S.; Palaga, T.; Kageyama, H.; Waditee-Sirisattha, R. Mycosporine-2- glycine exerts anti-inflammatory and antioxidant effects in lipopolysaccharide (LPS)- stimulated RAW 264.7 macrophages. *Arch. Biochem. Biophys.* **2019**, *662*, 33–39. [CrossRef] [PubMed]

123. Kumari, N.; Pandey, A.; Gupta, A.; Mishra, S.; Sinha, R.P. Characterization of UV-screening pigment scytonemin from cyanobacteria inhabiting diverse habitats of Varanasi, India. *Biologia* **2023**, *78*, 319–330. [CrossRef]

124. Bhatia, S.; Garg, A.; Sharma, K.; Kumar, S.; Sharma, A.; Purohit, A.P. Mycosporine and mycosporine-like amino acids: A paramount tool against ultra violet irradiation. *Phcog. Rev.* **2011**, *5*, 138–146. [CrossRef] [PubMed]

125. Siezen, R.J. Microbial sunscreens. *Microb. Biotechnol.* **2011**, *4*, 1–7. [CrossRef] [PubMed]

126. Schmid, B.D.; Schürch, C.; Zülli, F. Mycosporine-like amino acids from red algae protect against premature skin-aging. *Cosmetics* **2006**, *9*, 1–4.

127. Tarasuntisuk, S.; Patipong, T.; Hibino, T.; Waditee-Sirisattha, R.; Kageyama, H. Inhibitory effects of mycosporine-2-glycine isolated from a halotolerant cyanobacterium on protein glycation and collagenase activity. *Lett. Appl. Microbiol.* **2018**, *67*, 314–320. [CrossRef]

128. Kornhauser, A.; Coelho, S.G.; Hearing, V.J. Effects of cosmetic formulations containing hydroxyacids on sun-exposed skin: Current applications and future developments. *Dermatol. Res. Pract.* **2012**, *2012*, 710893. [CrossRef] [PubMed]

129. Matsumoto, G.I.; Nagashima, H. Occurrence of 3-hydroxy acids in microalgae and cyanobacteria and their geochemical significance. *Geochem. Cosmochim. Acta* **1984**, *48*, 1683–1687. [CrossRef]

130. Matsumoto, G.I.; Shioya, M.; Nagashima, H. Occurrence of 2-hydroxy acids in microalgae. *Phytochemistry* **1984**, *23*, 1421–1423. [CrossRef]

131. Pillaiyar, T.; Manickam, M.; Namasivayam, V. Skin whitening agents: Medicinal chemistry perspective of tyrosinase inhibitors. *J. Enzym. Inhib. Med. Chem.* **2017**, *32*, 403–425. [CrossRef]

132. Sung, H.J.; Khan, M.F.; Kim, Y.H. Recombinant lignin peroxidase-catalyzed decolorization of melanin using in-situ generated H_2O_2 for application in whitening cosmetics. *Int. J. Biol. Macromol.* **2019**, *136*, 20–26. [CrossRef] [PubMed]

133. Sahin, S.C. The Potential of *Arthrospira platensis* extract as a tyrosinase inhibitor for pharmaceutical or cosmetic applications. *South Afr. J. Bot.* **2018**, *119*, 236–243. [CrossRef]

134. Oh, G.; Ko, S.; Heo, S.; Nguyen, V.; Jung, W. A novel peptide purified from the fermented microalga Pavlova lutheri attenuates oxidative stress and melanogenesis in B16F10 melanoma cells. *Process Biochem.* **2015**, *50*, 1318–1326. [CrossRef]

135. Rao, A.R.; Sindhuja, H.N.; Dharmesh, S.M.; Sankar, K.U.; Sarada, R.; Ravishankar, G.A. Effective inhibition of skin cancer, tyrosinase, and antioxidative properties by astaxanthin and astaxanthin esters from the green alga *Haematococcus pluvialis*. *J. Agric. Food Chem.* **2013**, *61*, 3842–3851. [CrossRef] [PubMed]

136. De Jesus Raposo, M.F.; De Morais, R.M.S.C.; De Morais, A.M.M.B. Health applications of bioactive compounds from marine microalgae. *Life Sci.* **2013**, *93*, 479–486. [CrossRef]

137. Marrot, L.; Meunier, J.R. Skin DNA photodamage and its biological consequences. *J. Am. Acad. Dermatol.* **2008**, *58*, S139–S148. [CrossRef] [PubMed]

138. Chen, L.; Hu, J.Y.; Wang, S.Q. The role of antioxidants in photoprotection: A critical review. *J. Am. Acad. Dermatol.* **2012**, *67*, 1013–1024. [CrossRef]

139. Wang, S.Q.; Osterwalder, U.; Jung, K. Ex vivo evaluation of radical sun protection factor in popular sunscreens with antioxidants. *J. Am. Acad. Dermatol.* **2011**, *65*, 525–530. [CrossRef] [PubMed]

140. Matsui, M.S.; Hsia, A.; Miller, J.D.; Hanneman, K.; Scull, H.; Cooper, K.D.; Baron, E. Non-sunscreen photoprotection: Antioxidants add value to a sunscreen. *J. Investig. Dermatol. Symp.Proc.* **2009**, *14*, 56–59. [CrossRef]

141. Lin, F.H.; Lin, J.Y.; Gupta, R.D.; Tournas, J.A.; Burch, J.A.; Selim, M.A.; Monteiro-Riviere, N.A.; Grichnik, J.M.; Zielinski, J.; Pinnell, S.R. Ferulic acid stabilizes a solution of vitamins C and E and doubles its photoprotection of skin. *J. Investig. Dermatol.* **2005**, *125*, 826–832. [CrossRef] [PubMed]

142. Murray, J.C.; Burch, J.A.; Streilein, R.D.; Iannacchione, M.A.; Hall, R.P.; Pinnell, S.R. A topical antioxidant solution containing vitamins C and E stabilized by ferulic acid provides protection for human skin against damage caused by ultraviolet irradiation. *J. Am. Acad. Dermatol.* **2008**, *59*, 418–425. [CrossRef] [PubMed]

143. Kligman, A.M.; Grove, G.L.; Hirose, R.; Leyden, J.J. Topical tretinoin for photoaged skin. *J. Am. Acad. Dermatol.* **1986**, *15*, 836–859. [CrossRef]

144. Rosenthal, A.; Stoddard, M.; Chipps, L.; Herrmann, J. Skin cancer prevention: A review of current topical options complementary to sunscreens. *J. Eur. Acad. Dermatol. Venereol.* **2019**, *33*, 1261–1267. [CrossRef] [PubMed]

145. Dunaway, S.; Odin, R.; Zhou, L.; Ji, L.; Zhang, Y.; Kadekaro, A.L. Natural antioxidants: Multiple mechanisms to protect skin from solar radiation. *Front. Pharmacol.* **2018**, *9*, 392. [CrossRef] [PubMed]

146. Rezzani, R.; Rodella, L.F.; Favero, G.; Damiani, G.; Paganelli, C.; Reiter, R.J. Attenuation of ultraviolet A-induced alterations in NIH3T3 dermal fibroblasts by melatonin. *Br. J. Dermatol.* **2014**, *170*, 382–391. [CrossRef]

147. Janjetovic, Z.; Jarrett, S.G.; Lee, E.F.; Duprey, C.; Reiter, R.J.; Slominski, A.T. Melatonin and its metabolites protect human melanocytes against UVB-induced damage: Involvement of NRF2-mediated pathways. *Sci. Rep.* **2017**, *7*, 1274. [CrossRef] [PubMed]

148. Wei, H.; Cai, Q.; Rahn, R.O. Inhibition of UV light-and Fenton Reaction-induced oxidative DNA damage by the soybean isoflavone genistein. *Carcinogenesis* **1996**, *17*, 73–77. [CrossRef]

149. Kang, S.; Chung, J.H.; Lee, J.H.; Fisher, G.J.; Wan, Y.S.; Duell, E.A.; Voorhees, J.J. Topical N-acetyl cysteine and genistein prevent ultraviolet-light-induced signaling that leads to photoaging in human skin in vivo. *J. Investig. Dermatol.* **2003**, *120*, 835–841. [CrossRef]

150. Yeager, D.G.; Lim, H.W. What's new in photoprotection: A review of new concepts and controversies. *Dermatol. Clin.* **2019**, *37*, 149–157. [CrossRef] [PubMed]

151. Emanuele, E.; Spencer, J.M.; Braun, M. An experimental double-blind irradiation study of a novel topical product (TPF 50) compared to other topical products with DNA repair enzymes, antioxidants, and growth factors with sunscreens: Implications for preventing skin aging and cancer. *J. Drugs Dermatol.* **2014**, *13*, 309–314.

152. Antoniou, C.; Kosmadaki, M.G.; Stratigos, A.J.; Katsambas, A.D. Photoaging: Prevention and topical treatments. *Am. J. Clin. Dermatol.* **2010**, *11*, 95–102. [CrossRef]
153. Green, A.; Williams, G.; Nèale, R.; Hart, V.; Leslie, D.; Parsons, P.; Marks, G.C.; Gaffney, P.; Battistutta, D.; Frost, C.; et al. Daily sunscreen application and betacarotene supplementation in prevention of basal-cell and squamous-cell carcinomas of the skin: A randomized controlled trial. *Lancet* **1999**, *354*, 723–729. [CrossRef]
154. Darlington, S.; Williams, G.; Neale, R.; Frost, C.; Green, A. A randomized controlled trial to assess sunscreen application and beta carotene supplementation in the prevention of solar keratoses. *Arch. Dermatol.* **2003**, *139*, 451–455. [CrossRef] [PubMed]
155. Megna, M.; Lembo, S.; Balato, N.; Monfrecola, G. " Active" photoprotection: Sunscreens with DNA repair enzymes. *G. Ital. Dermatol. Venereol.* **2017**, *152*, 302–307. [CrossRef] [PubMed]
156. Haney, A.M.; Sanfilippo, J.E.; Garczarek, L.; Partensky, F.; Kehoe, D.M. Multiple photolyases protect the marine cyanobacterium *Synechococcus* from ultraviolet radiation. *Mbio* **2022**, *13*, 1511–1522. [CrossRef]
157. Zhang, M.; Wang, L.; Zhong, D. Photolyase: Dynamics and electron-transfer mechanisms of DNA repair. *Arch. Biochem. Biophys.* **2017**, *632*, 158–174. [CrossRef] [PubMed]
158. Pathak, J.; Singh, P.R.; Häder, D.P.; Sinha, R.P. UV-induced DNA damage and repair: A cyanobacterial perspective. *Plant Gene* **2019**, *19*, 100194. [CrossRef]
159. Dong, K.K.; Damaghi, N.; Picart, S.D.; Markova, N.G.; Obayashi, K.; Okano, Y.; Masaki, H.; Grether-Beck, S.; Krutmann, J.; Smiles, K.A.; et al. UV-induced DNA damage initiates release of MMP-1 in human skin. *Exp. Dermatol.* **2008**, *17*, 1037–1044. [CrossRef]
160. Liu, Z.; Wang, L.; Zhong, D. Dynamics and mechanisms of DNA repair by photolyase. *Phys. Chem. Chem. Phys.* **2015**, *17*, 11933–11949. [CrossRef]
161. Stege, H.; Roza, L.; Vink, A.A.; Grewe, M.; Ruzicka, T.; Grether-Beck, S.; Krutmann, J. Enzyme plus light therapy to repair DNA damage in ultraviolet-B-irradiated human skin. *Proc. Natl. Acad. Sci. USA* **2000**, *97*, 1790–1795. [CrossRef]
162. Wlaschek, M.; Briviba, K.; Stricklin, G.P.; Sies, H.; Scharffetter-Kochanek, K. Singlet oxygen may mediate the ultraviolet A-induced synthesis of interstitial collagenase. *J. Investig. Dermatol.* **1995**, *104*, 194–198. [CrossRef]
163. Wolf, P.; Müllegger, R.R.; Soyer, H.P.; Hofer, A.; Smolle, J.; Horn, M.; Cerroni, L.; Hofmann-Wellenhof, R.; Kerl, H.; Maier, H.; et al. Topical treatment with liposomes containing T4 endonuclease V protects human skin in vivo from ultraviolet-induced upregulation of interleukin-10 and tumor necrosis factor-α. *J. Investig. Dermatol.* **2000**, *114*, 149–156. [CrossRef]
164. Yarosh, D.; Klein, J.; O'Connor, A.; Hawk, J.; Rafal, E.; Wolf, P. Effect of topically applied T4 endonuclease V in liposomes on skin cancer in xeroderma pigmentosum: A randomized study. *Lancet* **2001**, *357*, 926–929. [CrossRef] [PubMed]
165. Cafardi, J.A.; Elmets, C.A. T4 endonuclease V: Review and application to dermatology. *Expert Opin. Biol. Ther.* **2008**, *8*, 829–838. [CrossRef] [PubMed]
166. Zattra, E.; Coleman, C.; Arad, S.; Helms, E.; Levine, D.; Bord, E.; Guillaume, A.; El-Hajahmad, M.; Zwart, E.; van Steeg, H.; et al. Polypodium leucotomos extract decreases UV-induced Cox-2 expression and inflammation, enhances DNA repair, and decreases mutagenesis in hairless mice. *Am. J. Pathol.* **2009**, *175*, 1952–1961. [CrossRef] [PubMed]
167. Shiota, S.; Nakayama, H. UV endonuclease of Micrococcus luteus, a cyclobutane pyrimidine dimer–DNA glycosylase/abasic lyase: Cloning and characterization of the gene. *Proc. Natl. Acad. Sci. USA* **1997**, *94*, 593–598. [CrossRef]
168. Wolf, P.; Cox, P.; Yarosh, D.B.; Kripke, M.L. Sunscreens and T4N5 liposomes differ in their ability to protect against ultraviolet-induced sunburn cell formation, alterations of dendritic epidermal cells, and local suppression of contact hypersensitivity. *J. Investig. Dermatol.* **1995**, *104*, 287–292. [CrossRef]
169. Carducci, M.; Pavone, P.S.; De Marco, G.; Lovati, S.; Altabas, V.; Altabas, K.; Emanuele, E. Comparative effects of sunscreens alone vs sunscreens plus DNA repair enzymes in patients with actinic keratosis: Clinical and molecular findings from a 6-month, randomized, clinical study. *J. Drugs Dermatol.* **2015**, *14*, 986–990.
170. Navarrete-Dechent, C.; Molgó, M. The use of a sunscreen containing DNA-photolyase in the treatment of patients with field cancerization and multiple actinic keratoses: A case-series. *Dermatol. Online J.* **2017**, *23*. [CrossRef]
171. Krutmann, J.; Hansen, P.M. Algenenzym Photolyase verbessert Schutz vor UVB-Schäden. *Pharm. Ztg.* **2004**, *149*, 50–53.
172. Puviani, M.; Barcella, A.; Milani, M. Efficacy of a photolyase-based device in the treatment of cancerization field in patients with actinic keratosis and non-melanoma skin cancer. *G. Ital. Dermatol. Venereol.* **2013**, *148*, 693–698.

MDPI

St. Alban-Anlage 66

4052 Basel

Switzerland

www.mdpi.com

Catalysts Editorial Office

E-mail: catalysts@mdpi.com

www.mdpi.com/journal/catalysts

9 783725 805792